Object Detection Based on Vision Sensors and Neural Network

Object Detection Based on Vision Sensors and Neural Network

Guest Editors
Man Qi
Matteo Dunnhofer

Basel • Beijing • Wuhan • Barcelona • Belgrade • Novi Sad • Cluj • Manchester

Guest Editors

Man Qi
Department of Computing
Canterbury Christ Church
University
Canterbury
UK

Matteo Dunnhofer
Machine Learning and
Perception Lab
University of Udine
Udine
Italy

Editorial Office
MDPI AG
Grosspeteranlage 5
4052 Basel, Switzerland

This is a reprint of the Special Issue, published open access by the journal *Sensors* (ISSN 1424-8220), freely accessible at: https://www.mdpi.com/journal/sensors/special_issues/7A3ZUD09QO.

For citation purposes, cite each article independently as indicated on the article page online and as indicated below:

Lastname, A.A.; Lastname, B.B. Article Title. *Journal Name* **Year**, *Volume Number*, Page Range.

ISBN 978-3-7258-3659-8 (Hbk)
ISBN 978-3-7258-3660-4 (PDF)
https://doi.org/10.3390/books978-3-7258-3660-4

© 2025 by the authors. Articles in this book are Open Access and distributed under the Creative Commons Attribution (CC BY) license. The book as a whole is distributed by MDPI under the terms and conditions of the Creative Commons Attribution-NonCommercial-NoDerivs (CC BY-NC-ND) license (https://creativecommons.org/licenses/by-nc-nd/4.0/).

Contents

About the Editors vii

My-Tham Dinh, Deok-Jai Choi and Guee-Sang Lee
DenseTextPVT: Pyramid Vision Transformer with Deep Multi-Scale Feature Refinement Network for Dense Text Detection
Reprinted from: *Sensors* 2023, 23, 5889, https://doi.org/10.3390/s23135889 1

Jhonghyun An
Traversable Region Detection and Tracking for a Sparse 3D Laser Scanner for Off-Road Environments Using Range Images
Reprinted from: *Sensors* 2023, 23, 5898, https://doi.org/10.3390/s23135898 15

Didier Ndayikengurukiye and Max Mignotte
CoSOV1Net: A Cone- and Spatial-Opponent Primary Visual Cortex-Inspired Neural Network for Lightweight Salient Object Detection
Reprinted from: *Sensors* 2023, 23, 6450, https://doi.org/10.3390/s23146450 32

Wei Wang, Weizhen Yang, Maozhen Li, Zipeng Zhang and Wenbin Du
A Novel Approach for Apple Freshness Prediction Based on Gas Sensor Array and Optimized Neural Network
Reprinted from: *Sensors* 2023, 23, 6476, https://doi.org/10.3390/s23146476 60

Shuaihui Wang, Fengyi Jiang and Boqian Xu
Swin Transformer-Based Edge Guidance Network for RGB-D Salient Object Detection
Reprinted from: *Sensors* 2023, 23, 8802, https://doi.org/10.3390/s23218802 73

Yanping Chen, Chong Deng, Qiang Sun, Zhize Wu, Le Zou, Guanhong Zhang, et al.
Lightweight Detection Methods for Insulator Self-Explosion Defects
Reprinted from: *Sensors* 2024, 24, 290, https://doi.org/10.3390/s24010290 86

José A. Rodríguez-Rodríguez, Ezequiel López-Rubio, Juan A. Ángel-Ruiz and Miguel A. Molina-Cabello
The Impact of Noise and Brightness on Object Detection Methods
Reprinted from: *Sensors* 2024, 24, 821, https://doi.org/10.3390/s24030821 102

Tian Zhou, Wu Xie, Huimin Zhang and Yong Fan
Simple Conditional Spatial Query Mask Deformable Detection Transformer: A Detection Approach for Multi-Style Strokes of Chinese Characters
Reprinted from: *Sensors* 2024, 24, 931, https://doi.org/10.3390/s24030931 124

Lu Zheng, Wenhan Long, Junchao Yi, Lu Liu and Ke Xu
Enhanced Knowledge Distillation for Advanced Recognition of Chinese Herbal Medicine
Reprinted from: *Sensors* 2024, 24, 1559, https://doi.org/10.3390/s24051559 144

Ana L. Afonso, Gil Lopes and A. Fernando Ribeiro
Lizard Body Temperature Acquisition and Lizard Recognition Using Artificial Intelligence
Reprinted from: *Sensors* 2024, 24, 4135, https://doi.org/10.3390/s24134135 162

Zexin Yan, Jie Fan, Zhongbo Li and Yongqiang Xie
Elevating Detection Performance in Optical Remote Sensing Image Object Detection: A Dual Strategy with Spatially Adaptive Angle-Aware Networks and Edge-Aware Skewed Bounding Box Loss Function
Reprinted from: *Sensors* 2024, 24, 5342, https://doi.org/10.3390/s24165342 186

Chunqing Liu, Fengliang Zhang, Yanchun Ni, Botao Ai, Siyan Zhu, Zezhou Zhao, et al.
Efficient Model Updating of a Prefabricated Tall Building by a DNN Method
Reprinted from: *Sensors* **2024**, *24*, 5557, https://doi.org/10.3390/s24175557 213

Ali Asghar Sharifi, Ali Zoljodi and Masoud Daneshtalab
TrajectoryNAS: A Neural Architecture Search for Trajectory Prediction
Reprinted from: *Sensors* **2024**, *24*, 5696, https://doi.org/10.3390/s24175696 234

Mirosław Łącki
Determining the Level of Threat in Maritime Navigation Based on the Detection of Small Floating Objects with Deep Neural Networks
Reprinted from: *Sensors* **2024**, *24*, 7505, https://doi.org/10.3390/s24237505 249

About the Editors

Man Qi

Dr. Man Qi is a Reader in Computing at Canterbury Christ Church University with over 30 years of teaching and research experience in computing. Dr Qi is a Fellow of British Computer Society (FBCS) and Fellow of Higher Education Academy (FHEA). Her research interests are Intelligent Computing and Applications, Human Computer Interaction, Internet of Things, and Cyber Security. She has engaged in EU, EPSRC, and other international collaborative projects. Dr Qi has over 80 research publications including over 40 peer-reviewed journal papers. She is an editorial board member for several international journals and has served as programme committee member/chair for over 50 international conferences.

Matteo Dunnhofer

Dr. Matteo Dunnhofer is a Marie Skłodowska-Curie Postdoctoral Fellow between the Department of Mathematics, Computer Science, and Physics of the University of Udine (Udine, Italy), and the Centre for Vision Research at York University (Toronto, Canada). From the University of Udine, he received a BSc and MSc in Computer Science in 2016 and 2018, and a PhD in Industrial and Information Engineering in 2022. In 2018, he was a visiting student at the Australian Centre for Robotic Vision hosted at the Queensland University of Technology. More recently, he has spent time as a visiting scientist at MIT and the University of Alcalá. His research focuses on the use of deep learning to tackle computer vision problems, especially for visual object tracking. He is also interested in applying deep learning techniques in the context of medical image analysis and video-based sport analytics. On these topics, he published several papers that appeared in international journals and conferences, and he organised workshops and tutorials, including the series of Workhops on Computer Vision for Winter Sports at WACV. In 2021, he was awarded for winning the Visual Object Tracking VOT2021 Long-term Challenge held at ICCV 2021. He also serves as reviewer for mant influential journals and conferences in computer vision, pattern recognition, and robotics. He was recognised as an outstanding reviewer by ECCV 2022. He is an associate editor for *The Visual Computer* journal.

Article

DenseTextPVT: Pyramid Vision Transformer with Deep Multi-Scale Feature Refinement Network for Dense Text Detection

My-Tham Dinh, Deok-Jai Choi and Guee-Sang Lee *

Department of Artificial Intelligence Convergence, Chonnam National University, 77 Yongbong-ro, Gwangju 500-757, Republic of Korea; thamdinh.dmt@gmail.com (M.-T.D.)
* Correspondence: gslee@jnu.ac.kr

Abstract: Detecting dense text in scene images is a challenging task due to the high variability, complexity, and overlapping of text areas. To adequately distinguish text instances with high density in scenes, we propose an efficient approach called DenseTextPVT. We first generated high-resolution features at different levels to enable accurate dense text detection, which is essential for dense prediction tasks. Additionally, to enhance the feature representation, we designed the Deep Multi-scale Feature Refinement Network (DMFRN), which effectively detects texts of varying sizes, shapes, and fonts, including small-scale texts. DenseTextPVT, then, is inspired by Pixel Aggregation (PA) similarity vector algorithms to cluster text pixels into correct text kernels in the post-processing step. In this way, our proposed method enhances the precision of text detection and effectively reduces overlapping between text regions under dense adjacent text in natural images. The comprehensive experiments indicate the effectiveness of our method on the TotalText, CTW1500, and ICDAR-2015 benchmark datasets in comparison to existing methods.

Keywords: scene text detection; pyramid vision transformer; dense adjacent text

Citation: Dinh, M.-T.; Choi, D.-J.; Lee, G.-S. DenseTextPVT: Pyramid Vision Transformer with Deep Multi-Scale Feature Refinement Network for Dense Text Detection. *Sensors* 2023, 23, 5889. https://doi.org/10.3390/s23135889

Academic Editor: Man Qi and Matteo Dunnhofer

Received: 12 May 2023
Revised: 15 June 2023
Accepted: 21 June 2023
Published: 25 June 2023

Copyright: © 2023 by the authors. Licensee MDPI, Basel, Switzerland. This article is an open access article distributed under the terms and conditions of the Creative Commons Attribution (CC BY) license (https://creativecommons.org/licenses/by/4.0/).

1. Introduction

Scene text detection has made significant progress in computer vision and plays a crucial role in various practical applications such as scene understanding, scene reading, and autonomous driving. The application of deep learning has led to remarkable achievements in detecting text in natural scenes [1–15].

Recent methods in scene text detection have extensively utilized deep neural networks (DNNs) to extract features and achieve impressive performance on benchmark datasets [16–18]. Despite these advancements, scene text detection remains a challenging task, primarily due to the irregular shapes, diverse scales, and high density of text instances in scenes (as illustrated in Figure 1). Existing methods like SegLink++ [13] and MSR [19] have shown effectiveness in handling text lines and accommodating variations in text line length. However, they have still faced difficulties in dealing with overlapping dense text regions, especially in small-scale texts. Following that, methods like PAN [1], TextSnake [20], and CT [12] aim to address overlap phenomena by expanding text regions from text kernels, but they fall short in achieving competitive results in scene text detection.

To overcome these challenges, our approach explores a multi-scale strategy with three different kernel filters and attention mechanisms, namely, Deep Multi-scale Feature Refinement Network (DMFRN). This method generates and fuses the multi-level features that provide comprehensive representations for scene text instances.

Moreover, this study is inspired by the merits of Transformer [21–26], which has been employed to eliminate the complex and understand spatial arrangement and contextual information in manually designed procedures of object detection. Transformer models like DETR [22] tackle the object detection task in a fully end-to-end manner, eliminating the need for complex handcrafted components such as anchor generation, region proposal networks, and non-maximum suppression. However, they are not capable of effectively extracting low-level visual features at a local level effectively, and they also struggle to detect small objects. Although ViT [24] employs a self-attention mechanism within Transformer to model the interactions between patches, enabling the model to capture both local and global contextual information, ViT has struggled to achieve pixel-level dense prediction.

Figure 1. Sample of inaccurate dense predictions in previous works.

In this work, we propose a solution to accurately predict dense text by employing the PvTv2 versatile backbone [26], which is designed to achieve high output resolution for dense prediction tasks in object detection while reducing resource consumption through a progressive shrinking pyramid. Unlike the original backbone, we added a channel attention module (CAM) and spatial attention module (SAM) between feature levels to effectively capture and leverage informative features in both the channel-wise and spatial dimensions. This work leads to enlarging the receptive fields and preserving high-resolution features, which is crucial for the dense prediction task.

To further enhance the quality of the feature representation, we incorporated a post-processing step based on PAN [1]. This step is designed to reduce the overlap between text regions. By applying this post-processing technique, we can improve the accuracy and clarity of the detected text regions, leading to more reliable results.

Our core contributions are as follows:

1. We propose an effective approach, called DenseTextPVT, which incorporates the advantages of dense prediction backbone in object detection tasks, Pyramid Vision Transformer (PvTv2) [26], with a channel attention module (CAM) [27] and spatial attention module (SAM) [27] to obtain high-resolution features that make our model well suited for dense text prediction in natural scene images.
2. We employed a Deep Multi-scale Feature Refinement Network (DMFRN) using three kernel filters simultaneously (3×3, 5×5, 7×7) with CBAM [27] at each feature. This allows for adaptive feature refinement, enabling our model to enrich feature representations with different scales, including small representations.

The paper consists of the following sections: Section 2 provides a summary of related works in scene text detection and Transformer. Section 3 describes the architecture of the proposed method in detail. Section 4 presents experimental results. Finally, Section 5 concludes the paper and discusses future work.

2. Related Work

2.1. Scene Text Detection

The regression-based method [8–11,15,28] directly adopts bounding boxes annotation regarding text as an object. He et al. [15] proposed a method for detecting multi-oriented text in scene images using a deep regression network. They utilized semantic segmentation at the pixel level to classify the text and directly calculated offsets between a pixel point and the corresponding box vertices to determine the text quadrangle. SegLink++ [13] presented an approach to detect dense and arbitrarily shaped text in scene images using a network that leverages instance-aware component grouping (ICG). EAST [8] predicted the multi-orientation of text lines or words within the full image directly by employing a fully convolutional network (FCN). FCE [28] formulated text contours in the Fourier domain and represented these arbitrarily shaped texts as compact signatures. Despite their ability to handle text instances with arbitrary shapes, they may struggle with text lines that are challenging to orient and tiny texts.

The segmentation-based method [1–3,6,7] mainly focuses on pixel-level feature representations [1–3,7,29], or segment-level [11,20] or contour-level segmentation [9,30,31]. Typically, these methods usually first segment text kernels and then cluster them into text instances via post-processing. For instance, PSENet [7] utilized a progressive scale algorithm to create a variety of kernels for each text instance and expand, bit by bit, the kernel to cover the entire text instance. Similarly, CT [12] predicted text instances by using text kernels and centripetal shifts, which were used to aggregate pixels, and then directing external text pixels towards the internal text kernels. PAN [1] implemented a clustering approach to precisely aggregate text pixels to exact text kernels based on the similarity vectors. DB++ [2] is an extension of the previous work on differentiable binarization (DB) [29], which incorporated the binarization process into a segmentation network for more accurate results. [32] employed an effective central text region mask and adjusted the expanding ratio from the central text region to the full text instance. However, the performance of these methods is heavily influenced by the quality of the segmentation accuracy.

2.2. Transformer

Transformer has become an increasingly popular topic of research in computer vision. Ref. [21] was the accredited father of Transformer, which was based solely on attention modules. Inspired by this architecture, refs. [21–26,33–35] utilized Transformer-based architecture to approach object detection as a problem of predicting sets. Transformer introduced a simple end-to-end framework that eliminated the need for intricate, hand-crafted anchor generation and post-processing steps. ViT [24] is a Transformer architecture specifically designed for computer vision tasks, and has demonstrated outstanding performance on image classification tasks by directly applying the Transformer to sequences of image patches. DeiT [25] was an extension of ViT that used a new distillation approach to train transformers more efficiently for image classification tasks. It required less data and computing resources than the original ViT model. PvTv2 [26], which was expanded from PVT [35], proposed a flexible backbone that could achieve high output resolution for various vision tasks, particularly dense prediction tasks, while also reducing time consumption by inheriting the advantages of both CNNs and Transformers.

In addition, ref. [33] utilized a Transformer-based architecture to address the problem of detecting multi-oriented texts in images using rotated bounding boxes, but it does not work well in curved text cases. Ref. [34] proposed an end-to-end trainable framework using Transformers (DETR) to predict polygon points or Bezier control points for determining the localization of text instances. Additionally, in [36], point coordinates were directly utilized to generate position queries and progressively updated while also enhancing the spatial awareness of non-local self-attention in the Transformer. Despite significant advancements, methods utilizing the Transformer approach have still faced challenges in accurately detecting small and dense adjacent texts.

Developing robust representations is crucial for a successful scene text detector, as it necessitates the learning of discriminative features that can detect accurately text regions. As previously noted, PvTv2 [26] has demonstrated great potential as a representation of dense prediction tasks in various image applications, such as image classification, object detection, and also semantic segmentation. In this study, we introduce DenseTextPVT, which employs the PvTv2 architecture to generate improved features for dense text in scene text detection.

3. Methodology
3.1. Overall Architecture

The overall framework of our proposed method is illustrated in Figure 2. Given a scene image I ($HxWx3$), we utilized a PvTv2 backbone to extract pyramid features according to four stages, F_1, F_2, F_3, and F_4, whose strides are 4, 8, 16, and 32 pixels following the input image I. To refine the feature information with high resolution, we used channel attention module (CAM) and spatial attention module (SAM) approaches at F_1, F_2 and F_3, F_4 features, respectively. Then, we employed a Deep Multi-scale Feature Refinement Network (DMFRN) with three irregular kernel filters, 3×3, 5×5, and 7×7, and applied CBAM [27] at each output feature to produce multi-level features, F_1^n, F_2^n, F_3^n, and F_4^n ($n = 3, 5, 7$), with rich information on text contents of various sizes. Afterward, to prepare for the prediction stage, we scaled up F_2^n, F_3^n, and F_4^n features into F_1^n size and concatenated them into a single robust feature map F, as shown in Figure 3. Finally, our detection stage was inspired by PAN post-processing [1], which is depicted in Figure 4. In this way, our method can determine which text pixels belong to the correct text kernels, helping us accurately discriminate and mitigate the overlap phenomena between dense text regions.

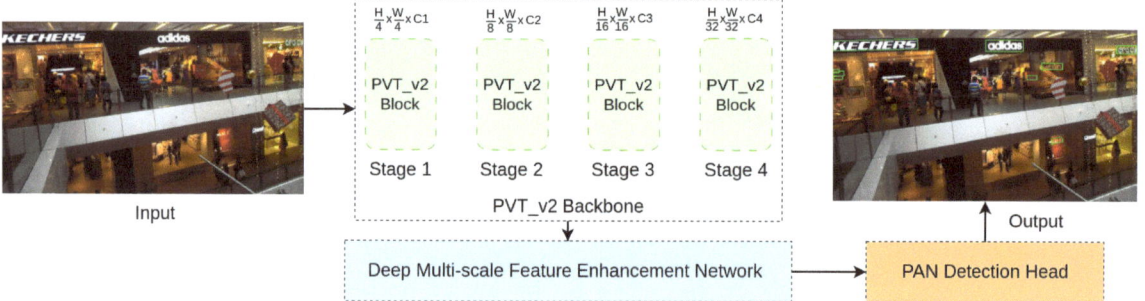

Figure 2. The overall framework of our DenseTextPVT approach.

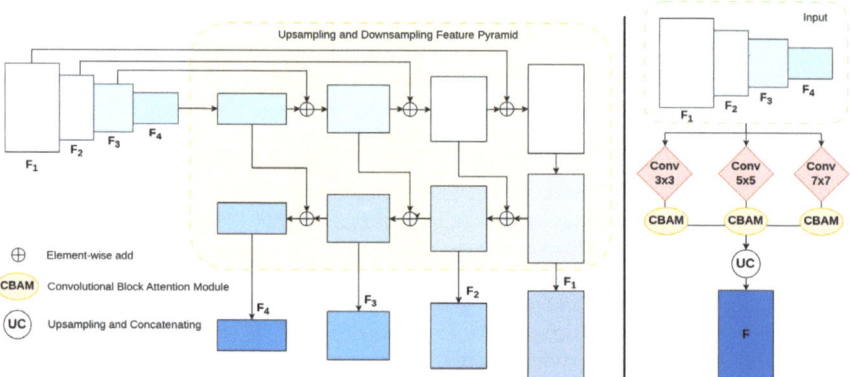

Figure 3. The detail of Deep Multi-scale Feature Refinement Network (DMFRN). The detail of each upsampling and downsampling feature pyramid enhancement (**left**), the overall DMFRN architecture (**right**).

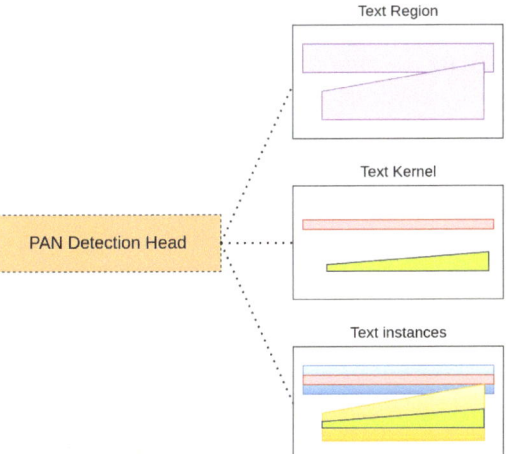

Figure 4. Pixel aggregation detection head.

3.2. PvTv2 Backbone

Different from convolutional neural networks such as ResNet or VGG, PvTv2 [26] serves as a versatile backbone specifically designed for various dense prediction tasks. This approach adopts the Transformer architecture and incorporates a progressive shrinking algorithm to generate feature maps of different scales using patch-embedding layers. Following the structure of [26], the algorithm consists of four pyramid stages, each comprising an overlapping patch-embedding layer and Transformer encoder layers L_i (where i represents the stage of the process).

In each stage, the input image I is divided into patches of size $\frac{H}{j} \times \frac{W}{j}$ (where j denotes the stride sizes: 4, 8, 16, and 32 pixels), as illustrated in Figure 5. These patches are then flattened and passed through a linear projection, resulting in embedded patches of size $\frac{H}{j} \times \frac{W}{j} \times C_i$. PvTv2 employs an Overlapping Patch-Embedding technique by enlarging the patch window size by half of its area and utilizing convolution with zero paddings to preserve resolution. In the Transformer encoder layer, to address the computational cost associated with the attention mechanism, the authors introduced a linear shifted row attention (linearSRA) as a replacement for the traditional multi-head attention. The SRA utilizes average pooling to reduce the spatial dimensions (H, W) to a fixed size (P, P). The linearSRA can be defined as follows, with P set to 7:

$$linearSRA = 2 \times H \times W \times P \times P \times C \quad (1)$$

In addition, PvTv2 introduces a 3×3 depth-wise convolution layer with a padding size of 1 between the first fully connected (FC) and GELU layer in the feed-forward network, as shown in Figure 5. This is to eliminate the fixed-size position encoding.

The construction of feature maps with different resolutions usually loses some details of context and texture structures. To make robust our algorithm, we used channel and spatial attention modules (CAM and SAM). In general, CAM [27] captures the most meaningful and relevant information for the extracted features F_i (i = 1, 2, 3, 4) through the following process: first, it performs average pooling and max pooling on the global context; next, it applies them to shared MLP; and finally, it merges feature vectors element-wise to generate a $1 \times 1 \times C$ feature map M_{CAM}.

$$M_{CAM} = \theta(MLP(AvgPool(F_i)) + MLP(MaxPool(F_i))) \quad (2)$$

where θ represents the Sigmoid function.

Similarly, SAM [27] is also designed to extract global contextual information. It first applies average pooling and max pooling operations along the channel axis, and then it concatenates the resulting feature maps to generate a $1 \times H \times W$ feature map M_{SAM} using a $conv^{7\times 7}$ convolutional filter.

$$M_{SAM} = \theta(conv^{7\times 7}(AvgPool(F_i), MaxPool(F_i))) \qquad (3)$$

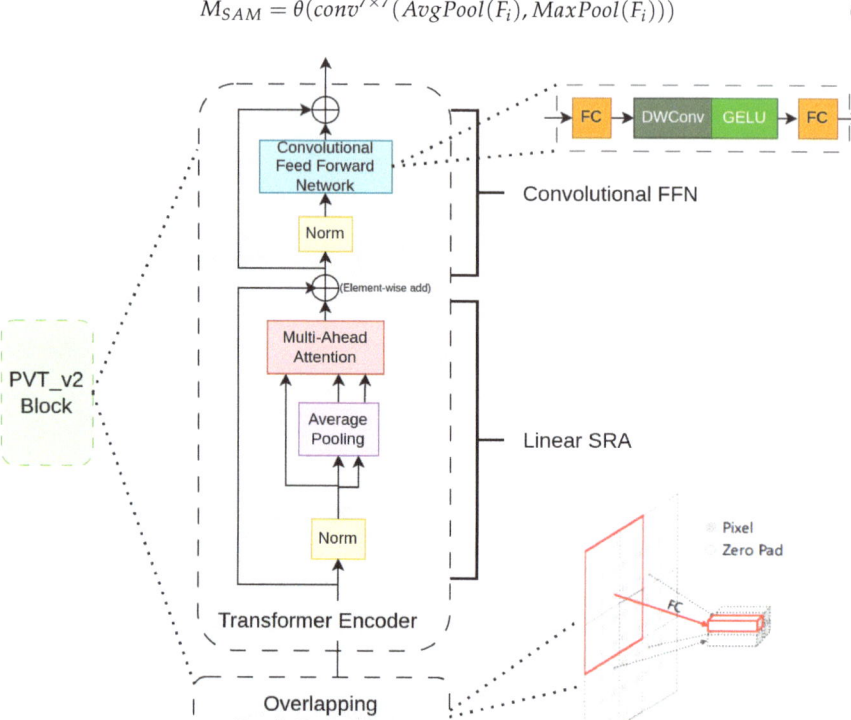

Figure 5. The details of PvTv2 Block. There are two main parts: overlapping patch embedding and Transformer encoder.

3.3. Deep Multi-Scale Feature Refinement Network

Typically, in a pyramid structure, high-level features contain rich semantic information but lack precise location details, while low-level features have more details but are filled with background noise. Combining multi-level features can lead to better feature maps. To do that, we exploit a DMFRN with different receptive fields to detect effectively small-scale and dense adjacent texts in images. The features extracted from PvTv2, denoted as F_1, F_2, F_3, and F_4, are fed as inputs to our DMFRN stage, which consists of three convolutional kernel filters with different sizes (3×3, 5×5, and 7×7). Each block in our DMFRN stage is a U-shaped module comprising two phases: upsampling and downsampling feature pyramids, which enhances the depth of the network. By simultaneously learning the irregular kernel sizes, our model can not only enlarge receptive fields but also capture multi-level information at varying levels of text in scene images. F_1^n, F_2^n, F_3^n, and F_4^n (n = 3, 5, 7) are generated by this process. Specifically, to enable the learning of relevant information in both the channel and spatial dimensions of the extracted features at each stage in the multi-scale process, we incorporate a convolutional block attention module (CBAM) [27] at each output feature, which is different from MFEN [36]. This work can boost the accuracy of the detection of dense and small text in images. Afterwards, we fuse features $\{F_1^3, F_2^3, F_3^3, F_4^3\}$, $\{F_1^5, F_2^5, F_3^5, F_4^5\}$, and $\{F_1^7, F_2^7, F_3^7, F_4^7\}$ via an element-wise sum operation, respectively, to generate F_1^f, F_2^f, F_3^f, and F_4^f. Finally, we use upsampling and a concatenating algorithm to fuse these features into a final enrich feature F.

$$F = Concate(F_1^f, F_2^f, F_3^f, F_4^f) \quad (4)$$

Then, we use F to make predictions by applying PAN [1] post-processing as in Figure 4. In this stage, we predict text instances by using similarity vectors to cluster correct text pixels with adequate text kernels.

3.4. Loss Function

The training loss L is the weighted sum of loss segmentation L_{seg} and loss detection L_{det}. To keep the weights among these losses balanced, we set it to 0.25 experimentally.

$$L = L_{seg} + 0.25 \times L_{det} \quad (5)$$

In detail, we adopt dice loss [37] to classify text/non-text in segmentation, which can be formulated as:

$$L_{seg} = \frac{1}{N} \sum_{k=1}^{N} (1 - \frac{2 \times (P_k \cap G_k)}{P_k^2 + G_k^2}) \quad (6)$$

where N denotes the number of text instance samples. P_i and G_i represent the prediction and ground truth of the kth text instances. The object containing text is labeled as 1 and non-text is labeled as 0.

Additionally, L_{det} represents the loss function of pixel aggregation (PA) in [1] that is applied to ensure that the text pixels are correctly associated with the appropriate text regions. This means that the distance between a text pixel and the kernel D_{pix_l,Ker_l} of the same lth text instance should be minimized.

$$D_{pix_l,Ker_l} = \begin{cases} \leq 6, & \text{if } pix \in (G_l - Ker_l) \\ > 6, & \text{otherwise} \end{cases} \quad (7)$$

where pix_l and Ker_l define the text pixel and text kernel of lth text sample. G_l is the ground truth of the lth text instance. The threshold of distance is set to 6 based on the PAN experiment.

4. Experiments and Results

4.1. Dataset

TotalText [16] comprises 1555 images, divided into 1255 training images and 300 testing images. It contains 11,459 text-bounding boxes, with 3936 and 971 instances of curved text in the training and testing sets, respectively. The number of annotated clockwise points varies for each text instance and is not fixed.

CTW1500 [17] contains 1000 training images and 500 testing images, each with long, dense, and curved text instances. There are 10,751 text instances in total. The scenes in the dataset are challenging and diverse, and environmental factors such as blur, low resolution, and perspective distortion are present in the images.

ICDAR 2015 [18] is a collection of incidental scene texts used in Challenge 4 on the website https://rrc.cvc.uab.es/ (accessed on 12 May 2023). The dataset contains 1000 natural images for the training process and 500 images for the testing set. It is a popular dataset for scene text detection and includes word-level text instances with multi-oriented texts, making it a useful resource for researchers in this field.

4.2. Implementations

During the pre-processing step, data augmentation techniques are applied for the training phase such as random crop, random rotation, random horizontal flip, and random scale, which help our model learn different scales and densities of features, leading to a better generalization ability during training and inference.

In the training phase, we only utilize the original training images of each dataset, as well as TotalText, CTW1500, and ICDAR 2015. The short side of the images is set to 640, 640, and 736 in the three datasets above, respectively. We use the PvTv2 backbone, which is a backbone for dense prediction, with strides of 4, 8, 16, and 32 pixels in input images. All the networks are optimized by the AdamW, https://pytorch.org/docs/stable/generated/torch.optim, accessed on 10 May 2023 [37] optimizer. Dice loss [38] and loss function in post-processing of PAN [1] are applied for optimization. Our model is implemented in Pytorch and trained scratch with a batch size of 4 on 1 GPU 2080Ti in 600 epochs for 150 k iterations. We use the "poly" learning rate strategy where the initial learning rate and power are set to 1×10^{-4} and 0.9, respectively.

During the inference stage, we set the batch size to 1 on 1 GPU and maintain the aspect ratio of the test images as in training phase. This ensures that the images are standardized and allows for consistent processing.

In scene text detection, regions of blurred text that are labeled as "DO NOT CARE" (###) in all datasets are commonly ignored. To address hard examples during training, online hard example mining (OHEM) [39] is utilized, with a negative–positive ratio typically set to 3. For ICDAR 2015, a minimal-area rectangle and polygon are fitted for each predicted text instance. The shrink ratio of the kernels is set to 0.7 on TotalText and CTW1500, and 0.5 on ICDAR 2015 to better fit the predicted text instance to the actual text region.

4.3. Evaluation Metrics

To assess the effectiveness of our proposed approach, we utilize standard metrics such as Precision (P), Recall (R), and F-measure (F). For this purpose, we consider a rectangular box containing text with a closed bounding box as True Positive (TP), while a rectangular box without any text inside is considered False Positive (FP). If there is text but no rectangular box, it is labeled as True Negative (TN), since our method failed to detect it.

In detail, Precision (P) is calculated as the ratio of the correctly identified words by our proposed method to the sum of correctly and incorrectly recognized words. It assesses the accuracy of the detected text regions. Recall (R) measures the ratio of the correct recognition to the total possible recognition at the word level. Briefly, it evaluates the ability of the method to identify all the text instances in the scene. We calculate these metrics both before and after restoration to showcase the effectiveness of our proposed approach in terms of restoring missing information, called F-measure (F). The higher the F-measure, the better the performance.

Moreover, we apply the Intersection over Union (IoU) ratio, which is used as a threshold for determining whether a predicted outcome is a True Positive (TP) or a False Positive (FP). In this paper, we set it to 0.5.

The equations of Precision, Recall, and F-measure are described below:

$$P = \frac{TP}{TP + FP} \tag{8}$$

$$R = \frac{TP}{TP + FN} \tag{9}$$

$$F = \frac{2 \times (P \times R)}{(P + R)} \tag{10}$$

4.4. Results

As presented in Tables 1–3, we compare our proposed DenseTextPVT approach with existing methods using three benchmark datasets: TotalText [16], CTW1500 [17], and ICDAR 2015 [18]. To evaluate the effectiveness of our method, we utilize the F-measure metric as in Equation (10). The results demonstrate the superior performance of our DenseTextPVT method when compared to previous algorithms.

Table 1. Quantitative detection results on TotalText. "-"/ "✓" means without/within training data. "P", "R", and "F" represent the Precision, Recall, and F-measure, respectively.

Method	Ext	P	R	F
EAST [8]	-	80.9	76.2	78.5
TextSnake [20]	✓	82.7	74.5	78.4
MSC [19]	✓	83.8	74.8	79.0
PSENet [7]	-	84.0	78.0	80.9
PAN [1]	-	88.0	79.4	83.5
TextRay [30]	-	83.5	77.9	80.6
SegLink++ [13]	✓	82.1	80.9	81.5
LOMO [14]	✓	87.6	79.3	83.3
SPCNet [40]	✓	83.0	82.8	82.9
PCR [10]	-	86.4	81.5	83.9
CRAFT [27]	✓	87.6	79.9	83.6
Ours_DenseTextPVT	-	89.4	80.1	84.7

Table 2. Quantitative detection results on CTW1500. "-"/ "✓" means without/within training data. "P", "R", and "F" represent the Precision, Recall, and F-measure, respectively.

Method	Ext	P	R	F
EAST [8]	-	78.7	49.1	60.4
PSENet [7]	-	80.6	75.6	78.0
PAN [1]	-	84.6	77.7	81.0
SegLink++ [13]	✓	82.8	79.8	81.3
LOMO [14]	✓	85.7	76.5	80.8
CT [12]	-	85.5	79.2	82.2
MSC [19]	✓	85.0	78.3	81.5
PCE [10]	-	85.3	79.8	82.4
TextRay [30]	-	82.8	80.4	81.6
DB [29]	✓	86.9	80.2	83.4
PAN [1]	✓	86.4	81.2	83.7
CRAFT [27]	✓	86.0	81.1	83.5
Xiufeng et al. [32]	✓	84.9	80.3	82.5
Ours_DenseTextPVT	-	88.3	79.8	83.9

Table 3. Quantitative detection results on ICDAR 2015. "-"/ "✓" means without/within training data. "P", "R", and "F" represent the Precision, Recall, and F-measure, respectively.

Method	Ext	P	R	F
EAST [8]	-	83.6	73.5	78.2
PSENet [7]	-	81.5	79.7	80.6
DPTNet-Tiny [41]	✓	90.3	77.4	83.3
LOMO [14]	✓	83.7	80.3	82.0
TextSnake [20]	✓	84.9	80.4	82.6
Xiufeng et al. [32]	-	85.8	79.7	82.6
MFEN [38]	-	84.5	79.7	82.0
SegLink++ [13]	✓	83.7	80.3	82.0
MSC [19]	✓	86.6	78.4	82.3
PAN [1]	-	82.9	77.8	80.3
PAN [1]	✓	84.0	81.9	82.9
Ours_DenseTextPVT	-	87.8	79.4	83.4

Our proposed method's effectiveness is demonstrated on the curved TotalText dataset (as shown in Table 1). Although the Recall (R) is lower compared to SegLink++ [13] and

SPCNet [40], our DenseTextPVT achieves significantly higher Precision (P) and F-measure (F) scores, 89.4% and 84.7%, respectively, without relying on any external dataset. The visualization in Figure 6 clearly illustrates that our DenseTextPVT is capable of accurately detecting dense curved texts.

Figure 6. The visualization samples on TotalText [16]. It is shown that our DenseTextPVT is capable of accurately detecting dense curved texts.

Similarly, our approach demonstrates strong performance on the long curved CTW1500 benchmark, achieving Precision (P) and F-measure (F) scores of 88.3% and 83.9%, respectively (as depicted in Table 2). While some algorithms, such as TextRay [30], DB [29], PAN [1], CRAFT [27], and Xiufeng et al. [32], have slightly higher Recall (R) scores, our approach outperforms the existing algorithms in terms of overall performance. Additionally, Figure 7 provides visual evidence that our proposed method accurately locates not only long curved texts but also dense adjacent text instances.

When examining the results on the ICDAR 2015 dataset (as presented in Table 3), it is observed that our DenseTextPVT does not achieve the highest Precision score, such as DPTNet-Tiny [41,42] with a score of 90.3%. There is also a slight variation in the Recall score compared to algorithms like LOMO [14], MFEN [36], TextSnake [20], Xiufeng et al. [32], SegLink++ [13], and PAN [1]. However, our proposed algorithm demonstrates an impressive overall performance with an F-measure of 83.4% when trained from scratch. The visualization in Figure 8 demonstrates the effectiveness of our method in detecting dense adjacent scene texts with multiple orientations.

Figure 7. Several visualization results on long curved text lines on CTW1500 [17]. It demonstrates the accurate localization of long curved texts with dense adjacent information by our proposed method.

Figure 8. Samples demonstrate that our DenseTextPVT algorithm is capable of effectively detecting dense multi-oriented text in scene images on ICDAR 2015 [18].

5. Conclusions

In this study, we introduced a new method, namely, DenseTextPVT, for detecting dense adjacent scene text. Our method manipulates the PvTv2 backbone with the combination of channel and spatial attention module for dense prediction, and exploits a Deep Multi-scale Feature Refinement Network to efficiently learn multi-level feature information. Afterwards, we inherit a post-processing technique in PAN to reduce overlap phenomena among text regions. Our results outperform state-of-the-art methods on several popular benchmark datasets, achieving superior F-measure scores of 84.7% on TotalText, 83.9% on CTW1500, and 83.4% on ICDAR 2015.

In the future, we plan to explore the possibility of an end-to-end framework for dense adjacent text detection. Moreover, we aim to investigate the potential of using the progressive scale expansion algorithm for segmentation mask in detection tasks, especially in benchmarks with a high density of object instances.

Author Contributions: Conceptualization, M.-T.D.; Methodology, M.-T.D.; Writing—review and editing, M.-T.D.; Supervision, D.-J.C. and G.-S.L.; Project administration, G.-S.L.; Funding acquisition, G.-S.L. All authors have read and agreed to the published version of the manuscript.

Funding: This research was supported by Basic Science Research Program through the National Research Foundation of Korea (NRF) funded by the Ministry of Education (NRF-2018R1D1A3B05049058).

Institutional Review Board Statement: Not applicable.

Informed Consent Statement: Not applicable.

Data Availability Statement: Not applicable.

Conflicts of Interest: The authors declare no conflict of interest.

Abbreviations

The following abbreviations are used in this manuscript:

PvTv2	Pyramid Vision Transformer
CAM	Channel Attention Module
SAM	Spatial Attention Module
CBAM	Convolutional Block Attention
DNNs	Deep Neural Networks
DMFRN	Deep Multi-scale Feature Refinement Network
PA	Pixel Aggregation
PAN	Pixel Aggregation Network
LinearSRA	Linear Shifted Row Attention
FC	Fully Connected
FFN	Feed Forward Network
P	Precision
R	Recall
F	F-measure

References

1. Wang, W.; Xie, E.; Song, X.; Zang, Y.; Wang, W.; Lu, T.; Yu, G.; Shen, C. Efficient and accurate arbitrary-shaped text detection with pixel aggregation network. In Proceedings of the IEEE/CVF International Conference on Computer Vision, Long Beach, CA, USA, 19–20 June 2019; pp. 8440–8449.
2. Liao, M.; Wan, Z.; Yao, C.; Chen, K.; Bai, X. Real-time scene text detection with differentiable binarization and adaptive scale fusion. *IEEE Trans. Pattern Anal. Mach. Intell.* **2022**, *45*, 919–931. [CrossRef] [PubMed]
3. Zhang, S.X.; Zhu, X.; Chen, L.; Hou, J.B.; Yin, X.C. Arbitrary Shape Text Detection via Segmentation with Probability Map. *IEEE Trans. Pattern Anal. Mach. Intell.* **2022**, *45*, 2736–2750. [CrossRef] [PubMed]
4. Tang, J.; Zhang, W.; Liu, H.; Yang, M.; Jiang, B.; Hu, G.; Bai, X. Few Could Be Better Than All: Feature Sampling and Grouping for Scene Text Detection. In Proceedings of the IEEE/CVF Conference on Computer Vision and Pattern Recognition, Virtual, 20–25 June 2022; pp. 4563–4572.
5. Yin, X.-C.; Yin, X.; Huang, K.; Hao, H.-W. Robust text detection in natural scene images. *IEEE Trans. Pattern Anal. Mach. Intell.* **2013**, *36*, 970–983.
6. Chen, Z.; Wang, J.; Wang, W.; Chen, G.; Xie, E.; Luo, P.; Lu, T. FAST: Searching for a Faster Arbitrarily-Shaped Text Detector with Minimalist Kernel Representation. *arXiv* **2021**, arXiv:2111.02394.
7. Wang, W.; Xie, E.; Li, X.; Hou, W.; Lu, T.; Yu, G.; Shao, S. Shape robust text detection with progressive scale expansion network. In Proceedings of the IEEE/CVF Conference on Computer Vision and Pattern Recognition, Long Beach, CA, USA, 19–20 June 2019; pp. 9336–9345.
8. Zhou, X.; Yao, C.; Wen, H.; Wang, Y.; Zhou, S.; He, W.; Liang, J. East: An efficient and accurate scene text detector. In Proceedings of the IEEE Conference on Computer Vision and Pattern Recognition, Honolulu, HI, USA, 21–26 July 2017; pp. 5551–5560.
9. Zhou, B.; Khosla, A.; Lapedriza, A.; Oliva, A.; Torralba, A. Learning deep features for discriminative localization. In Proceedings of the IEEE Conference on Computer Vision and Pattern Recognition, Las Vegas, NV, USA, 26 June–1 July 2016; pp. 2921–2929.
10. Dai, P.; Zhang, S.; Zhang, H.; Cao, X. Progressive contour regression for arbitrary-shape scene text detection. In Proceedings of the IEEE/CVF Conference on Computer Vision and Pattern Recognition, Virtual, 20–25 June 2021; pp. 7393–7402.

11. Baek, Y.; Lee, B.; Han, D.; Yun, S.; Lee, H. Character region awareness for text detection. In Proceedings of the IEEE/CVF Conference on Computer Vision and Pattern Recognition, Long Beach, CA, USA, 15–20 June 2019; pp. 9365–9374.
12. Sheng, T.; Chen, J.; Lian, Z. Centripetaltext: An efficient text instance representation for scene text detection. *Adv. Neural Inf. Process. Syst.* **2021**, *34*, 335–346.
13. Shi, B.; Xiang, B.; Serge, B. Detecting oriented text in natural images by linking segments. In Proceedings of the IEEE Conference on Computer Vision and Pattern Recognition, Honolulu, HI, USA, 21–26 July 2017.
14. Zhang, C.; Borong, L.; Zuming, H.; Mengyi, E.; Junyu, H.; Errui, D.; Xinghao, D. Look more than once: An accurate detector for text of arbitrary shapes. In Proceedings of the IEEE/CVF Conference on Computer Vision and Pattern Recognition, Long Beach, CA, USA, 15–20 June 2019; pp. 10552–10561.
15. He, W.; Zhang, X.-Y.; Yin, F.; Liu, C.-L. Deep direct regression for multi-oriented scene text detection. In Proceedings of the 2017 IEEE International Conference on Computer Vision (ICCV), Venice, Italy, 22–29 October 2017; pp. 745–753.
16. Kheng, C.C.; Chan, C.S. TotalText: A comprehensive dataset for scene text detection and recognition. In *Proceedings of the 2017 14th IAPR International Conference on Document Analysis and Recognition (ICDAR), Kyoto, Japan, 9–15 November 2017*; IEEE: Piscataway, NJ, USA, 2017; Volume 1.
17. Liu, Y.; Jin, L.; Zhang, S.; Zhang, S. Detecting curve text in the wild: New dataset and new solution. *arXiv* **2017**, arXiv:1712.02170.
18. Karatzas, D.; Gomez-Bigorda, L.; Nicolaou, A.; Ghosh, S.; Bagdanov, A.; Iwamura, M.; Matas, J.; Neumann, L.; Chandrasekhar, V.R.; Lu, S.; et al. ICDAR 2015 competition on robust reading. In Proceedings of the 2015 13th International Conference on Document Analysis and Recognition (ICDAR), Tunis, Tunisia, 23–26 August 2015; pp. 1156–1160.
19. Xue, C.; Shijian, L.; Wei, Z. MSR: Multi-scale shape regression for scene text detection. *arXiv* **2019**, arXiv:1901.02596.
20. Long, S.; Jiaqiang, R.; Wenjie, Z.; Xin, H.; Wenhao, W.; Cong, Y. Textsnake: A flexible representation for detecting text of arbitrary shapes. In Proceedings of the European Conference on Computer Vision (ECCV), Munich, Germany, 8–14 September 2018; pp. 20–36.
21. Vaswani, A.; Shazeer, N.; Parmar, N.; Uszkoreit, J.; Jones, L.; Gomez, A.N.; Kaiser, Ł.; Polosukhin, I. Attention is all you need. *Adv. Neural Inf. Process. Syst.* **2017**, *130*, 5998–6008.
22. Carion, N.; Massa, F.; Synnaeve, G.; Usunier, N.; Kirillov, A.; Zagoruyko, S. End-to-end object detection with transformers. In *Computer Vision–ECCV 2020: 16th European Conference, Glasgow, UK, 23–28 August 2020*; Springer International Publishing: Berlin/Heidelberg, Germany, 2020; pp. 213–229.
23. Ze, L.; Lin, Y.; Cao, Y.; Hu, H.; Wei, Y.; Zhang, Z.; Lin, S.; Guo, B. Swin transformer: Hierarchical vision transformer using shifted windows. In Proceedings of the IEEE/CVF Conference on Computer Vision and Pattern Recognition, Virtual, 20–25 June 2021; pp. 10012–10022.
24. Dosovitskiy, A.; Beyer, L.; Kolesnikov, A.; Weissenborn, D.; Zhai, X.; Unterthiner, T.; Dehghani, M.; Minderer, M.; Heigold, G.; Gelly, S.; et al. An image is worth 16 × 16 words: Transformers for image recognition at scale. In Proceedings of the International Conference on Learning Representations, Virtual Event, 3–7 May 2021.
25. Hugo, T.; Cord, M.; Douze, M.; Massa, F.; Sablayrolles, A.; Jégou, H. Training data-efficient image transformers & distillation through attention. *Int. Conf. Mach. Learn.* **2021**, *139*, 10347–10357.
26. Wang, W.; Xie, E.; Li, X.; Fan, D.; Song, K.; Liang, D.; Lu, T.; Luo, P.; Shao, L. Pvt v2: Improved baselines with pyramid vision transformer. *Comput. Vis. Media* **2022**, *8*, 415–424. [CrossRef]
27. Woo, S.; Park, J.; Lee, J.-Y.; Kweon, I.S. Cbam: Convolutional block attention module. In Proceedings of the European Conference on Computer Vision (ECCV), Munich, Germany, 8–14 September 2018; pp. 3–19.
28. Zhu, Y.; Chen, J.; Liang, C.; Kuang, Z.; Jin, L.; Zhang, W. Fourier contour embedding for arbitrary-shaped text detection. In Proceedings of the IEEE/CVF Conference on Computer Vision and Pattern Recognition, Virtual, 20–25 June 2021; pp. 3123–3131.
29. Liao, M.; Wan, Z.; Yao, C.; Chen, K.; Bai, X. Real-time scene text detection with differentiable binarization. In Proceedings of the AAAI Conference on Artificial Intelligence, New York, NY, USA, 7–12 February 2020; Volume 34, pp. 11474–11481.
30. Wang, F.; Chen, Y.; Wu, F.; Li, X. Textray: Contour-based geometric modeling for arbitrary-shaped scene text detection. In Proceedings of the 28th ACM International Conference on Multimedia, Seattle, WA, USA, 12–16 October 2020; pp. 111–119.
31. Dang, Q.-V.; Lee, G.-S. Document image binarization with stroke boundary feature guided network. *IEEE Access* **2021**, *9*, 36924–36936. [CrossRef]
32. Jiang, X.; Xu, S.; Zhang, S.; Cao, S. Arbitrary-shaped text detection with adaptive text region representation. *IEEE Access* **2020**, *8*, 102106–102118. [CrossRef]
33. Zobeir, R.; Naiel, M.A.; Younes, G.; Wardell, S.; Zelek, J.S. Transformer-based text detection in the wild. In Proceedings of the IEEE/CVF Conference on Computer Vision and Pattern Recognition, Virtual, 20–25 June 2021; pp. 3162–3171.
34. Zobeir, R.; Younes, G.; Zelek, J. Arbitrary shape text detection using transformers. In Proceedings of the 2022 26th International Conference on Pattern Recognition (ICPR), Montreal, QC, Canada, 21–25 August 2022; pp. 3238–3245.
35. Wang, W.; Xie, E.; Li, X.; Fan, D.; Song, K.; Liang, D.; Lu, T.; Luo, P.; Shao, L. Pyramid vision transformer: A versatile backbone for dense prediction without convolutions. In Proceedings of the IEEE/CVF Conference on Computer Vision and Pattern Recognition, Virtual, 20–25 June 2021; pp. 568–578.
36. Dinh, M.-T.; Lee, G.-S. Arbitrary-shaped Scene Text Detection based on Multi-scale Feature Enhancement Network. In Proceedings of the Korean Information Science Society Conference, Jeju, Korea, 29 June–1 July 2022.

37. Sudre, C.H.; Li, W.; Vercauteren, T.; Ourselin, S.; Cardoso, M.J. Generalised dice overlap as a deep learning loss function for highly unbalanced segmentations. In *Deep Learning in Medical Image Analysis and Multimodal Learning for Clinical Decision Support: Third International Workshop, DLMIA 2017, and 7th International Workshop, ML-CDS 2017, Held in Conjunction with MICCAI 2017, Québec City, QC, Canada, 14 September 2017*; Springer International Publishing: Berlin/Heidelberg, Germany, 2017; pp. 240–248.
38. Loshchilov, I.; Hutter, F. Decoupled weight decay regularization. *arXiv* **2017**, arXiv:1711.05101.
39. Shrivastava, A.; Gupta, A.; Girshick, R. Training region-based object detectors with online hard example mining. In Proceedings of the IEEE Conference on Computer Vision and Pattern Recognition, Las Vegas, NV, USA, 26 June–1 July 2016; pp. 761–769.
40. Enze, X.; Zang, Y.; Shao, S.; Yu, G.; Yao, C.; Li, G. Scene text detection with supervised pyramid context network. In Proceedings of the AAAI Conference on Artificial Intelligence, Honolulu, HI, USA, 27 January–1 February 2019; Volume 33, pp. 9038–9045.
41. Lin, J.; Jiang, J.; Yan, Y.; Guo, C.; Wang, H.; Liu, W.; Wang, H. DPTNet: A Dual-Path Transformer Architecture for Scene Text Detection. *arXiv* **2022**, arXiv:2208.09878.
42. Deng, D.; Liu, H.; Li, X.; Cai, D. Pixellink: Detecting scene text via instance segmentation. In Proceedings of the AAAI Conference on Artificial Intelligence, New Orleans, LA, USA, 2–3 February 2018.

Disclaimer/Publisher's Note: The statements, opinions and data contained in all publications are solely those of the individual author(s) and contributor(s) and not of MDPI and/or the editor(s). MDPI and/or the editor(s) disclaim responsibility for any injury to people or property resulting from any ideas, methods, instructions or products referred to in the content.

Article

Traversable Region Detection and Tracking for a Sparse 3D Laser Scanner for Off-Road Environments Using Range Images

Jhonghyun An

School of Computing, Gachon University, Seongnam-si 1332, Gyeonggi-do, Republic of Korea; jhonghyun@gachon.ac.kr

Abstract: This study proposes a method for detecting and tracking traversable regions in off-road conditions for unmanned ground vehicles (UGVs). Off-road conditions, such as rough terrain or fields, present significant challenges for UGV navigation, and detecting and tracking traversable regions is essential to ensure safe and efficient operation. Using a 3D laser scanner and range-image-based approach, a method is proposed for detecting traversable regions under off-road conditions; this is followed by a Bayesian fusion algorithm for tracking the traversable regions in consecutive frames. Our range-image-based traversable-region-detection approach enables efficient processing of point cloud data from a 3D laser scanner, allowing the identification of traversable areas that are safe for the unmanned ground vehicle to drive on. The effectiveness of the proposed method was demonstrated using real-world data collected during UGV operations on rough terrain, highlighting its potential as a solution for improving UGV navigation capabilities in challenging environments.

Keywords: 3D; laser scanner; LIDAR; traversability; traversable region; detection; tracking; autonomous driving; unmanned ground vehicle (UGV); off-road; range image; Bayesian fusion

1. Introduction

Unmanned ground vehicles (UGVs) are mission-critical assets designed to operate in hazardous or harsh environments, where human intervention may be impractical, too dangerous, or infeasible, such as complex and hostile environments, characterized by rugged terrain, extreme weather conditions, or enemy threats. They are essential for various military operations, including reconnaissance, surveillance, target acquisition, and weapon delivery.

The detection of clear road boundaries and relatively flat surfaces is critical for commercial autonomous vehicles, which primarily operate on roads. Road curbs, lanes, and other structures are typically used to determine traversable areas [1–7]. Elevation mapping is also used to detect flat surfaces and estimate the location of vehicles relative to the ground [8–17]. Obstacle detection is another critical aspect, and previous methods have focused on identifying obstacles, to determine free spaces that are suitable for vehicle traversal [18–22]. These methods identify the areas where no physical obstacles directly impede the progress of the ego vehicle as traversable [23–30].

As noted in [31], the operating environment of a UGV is significantly different from that of a commercial autonomous vehicle. In such an environment, road boundaries are not always clear, and surfaces are often uneven under off-road conditions, making determining traversable regions challenging. In addition, the terrain is not paved, and road elevations can vary significantly. In contrast to the case in commercial autonomous driving, detecting structures such as road boundaries or relatively flat terrain to identify traversable regions is not always possible for a UGV. For a UGV, traversable areas refer to locations where the vehicle can move from any starting point to a target destination, without encountering obstacles or topographical restrictions. Therefore, a different approach is required to detect traversable regions for UGVs compared to that for commercial autonomous vehicles.

Citation: An, J. Traversable Region Detection and Tracking for a Sparse 3D Laser Scanner for Off-Road Environments Using Range Images. *Sensors* **2023**, *23*, 5898. https://doi.org/10.3390/s23135898

Academic Editors: Man Qi and Matteo Dunnhofer

Received: 16 May 2023
Revised: 22 June 2023
Accepted: 24 June 2023
Published: 25 June 2023

Copyright: © 2023 by the author. Licensee MDPI, Basel, Switzerland. This article is an open access article distributed under the terms and conditions of the Creative Commons Attribution (CC BY) license (https:// creativecommons.org/licenses/by/ 4.0/).

Traditionally, image sensors have been the primary type of sensor used to detect traversable regions. However, with the recent development of deep learning models, recognizing various scenes based on texture and other features has become possible [32–37]. These methods are a type of image analysis in which pixels corresponding to a particular object are classified into classes of objects and the target object is divided into meaningful units. In this approach, ground areas are classified based on their texture, such as mud, grass, or bushes. Traversable regions are then recognized as specific classes, based on the results of the classification process. However, these methods only identify areas that contrast with the surrounding environment in color images and ground texture. They do not provide information about whether driving in these areas is possible. In addition, this is limited in natural environments, where the illumination intensity changes rapidly due to the limitations of image sensors.

To address the limitations of current approaches to determining traversable regions, an alternative methodology employing 3D laser scanner technology has been proposed [38–40]. The use of 3D laser scanners enables highly precise and accurate measurements of the surrounding environment, including the ground surface, without being affected by texture or color. By analyzing the resulting point cloud data, specific criteria such as surface roughness, slope, or height differences can be utilized to identify traversable regions. This approach provides a reliable and efficient means of determining traversable regions, particularly in off-road conditions where road boundaries are not clearly defined and surfaces are uneven.

Therefore, this paper proposes a real-time traversable region-detection method using a 3D laser scanner. For real-time processing, a method for converting the 3D point cloud data into 2D images is used [12,13]. Previous studies have demonstrated the effectiveness of these approaches for accurately detecting traversable regions for paved roads in real-time processing. However, for unpaved and rough terrain, a different approach is necessary [41]. In this context, [41] processed point cloud data into a 2D range image and generated, not only vertical angle images, but also horizontal angle images, to detect the traversable regions for ego vehicles. Consequently, this method exhibited a commendable performance even in open-field environments. However, in environments where the vehicle's pose undergoes significant changes due to the terrain, relying on a single frame for detecting traversable regions is inadequate. To overcome this challenge, a traversable region tracking method is proposed, which accumulates the detection results from previous frames. The confidence value of each pixel in the range image is leveraged to model its traversability. These confidences are then accumulated over consecutive frames using the Bayesian fusion method [42,43]. The contributions of this work can be summarized as follows:

- An effective traversable-region-detection method using a 3D laser scanner is proposed. To deal with a large amount of 3D point-cloud data, we used range images with each pixel indicating the range data of a specific space. Then, each pixel and the adjacent pixels are searched based on the vertical and horizontal inclination angles of the ground;
- A traversable-region-tracking algorithm was developed to integrate the previous detection results, to prevent detrimental effects from an unexpected pose of the vehicle. By modeling the range data of each pixel as a probability value, the traversability of the previous and current pixels in the traversable region detection results can be fused using the Bayesian fusion method.

The rest of this paper is structured as follows: In Section 2, we discuss related works in the field. Section 3 outlines the theoretical formulation of the proposed method. Moving on to Section 4, we describe the dataset configuration and the data logging system, and present the experimental results. Finally, in Section 5, we provide a summary of the paper.

2. Related Work

Three-dimensional laser scanner technology has been the subject of considerable attention in relation to autonomous vehicles. These scanners are widely used for environmental recognition purposes, such as object detection, map building, and route finding. In particular, the problems associated with map building can be divided into structure detection problems, such as surrounding buildings, and traversable region detection problems. When a 3D laser scanner is used for traversable region detection, many methods are available, depending on how the raw data are processed.

Thrun et al. [44] introduced a grid-based approach that divides grid cells into the ground and non-ground cells based on the height differences between points inside the cell. Moosmann et al. [8] used raw data to create a graph obtained from triangulation and utilized the concept of local convexity between two neighboring nodes in the graph. If the center point of the two connected nodes lies below the surface of the points, these points are clustered on the same ground. Himmelsbach et al. [9] represented all raw point-cloud data in a 2D x-y plane and divided this into discrete pieces of a circle. They then created each piece as a bin and identified inliers as traversable regions. Douillard et al. [45] proposed GP-INSAC, a Gaussian-process-based iterative method that classifies all raw points as the ground, with small variations in height relative to the mean of the Gaussian distribution. Chen et al. [11] presented individual raw scan data instances as a circular polar grid divided into segments. To distinguish the ground points, Chen applied a 1D Gaussian-process-based regression method for each segment, similarly to with a 1D bin. Babahajiani et al. [15] and Lai and Fox [46] used prior ground knowledge to dictate a set of all points and applied common plane fitting techniques, such as the random sample consensus (RANSAC) algorithm. Zermas et al. [16] proposed a multi-model plane fitting algorithm that divides raw data into a number of segments along the horizontal direction. The performances of these methods in terms of detecting traversable regions using all points have been verified for various specializations over many years. In addition, for improved time efficiency, computational optimization has been performed, according to the driving environment. However, searching for all the points remains inefficient with regard to time. In addition, certain model-based regression methods cannot sufficiently represent the actual surface, because the ground point does not form a perfect plane, and significant noise is generated over long distances because of raw data from 3D laser scanner measurements.

Alternatively, the raw 3D point data of the laser scanner can be projected onto a cylinder whose axis is the scanner's axis of rotation, as opposed to using the raw data in isolation. This projection creates a range image in which the pixel value corresponds to the distance measurement. Basic research work on this aspect was conducted by Hoover et al. [47], and subsequent key approaches to local surface fitting and clustering are still being applied today. Based on this method, Bogoslavsky and Stachniss [12,13] proposed an efficient ground-search method based on range images. The ground slope is calculated using the distance between each pixel in the range image. If there is a similar slope between each pixel, the pixel is detected as the ground. However, this method assumes that the ground is a well-paved flat area in a city and it employs a means of fixing changes in the slope values. In addition, because detection is performed every moment, the information detected in the previous frame is not used.

In addition, image sensors are commonly used for detecting traversable regions. Recent advancements in deep learning models have made it possible to recognize different scenes based on texture and other features [32–37]. These methods involve image analysis techniques that classify pixels into object classes and the target objects into meaningful units. S. Palazzo et al. [35] introduced a deep-learning-based approach that estimates and predicts the traversability score of different routes captured by an onboard RGB camera. S. Hosseinpoor et al. [36] presented a method based on semantic segmentation, where they adapted Deeplabv3+ using only an RGB camera. They fine-tuned a pretrained network originally trained on Cityscapes with their own dataset. T. Leung et al. [37] proposed a hybrid framework for analyzing traversability that utilizes both RGB cameras and LiDAR.

An RGB camera is used to gather semantic information, identifying different types of terrain, such as concrete or grass. Meanwhile, the LiDAR sensor provides geometric information, by creating an elevation map of the surrounding terrain.

3. Proposed Method

The term "traversable regions" refers to ground areas that allow a specific vehicle to move toward a target point without any obstacles. However, unmanned ground vehicles (UGVs) face more complex challenges. These include navigating through plants and grassy areas that can obstruct their movement, as well as the absence of clear structures that help distinguish between drivable and non-drivable terrain. Additionally, the presence of varying slopes makes it difficult to visually identify traversable areas. Hence, it is crucial to develop a new method for identifying traversable areas that are specifically tailored for UGVs, distinctly from the approaches used for commercial vehicles. Therefore, this paper presents a new method that uses a 3D laser scanner as the primary sensor for detecting and tracking traversable regions, as shown in Figure 1.

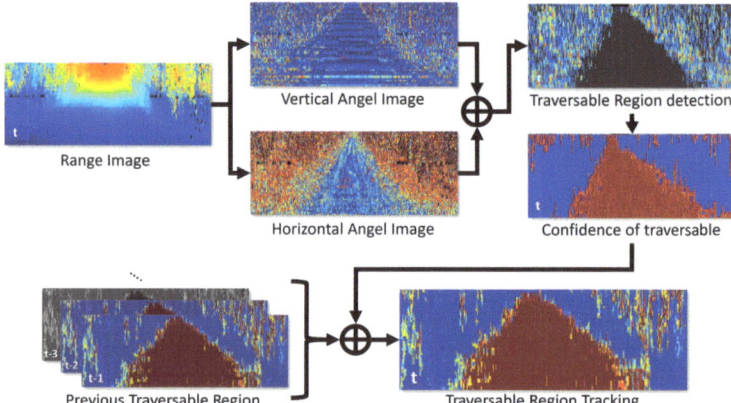

Figure 1. Illustration of the proposed method. In the first step, vertical and horizontal angle images are generated from the range image. Subsequently, the traversable region is detected, and the pixel-wise confidence is calculated based on this. Using this information, the traversable region is tracked using the proposed Bayesian fusion method.

3.1. Range-Image-Based Traversable Region Detection

3.1.1. Range Image

A 3D laser scanner is a device that measures the spatial information of objects or environments using laser beams, while rotating 360 degrees. It emits laser beams onto the target surfaces and measures the time it takes for the laser to bounce back to the scanner, allowing for the calculation of distances. Moreover, by maintaining a consistent vertical angle, the scanner ensures that the measurements are taken from the same reference plane throughout the entire rotation, as shown in Figure 2. These planes are called layers, i.e., 16, 32, 64, or 128 for Ouster scanners. As the scanner rotates, it emits laser beams and records the reflections from the surrounding objects at various angles. Therefore, the point cloud set $\mathbf{Z}_t$ measured at time t is defined as follows:

$$\mathbf{Z}_t = \left\{ z_t^1, z_t^2, z_t^3 \ldots z_t^P \right\} \quad (1)$$

$$z_t^k = \left(r_t^k, \theta_t^k, \ell_t^k \right) \quad (2)$$

where the total number of measurement points is denoted by P; and r_t^k, θ_t^k, and ℓ_t^k denote the kth distance, bearing, and layer at time t of the measurements, respectively. As such, a laser scanner provides individual layers per laser beam as raw data, along with timestamps and bearings. This facilitates the direct conversion of the raw point cloud into range images. Therefore, the range image I is a function.

$$I : U_I \to [0, r_{max}] \tag{3}$$

$$U_I = [[0; m-1] \times [0; n-1]] \tag{4}$$

are the pixels of the range image, where m is the number of rows in the range image defined by the number of layers in the vertical direction and n is the number of columns given by the range readings per 360-degree revolution of the scanner, i.e., $n = 360/\Delta\theta_h$ respectively. $\Delta\theta_h$ is the horizontal resolution of the laser scanner. Therefore, a specific pixel $I(i,j)$ has a distance value $r_{i,j}$ corresponding to the space. Therefore, the amount of data to be processed is reduced by the resolution of the range image in the form of millions of point clouds.

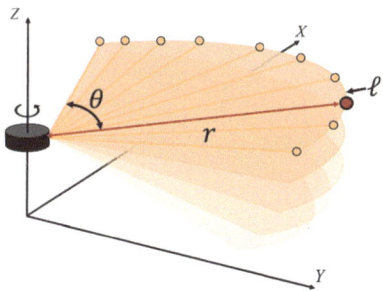

Figure 2. Schematic of a mechanical pulsed-time-of-flight (ToF) laser scanner.

3.1.2. Traversable Region Detection

Using the range image, we detect traversable regions by considering the vertical and horizontal inclinations of the ground. Vertical inclination refers to the slope of a path that enables the ego vehicle to move in its direction of travel, whereas horizontal inclination refers to the slope of a path that allows lateral movement based on the vehicle's direction of travel. This enables the identification of flat ground suitable for vehicle movement. To achieve this, we first calculate the angles between consecutive rows in the range image, as follows:

$$\alpha_{r,c} = \text{atan2}(\Delta x_\alpha, \Delta z_\alpha), \tag{5}$$

where

$$\begin{aligned}\Delta x_\alpha &= |I_{r-1,c} \sin\theta_{r-1} - I_{r,c} \sin\theta_r|, \\ \Delta z_\alpha &= |I_{r-1,c} \cos\theta_{r-1} - I_{r,c} \cos\theta_r|.\end{aligned} \tag{6}$$

Additionally, to determine the horizontal inclination of the ground, the angles between consecutive columns are calculated using the range image, as follows:

$$\beta_{r,c} = \text{atan2}(\Delta x_\beta, \Delta y_\beta), \tag{7}$$

where

$$\begin{aligned}\Delta x_\beta &= |I_{r,c} \cos\Delta\theta_c - I_{r,c-1}|, \\ \Delta y_\beta &= |I_{r,c} \sin\Delta\theta_c|.\end{aligned} \tag{8}$$

Consequently, we can treat all stacks of vertical and horizontal inclination angles as range images, so we define them as angle images M_α and M_β:

$$M : U_M \to \left[-\frac{pi}{2}, +\frac{pi}{2}\right], \tag{9}$$

where
$$U_{M_\alpha} = [[0; m-2] \times [0; n-1]], \qquad (10)$$

$$U_{M_\beta} = [[0; m-1] \times [0; n-2]]. \qquad (11)$$

Hence, Figure 3 illustrates the alpha angle and beta angle used for calculating the vertical and horizontal angle images. However, a mechanically rotating-type 3D laser scanner generates a significant number of outliers in the range measurements, which can affect the calculation of the angle α. The alpha angle represents the slope of the ground at the sensor's location in the direction of the vehicle's movement. As a result, it is heavily influenced by the vehicle's attitude. This is discussed in more detail in the work by Leonard et al. [48]. To address this issue, Bogoslavsky et al. applied a Savitzky–Golay filter to every column of the vertical angle range images, to fit a local polynomial of a given window size to the data [13]. This filter is a smoothing method that constructs a polynomial regression model for a short signal interval within a window, as applied to a continuous signal [49]. However, column-wise smoothing using a Savitzky–Golay filter is only suitable for flat paved roads and not unpaved roads with varying heights and curves. Therefore, as shown in Figure 4, the angle image exhibits salt-and-pepper noise owing to the condition of unspecified road surfaces.

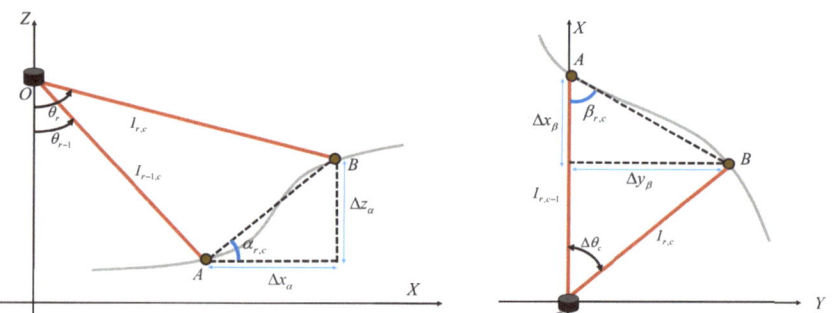

Figure 3. This figure illustrates the α and β angles used to compute the vertical and horizontal angle images. The red lines represent adjacent laser beams. The x-axis denotes the vehicle's forward direction, and the z-axis represents the vertical direction perpendicular to the ground.

In order to remove outliers from the range measurements obtained by the 3D laser, a median filter was employed, as depicted in Figure 4. To account for a specific pixel and its surrounding pixels in the vertical and horizontal directions, a 5×5 kernel was used. The actual and range images are shown in the first row, while the second row shows the vertical and horizontal angle images calculated based on the angle difference between consecutive range images in the same frame, respectively. The images obtained after noise removal using a median filter are shown in the third row. Vertical angle images highlight vertical obstacles, such as vehicles, whereas horizontal angle images provide details regarding the road surface conditions, for distinguishing horizontal obstacles, such as curbs.

Finally, we propose the detection of traversable regions using the noise-removed vertical and horizontal angle images. To achieve this, the breadth-first search (BFS) method is employed from the lowest row of the range image, which is considered the ground closest to the ego vehicle, and the adjacent pixels near the four neighborhood pixels are searched. However, when calculating the angle difference between a specific pixel (i, j) and its adjacent pixel $(i \pm 1, j \pm 1)$, both the vertical angle image M_α within a difference $\Delta \alpha$ and horizontal angle image M_β within a difference $\Delta \beta$ are considered, to determine if they fall within a specific angle range. Using this process, the pixels with small differences between the vertical and horizontal angles are identified as the traversable regions. Therefore, the result of traversable region detection is M_T.

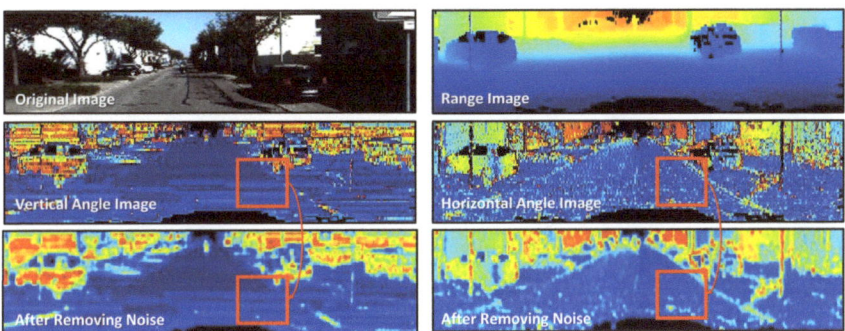

Figure 4. Outlier removal results in the range measurements of a 3D laser using a median filter. the second row shows the vertical and horizontal angle images, respectively. The images obtained after noise removal using a median filter are shown in the third row.

3.2. Probabilistic Traversable Region Tracking

In this section, we propose a probabilistic traversable region tracking method that utilizes Bayesian fusion. First, the confidence of the traversable region detected in the current frame is calculated in pixel units of the range image. Then, the confidence of each pixel is converted into class information, indicating whether the traverse is possible or not; this is also calculated in the pixel units of the range image. Finally, by applying Bayesian fusion to the accumulated class information, the traversable area can be tracked for each individual pixel within the range image, as shown in Figure 5.

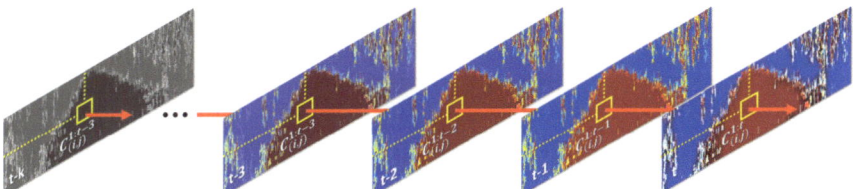

Figure 5. This figure presents an illustration of the pixel−wise tracking of the traversable region based on Bayesian fusion. It showcases the updated result of $\mathbf{C}_{i,j}^{1:t}$ for the pixel (i, j) over time. $\mathbf{C}_{i,j}^{1:t}$ denotes the sequence of the traversable region tracking results in a range image up to time t.

3.2.1. Confidence of Traversability

In the previous section, we detected the traversable region based on an angle image. However, to convert a traversable region in a single frame into a binary class of pixel units, the region must be converted into a probabilistic expression based on a specific value. The probability that a pixel belongs to a traversable region depends on the slope of the ground. The higher the slope of the ground, the lower the probability that it is a traversable region, and vice versa. Therefore, the confidence that a traverse is possible is calculated based on the difference in the height of the ground between a specific pixel and its surrounding pixels. To calculate the confidence of the traversable region for each pixel in the range image in the current frame, the following steps are performed:

$$p(M(i,j)) = 1 - \frac{1}{1 + e^{-k(x-\varphi)}}, \tag{12}$$

where x denotes the average angular difference between a particular pixel i, j in the range image and an adjacent pixel $i \pm 1, j \pm 1$. In addition, φ and k are hyperparameters, where φ is the midpoint for slope determination and k is the logistic growth rate of the range of drivable inclination. Therefore, when the difference between a specific pixel (i, j) and its adjacent pixels $(i \pm 1, j \pm 1)$ in the vertical and horizontal angle image M_α, M_β is within a certain angle range $\Delta\alpha$ and $\Delta\beta$, the ground confidence is modeled as increasing. In this study, all the pixels in the range images were defined as binary classes, i.e., either traversable or non-traversable regions.

$$c_{i,j} = \{T, NT\}. \tag{13}$$

Therefore, the class confidence of the traversable region predicted by the current range image can be expressed as

$$p(c = k|M(i,j)) = p_k, \text{ where } \sum_{k \in \{T, NT\}} p_k = 1, \tag{14}$$

where c denotes the class of traversability, k is one of the possible classes that c can take, and $M(i,j)$ is a specific pixel in the range image. Finally, the class is predicted using

$$class(M(i,j)) = \arg\max_{k \in \{T, NT\}} p_k. \tag{15}$$

3.2.2. Bayesian Fusion in a Sequence

In this section, we propose a Bayesian fusion method for tracking the class information in a sequence of range images. $\mathbf{C}_{i,j}^{1:t}$ is a sequence of the traversable region classes detected in a range image at the time up to t. We determine the class of the traversability sequence of the specific pixel of $\mathbf{C}_{i,j}^{1:t}$ using maximum a posteriori (MAP) estimation. That is, the class of the sequence of the traversability is predicted using

$$class\left(\mathbf{C}_{i,j}^{1:t}\right) = \arg\max_{k \in \{T, N\}} p\left(c_{i,j} = k \middle| \mathbf{C}_{i,j}^{1:t}\right). \tag{16}$$

Assuming that t denotes the current time, $\mathbf{C}_{i,j}^{1:t}$ is divided into the current measurement $\mathbf{C}_{i,j}^{t}$ and all previous measurements $\mathbf{C}_{i,j}^{1:t-1}$ based on the Markov property,

$$p\left(c = k|\mathbf{C}^{1:t}\right) = p\left(c = k|\mathbf{C}^{t}, \mathbf{C}^{1:t-1}\right). \tag{17}$$

This indicates that the measurements up to time $t-1$ have no impact on the measurements at time t. This is because the traversability has been accumulated for each individual pixel within the range image using the data up to time $t-1$. However, it is important to consider that when the traversable probability from time $t-1$ is propagated to time t, it influences the pixel-level probability values. Therefore, the probability of a traversable region in a specific pixel of a range image can be rewritten as follows using the Bayes rule:

$$p\left(c = k|\mathbf{C}^{1:t}\right) = \frac{p\left(\mathbf{C}^{t}|c = k, \mathbf{C}^{1:t-1}\right)p\left(c = k|\mathbf{C}^{1:t-1}\right)}{p(\mathbf{C}^{t}|\mathbf{C}^{1:t-1})}. \tag{18}$$

This is based on the Bayes rule, $p(x|y) = p(y|x) \cdot p(x)/P(y)$. As the previous measurements $\mathbf{C}^{1:t-1}$ do not affect the current measurement $\mathbf{C}^{t}$ and the current measurement is conditioned on $c = k$, we can obtain $p(\mathbf{C}^{t}|c = k, \mathbf{C}^{1:t-1})$ as $p(\mathbf{C}^{t}|c = k)$. For the sake of simplicity, the subscripts i and j, which indicate the coordinates of a specific pixel, are omitted in the following formulas. Subsequently, by applying the Bayes rule,

$$p\left(c = k|\mathbf{C}^{1:t}\right) = \frac{p(c = k|\mathbf{C}^{t})p(\mathbf{C}^{t})}{p(c = k)} \frac{p\left(c = k|\mathbf{C}^{1:t-1}\right)}{p(\mathbf{C}^{t}|\mathbf{C}^{1:t-1})}. \tag{19}$$

At this time, the sum of the class confidence values of a sequence for the binary class is 1; thus, dividing Equation (19) by the sum of the class confidence values of a sequence is mathematically equivalent to

$$p\left(c = k | \mathbf{C}^{1:t}\right) = \frac{p\left(c = k | \mathbf{C}^{1:t}\right)}{\sum_{k' \in \{T,N\}} p(c = k' | \mathbf{C}^{1:t})}. \tag{20}$$

Substituting Equation (19) into Equation (20) yields

$$p\left(c = k | \mathbf{C}^{1:t}\right) = \frac{\frac{p(c=k|\mathbf{C}^t)p(\mathbf{C}^t)}{p(c=k)} \frac{p(c=k|\mathbf{C}^{1:t-1})}{p(\mathbf{C}^t|\mathbf{C}^{1:t-1})}}{\sum_{k' \in \{T,N\}} \frac{p(c=k'|\mathbf{C}^t)p(\mathbf{C}^t)}{p(c=k')} \frac{p(c=k'|\mathbf{C}^{1:t-1})}{p(\mathbf{C}^t|\mathbf{C}^{1:t-1})}}. \tag{21}$$

By canceling each term, we obtain the following equation:

$$p\left(c = k | \mathbf{C}^{1:t}\right) = \frac{p(c = k | \mathbf{C}^{1:t-1}) p(c = k | \mathbf{C}^t) p(c = k')}{\sum_{k' \in \{T,N\}} \{p(c = k' | \mathbf{C}^{1:t-1}) p(c = k' | \mathbf{C}^t) p(c = k)\}}. \tag{22}$$

From Equation (22), we can update the sequence confidence $p(c = k | \mathbf{C}^{1:t})$ at time t from the previous sequence confidence $p(c = k | \mathbf{C}^{1:t-1})$ at time $t - 1$, and the current confidence of traversability $p(c = k | \mathbf{C}^t)$ directly. Thus, we do not need to retain all previous frames. Additionally, $p(c = k)$ is the initial confidence of traversability. The overall process of traversable region detection and tracking is summarized in Algorithm 1.

Algorithm 1 Traversable Region Detection and Tracking

Input: 3D point cloud $\mathbf{Z}_t$ and previous Traversable Region $\mathbf{C}^{1:t-1}$
Output: Traversable Region Probability $\mathbf{C}^{1:t}$
for every frame t **do**
   01: $\mathbf{I}_t \leftarrow$ Make Range Image($\mathbf{Z}_t$)
   02: $M_\alpha \leftarrow$ Make Vertical Angle Image ($\mathbf{I}_t$)
   03: $M_\beta \leftarrow$ Make Horizontal Angle Image($\mathbf{I}_t$)
   04: $M_T \leftarrow$ Traversable Region Detection(M_α, M_β)
   05: $\mathbf{C}^t \leftarrow$ Traversable Confidence ($\mathbf{I}_t, M_T$)
   06: $\mathbf{C}^{1:t} \leftarrow$ Tracking Traversable Region $(\mathbf{C}^t, \mathbf{C}^{1:t-1})$
end for
---

4. Experiment

4.1. Experiment Environment

We collected a unique dataset by conducting experiments on actual terrain. The total length of the track was approximately 1.2 km, and it contained numerous irregular slopes, thus providing a suitable environment for verifying the reliability of the proposed algorithm under various slopes and road surface conditions for a UGV. The maximum difference in the pitch angle was approximately 20 degrees, and the maximum difference in the roll angle was approximately 10 degrees, as shown in Figure 6. Based on these pitch and roll angle differences, we prepared three scenarios (routes A, B, and C) to evaluate the proposed method, using 1500 frames.

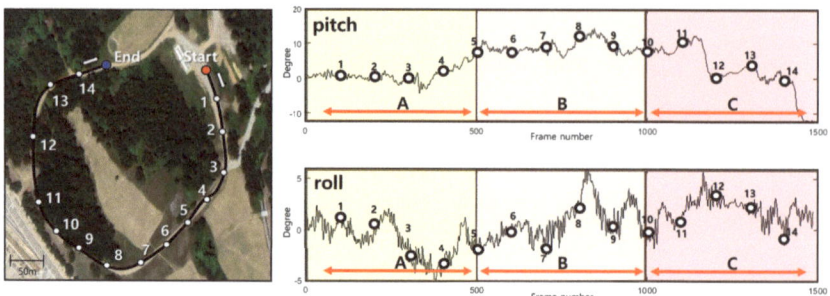

Figure 6. Experimental environment with actual rough terrain. (**Left**) Satellite image of the experimental route. (**Top**) Pitch angle variation with the frame number. (**Bottom**) Roll angle variation with the frame number.

Our experimental driving platform was equipped with a 3D laser scanner, Velodyne HDL-64E sensor, and high-precision positioning system, NovAtel OEMV-2 receiver, with a Honeywell HMR3500, as shown in Figure 7. Additionally, experimental data for performance evaluation were collected by measuring a 3D point cloud using LiDAR, while simultaneously recording the vehicle's motion information with GPS. This enabled us to collect precise route information for the driving platform in the rough terrain considered in this study.

Figure 7. Our experimental driving platform was equipped with a high-end 3D laser scanner, a high-precision positioning system, and a front camera.

4.2. Data Annotation

Defining traversable areas in places without road structures, such as road boundaries or lanes, can be challenging. In addition, even under similar road conditions, certain areas may prove difficult for vehicles to traverse. Consequently, there may be variations in the definition of traversable areas in such locations. However, annotating the ground truth based on multiple criteria may not result in good data, and evaluating the algorithm's performance may prove difficult. To ensure consistency in annotation, this study established a unified definition of a traversable area, as an area where other vehicles have traveled or where there are visible traces of such movement that differentiate it from other areas. Data were collected by multiple annotators using this definition, although different preferences might have resulted in different traversable regions.

To collect data in point units, over 100,000 points per frame must be collected. However, the manual labeling of each point is time-consuming. Hence, in this study, a pixel-wise annotation was conducted on the range image derived from the raw 3D point cloud, as shown in Figure 8. To label the traversable regions, the annotators needed to observe the variations in the range image across consecutive frames, while referring to the corresponding actual driving images. The identified traversable areas were then represented by polygons, as shown in the figure. To ensure accuracy, data were collected by multiple personnel in the same frame, and the intersection of the areas designated by each annotator was used as the final ground-truth dataset of the traversable areas.

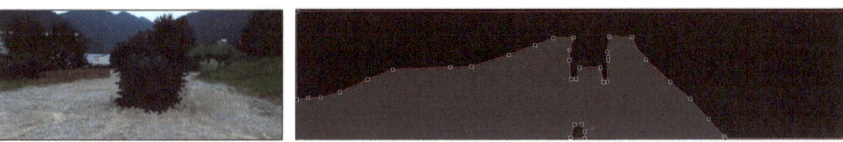

Figure 8. This figure shows the data annotation method. The data annotation is performed at the pixel level, marking traversable and non-traversable regions. The gray area represents the traversable region, while the black area indicates the non-traversable region.

4.3. Evaluation Metrics

Labeling every point to evaluate an algorithm is time-consuming, and visualizing the traversable region in the form of a point cloud is challenging. Therefore, in this study, the ground-truth data were collected in a pixel-wise manner in the range image, as shown in Figure 8. At this point, the balance between traversable and non-traversable regions within a single-range image may vary depending on the scenario. However, these imbalanced issues in a single-range image are common in real-world situations. Figure 9 illustrates that the traversable region (represented by the gray color) occupies a smaller proportion in the range image compared to the non-traversable region (depicted in black color). To address this issue and properly evaluate both the proposed and previous methods, two evaluation metrics were used: the Jaccard index (also known as intersection-over-union (IOU)) and the Dice coefficient (also known as F1 score).

Figure 9. The range image, converted from a raw 3D point cloud, was annotated pixel-wise to distinguish between traversable and non-traversable regions. The images on the left show the actual environment of the rough terrain, while those on the right represent the annotation results for the traversable regions.

The Jaccard index is one of the most commonly used metrics in semantic segmentation and is an effective indicator. It can be calculated as the area of overlap between the predicted union of the predicted segmentation and the ground truth:

$$IoU = \frac{TP}{TP + FP + FN}. \tag{23}$$

The Dice coefficient is a statistical measure employed to assess the similarity between two samples. It is commonly used as a metric to quantify the overlap between the predicted and ground-truth segmentations.

$$Dice = \frac{2TP}{2TP + FP + FN}. \qquad (24)$$

4.4. Quantitative Result

To evaluate the performance of the proposed method in detecting and tracking traversable regions, we conducted a comparative evaluation with three real-time computational methods using the entire dataset. The first method detects drivable areas using an elevation-based grid map [38], in which the slopes between cells are computed using height information from a 3D laser scanner. This approach is similar to the proposed method, as it projects 3D spatial information onto a 2D grid map. The second method detects drivable areas using a range image [13], which is more effective for dealing with sparse 3D laser scanner measurements, and contains 2.5D information. This method detects traversable regions by calculating the slope of the pixels in the column direction of the range image. The third method, based on the author's previous research, is a specialized approach designed solely for detecting traversable regions in challenging and uneven terrains [41].

The comparison results are presented in Figure 10 in the form of a range image. In the range image, each color represents an object perpendicular to the ground in different successive shades of red, with objects parallel to the ground shown in successive shades of blue. Therefore, blue indicates flat ground over which the UGV can be driven. Conversely, red indicates an impassable obstacle or non-flat ground over which the UGV cannot be driven, while gray indicates areas judged to be drivable or ground truths collected manually. The results presented in Table 1 correspond to the experiments conducted on datasets where data collection and ground truth annotation were performed. We established three types of driving paths along an approximately 1.2 km route and organized the experimental findings accordingly. Table 1 provides the conclusions in terms of the IOU and Dice values between the ground truth and predictions for each of the three path types.

Table 1. Quantitative Results of Traversable Region Detection.

Method	Route A		Route B		Route C	
	Iou	Dice	Iou	Dice	Iou	Dice
Elevation Map [38]	0.5090	0.6726	0.2004	0.3291	0.1610	0.2765
Range Image [13]	0.5617	0.7165	0.2069	0.3399	0.2997	0.4563
Detection only [41]	0.6509	0.7870	0.2773	0.4259	0.4816	0.6461
Proposed method	0.6701	0.7971	0.2871	0.4269	0.4826	0.6471

The top row of the figure presents the ground truth, which was obtained by manually collecting pixel-wise images for each frame along different routes. The leftmost image corresponds to route A, and it can be observed that this route offers a relatively wide traversable area with minimal changes in the pitch angle. Most areas in this route are deemed traversable, with only a few bushes on the left and right sides serving as obstacles. The middle image corresponds to route B, which has a gentle and flat slope on the left side of the route. In the ground truth, the flat land on the left side is labeled as non-traversable, because the annotator has prior information. In fields or rough terrain, areas that are likely to be traversable often exhibit traces of previous vehicle paths, which the annotator can identify. Based on this information, only the central areas were labeled as traversable regions, while the flat land on the left side was labeled as non-traversable. However, some methods detected the flat left side of the route as a traversable region, as shown by the results. Finally, the rightmost image represents a narrow mountain road bordered by trees on both sides. Despite significant variations in pitch and roll angles, there is no vegetation present for this terrain, except for the trees. This terrain can be traversed by pedestrians, but the areas where vehicles can travel are extremely limited.

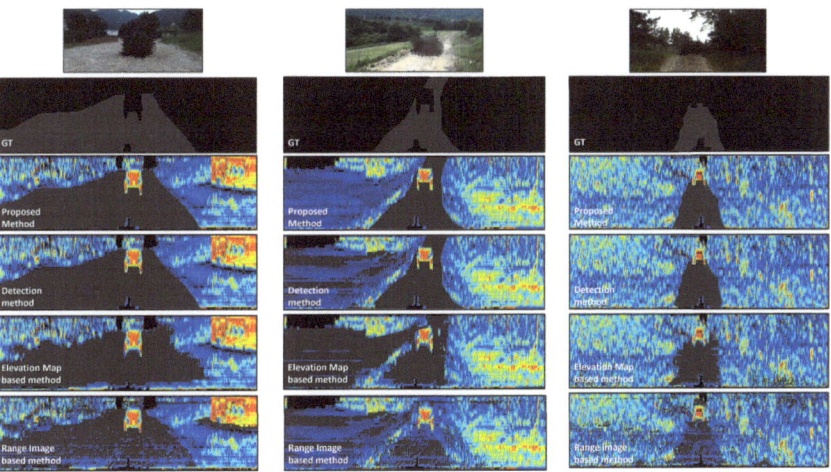

Figure 10. This figure showcases the experimental results. The first row exhibits the actual images, while the second row presents the labeled ground truth (GT). The experimental results are displayed as range images obtained from the raw 3D point cloud. The third row illustrates the approach proposed in this paper. Furthermore, from left to right in each column, they respectively represent the route segments A, B, and C. A comparative analysis with the other results demonstrated that the method proposed in this paper delivered improved outcomes.

The results figure shows the outcomes for each method in the third row to the sixth row. The third row presents the results of our proposed method. The fourth row displays the results of a previous study that focused on detection without tracking. The fifth row represents the results of an elevation-map-based approach, while the sixth row shows the results of a range-image-based approach. As we can see from the experimental results, our proposed method performed the best. It achieved the most accurate detection of traversable regions for route A, which had minimal changes in vehicle pitch and roll angles and a wide traversable area. Specifically, our proposed method achieved an IoU score of 0.6701 and a Dice coefficient of 0.7971, indicating its superior performance. However, the other methods also exhibited a similar performance. In particular, for route A, which had a relatively flat terrain and minimal variations in slope, all methods demonstrated a comparable performance. The range-image-based approach, in particular, achieved a relatively high performance, with an IoU score of 0.5617 and a Dice coefficient of 0.7165. This was because the range-image-based approach assumed that the traversable areas were relatively flat. However, this method tended to mistakenly identify gentle slopes as traversable areas, which could pose a risk to the vehicle.

For route B, all methods had a lower performance overall. The other methods achieved IoU scores in the range of 0.2 and Dice coefficients around 0.3 to 0.4. However, our proposed method in this study showed an improved performance by approximately 20–30% compared to the other algorithms, with IoU and Dice coefficients of 0.2871 and 0.4269, respectively. The lower performance of all methods can be attributed to the discrepancy between the region labeled as traversable in the ground truth and the region identified as traversable based on the algorithms. This inconsistency arose because the flat land on the left side was labeled as non-traversable in the ground truth by the annotators, due to the given information that other vehicles had already passed through. Despite this discrepancy, our proposed method demonstrated a better performance than the other algorithms, because it tracked the drivable areas based on the vehicle's trajectory. This allowed it to achieve a superior performance, as shown in the figures.

For route C, which had a narrow traversable area and many trees, the proposed method demonstrated a superior performance, with an IoU score of 0.4826 and Dice coefficient of 0.6471, compared to the other methods. Unlike routes A and B, the terrain characteristics of this route allowed us to visually identify areas that were passable. As a result, the traversable detection results of the algorithm were closer to the ground truth compared to the other routes. However, there was a potential misinterpretation of the grass existing among the trees as traversable. This was because the grass appears relatively flat compared to the trees, leading the algorithm to incorrectly detect some areas as passable. This problem was particularly pronounced with the range image-based method, which detected the entire range image as traversable.

4.5. Computation Time

In a subsequent experiment, we compared the real-time performance of each algorithm, considering both detection performance and operational speed as crucial factors. The run times of the algorithms were evaluated for all frames on a desktop computer with an i8-8700 3.20 GHz CPU, using only a single core of the CPU. The processing time was measured from the input of the raw point cloud data until the determination of the traversable region. As shown in Figure 11, the proposed method, the elevation-map-based method, the detection-only method, and the range-image-based method exhibited average calculation times of 2.106 ms, 19.445 ms, 2.316 ms, and 2.028 ms, respectively, for all frames. Both the range-image-based method and the detection-only method used range images, resulting in a calculation time of approximately 2 ms, since not all laser points were directly utilized. However, the proposed method required slightly more computation time compared to the range-image-based method, due to conducting searches in both the column and row directions within the range images. On the other hand, the elevation-map-based method searched through all points and required more time compared to the proposed method. Overall, these results indicated that the proposed method operated faster than the sensor's measurement period, ensuring a real-time performance capability.

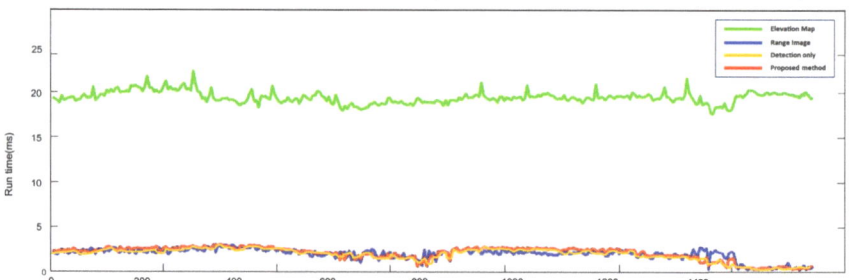

Figure 11. This figure represents the comparison results for computation time. The method proposed in this paper is indicated by the red line. It required a similar computation time as the other methods that utilized range images, but demonstrated a significantly higher efficiency compared to the methods using elevation maps.

5. Conclusions

In conclusion, this paper presents a novel approach for detecting and tracking traversable regions using 3D laser scanners in off-road conditions. To enhance the computational efficiency, the raw data from the laser scanner, which consists of millions of data points, are processed as range images that contain distance information. Unlike previous methods that primarily focused on flat roads, our proposed approach leverages both vertical and horizontal information from range images, to robustly detect traversable regions on uneven off-road terrain. Additionally, we introduced a sequence tracking method that incorporates

Bayesian fusion to integrate detection results from previous frames, ensuring resilience against abrupt changes in vehicle posture. To assess the performance of our method, we collected data while driving on an actual mountain road and obtained multiple annotations of the traversable regions in the range images. The experimental results provided compelling evidence of the effectiveness of our proposed method in real-world driving scenarios.

Funding: This research was supported by a National Research Foundation of Korea (NRF) grant funded by the Korea government (MSIT) (No. RS-2022-00165870) and also supported by the Gachon University research fund of 2021 (202110050001) and fund of 2023 (202300700001).

Institutional Review Board Statement: Not applicable.

Informed Consent Statement: Not applicable.

Data Availability Statement: Restrictions apply to the availability of these data. Data was obtained from Agency for Defense Development.

Conflicts of Interest: The author declares no conflict of interest.

References

1. Homm, F.; Kaempchen, N.; Burschka, D. Fusion of laserscannner and video based lanemarking detection for robust lateral vehicle control and lane change maneuvers. In Proceedings of the IEEE Intelligent Vehicles Symposium, Baden-Baden, Germany, 5–9 June 2011; pp. 969–974.
2. Guan, H.; Li, J.; Yu, Y.; Wang, C.; Chapman, M.; Yang, B. Using mobile laser scanning data for automated extraction of road markings. *ISPRS J. Photogramm. Remote Sens.* **2014**, *87*, 93–107. [CrossRef]
3. Kumar, P.; McElhinney, C.P.; Lewis, P.; McCarthy, T. Automated road markings extraction from mobile laser scanning data. *Int. J. Appl. Earth Obs. Geoinf.* **2014**, *32*, 125–137. [CrossRef]
4. Hata, A.Y.; Osorio, F.S.; Wolf, D.F. Robust curb detection and vehicle localization in urban environments. In Proceedings of the IEEE Intelligent Vehicles Symposium, Dearborn, MI, USA, 8–11 June 2014; pp. 1257–1262. [CrossRef]
5. Guan, H.; Li, J.; Member, S.; Yu, Y.; Chapman, M.; Wang, C. Automated Road Information Extraction From Mobile Laser Scanning Data. *IEEE Trans. Intell. Transp. Syst.* **2015**, *16*, 194–205. [CrossRef]
6. Zhang, Y.; Wang, J.; Wang, X.; Li, C.; Wang, L. 3D LIDAR-Based Intersection Recognition and Road Boundary Detection Method for Unmanned Ground Vehicle. In Proceedings of the IEEE Conference on Intelligent Transportation Systems, Gran Canaria, Spain, 15–18 September 2015. [CrossRef]
7. Han, J.; Kim, D.; Lee, M.; Sunwoo, M. Enhanced Road Boundary and Obstacle Detection Using a Downward-Looking LIDAR Sensor. *IEEE Trans. Veh. Technol.* **2012**, *61*, 971–985. [CrossRef]
8. Moosmann, F.; Pink, O.; Stiller, C. Segmentation of 3D lidar data in non-flat urban environments using a local convexity criterion. In Proceedings of the IEEE Intelligent Vehicles Symposium, Xi'an, China, 3–5 June 2009; pp. 215–220. [CrossRef]
9. Himmelsbach, M.; von Hundelshausen, F.; Wuensche, H. Fast segmentation of 3D point clouds for ground vehicles. In Proceedings of the IEEE Intelligent Vehicles Symposium, La Jolla, CA, USA, 21–24 June 2010; pp. 560–565. [CrossRef]
10. Choe, Y.; Ahn, S.; Chung, M.J. Fast point cloud segmentation for an intelligent vehicle using sweeping 2D laser scanners. In Proceedings of the International Conference on Ubiquitous Robots and Ambient Intelligence, Daejeon, Republic of Korea, 26–29 November 2012; pp. 38–43. [CrossRef]
11. Chen, T.; Dai, B.; Wang, R.; Liu, D. Gaussian-Process-Based Real-Time Ground Segmentation for Autonomous Land Vehicles. *J. Intell. Robot. Syst.* **2014**, *76*, 563–582. [CrossRef]
12. Bogoslavskyi, I.; Stachniss, C. Fast Range Image-Based Segmentation of Sparse 3D Laser Scans for Online Operation. In Proceedings of the IEEE/RSJ International Conference on Intelligent Robots and Systems, Daejeon, Republic of Korea, 9–14 October 2016.
13. Bogoslavskyi, I.; Stachniss, C. Efficient Online Segmentation for Sparse 3D Laser Scans. *J. Photogramm. Remote Sens. Geoinf. Sci.* **2017**, *85*, 41–52. [CrossRef]
14. Caltagirone, L.; Scheidegger, S.; Svensson, L.; Wahde, M. Fast LIDAR-based road detection using fully convolutional neural networks. In Proceedings of the IEEE Intelligent Vehicles Symposium, Los Angeles, CA, USA, 11–14 June 2017; pp. 1019–1024.
15. Babahajiani, P.; Fan, L.; Kämäräinen, J.K.; Gabbouj, M. Urban 3D segmentation and modelling from street view images and LiDAR point clouds. *Mach. Vis. Appl.* **2017**, *28*, 679–694. [CrossRef]
16. Zermas, D.; Izzat, I.; Papanikolopoulos, N. Fast segmentation of 3D point clouds: A paradigm on LiDAR data for autonomous vehicle applications. In Proceedings of the IEEE International Conference on Robotics and Automation, Singapore, 29 May–3 June 2017; pp. 5067–5073. [CrossRef]
17. Shin, M.O.; Oh, G.M.; Kim, S.W.; Seo, S.W. Real-time and accurate segmentation of 3-D point clouds based on gaussian process regression. *IEEE Trans. Intell. Transp. Syst.* **2017**, *18*, 3363–3377. [CrossRef]

18. Weiss, T.; Schiele, B.; Dietmayer, K. Robust Driving Path Detection in Urban and Highway Scenarios Using a Laser Scanner and Online Occupancy Grids. In Proceedings of the IEEE Intelligent Vehicles Symposium, Istanbul, Turkey, 13–15 June 2007; pp. 184–189. [CrossRef]
19. Homm, F.; Kaempchen, N.; Ota, J.; Burschka, D. Efficient occupancy grid computation on the GPU with lidar and radar for road boundary detection. In Proceedings of the IEEE Intelligent Vehicles Symposium, La Jolla, CA, USA, 21–24 June 2010; pp. 1006–1013. [CrossRef]
20. An, J.; Choi, B.; Sim, K.B.; Kim, E. Novel Intersection Type Recognition for Autonomous Vehicles Using a Multi-Layer Laser Scanner. *Sensors* **2016**, *16*, 1123. [CrossRef]
21. Jungnickel, R.; Michael, K.; Korf, F. Efficient Automotive Grid Maps using a Sensor Ray based Refinement Process. In Proceedings of the IEEE Intelligent Vehicles Symposium, Gotenburg, Sweden, 19–22 June 2016.
22. Eraqi, H.M.; Honer, J.; Zuther, S. Static Free Space Detection with Laser Scanner using Occupancy Grid Maps. In Proceedings of the IEEE Conference on Intelligent Transportation Systems, Maui, HI, USA, 4–7 November 2018.
23. Wang, C.C.; Thorpe, C.; Thrun, S. Online simultaneous localization and mapping with detection and tracking of moving objects: Theory and results from a ground vehicle in crowded urban areas. In Proceedings of the IEEE International Conference on Robotics and Automation, Taipei, Taiwan, 14–19 September 2003; Volume 1, pp. 842–849. [CrossRef]
24. Wang, C.C.; Thorpe, C.; Thrun, S.; Durrant-whyte, H. Simultaneous Localization, Mapping and Moving Object Tracking Moving Object Tracking. *Int. J. Robot. Res.* **2007**, *26*, 889–916. [CrossRef]
25. Vu, T.D.; Aycard, O.; Appenrodt, N. Online Localization and Mapping with Moving Object Tracking in Dynamic Outdoor Environments. In Proceedings of the IEEE Intelligent Vehicles Symposium, Istanbul, Turkey, 13–15 June 2007; pp. 190–195. [CrossRef]
26. Vu, T.D.; Burlet, J.; Aycard, O. Grid-based localization and local mapping with moving object detection and tracking. *Inf. Fusion* **2011**, *12*, 58–69. [CrossRef]
27. Steyer, S.; Tanzmeister, G.; Wollherr, D. Object tracking based on evidential dynamic occupancy grids in urban environments. In Proceedings of the IEEE Intelligent Vehicles Symposium, Los Angeles, CA, USA, 11–14 June 2017; pp. 1064–1070. [CrossRef]
28. Hoermann, S.; Bach, M.; Dietmayer, K. Dynamic Occupancy Grid Prediction for Urban Autonomous Driving: A Deep Learning Approach with Fully Automatic Labeling. *arXiv* **2017**, arXiv:1705.08781v2.
29. Gies, F.; Danzer, A.; Dietmayer, K. Environment Perception Framework Fusing Multi-Object Tracking, Dynamic Occupancy Grid Maps and Digital Maps. In Proceedings of the IEEE Conference on Intelligent Transportation Systems, Maui, HI, USA, 4–7 November 2018; pp. 3859–3865.
30. Hoermann, S.; Henzler, P.; Bach, M.; Dietmayer, K. Object Detection on Dynamic Occupancy Grid Maps Using Deep Learning and Automatic Label Generation. In Proceedings of the IEEE Intelligent Vehicles Symposium, Changshu, China, 26–30 June 2018; pp. 190–195. [CrossRef]
31. Papadakis, P. Terrain traversability analysis methods for unmanned ground vehicles: A survey. *Eng. Appl. Artif. Intell.* **2013**, *26*, 1373–1385. [CrossRef]
32. Yang, S.; Xiang, Z.; Wu, J.; Wang, X.; Sun, H.; Xin, J.; Zheng, N. Efficient Rectangle Fitting of Sparse Laser Data for Robust On-Road Object Detection. In Proceedings of the IEEE Intelligent Vehicles Symposium, Suzhou, China, 26–30 June 2018.
33. Jiang, P.; Osteen, P.; Wigness, M.; Saripalli, S. RELLIS-3D Dataset: Data, Benchmarks and Analysis. *arXiv* **2020**, arXiv:2011.12954.
34. Viswanath, K.; Singh, K.; Jiang, P.; Sujit, P.B.; Saripalli, S. OFFSEG: A Semantic Segmentation Framework For Off-Road Driving. *arXiv* **2021**, arXiv:2103.12417.
35. Palazzo, S.; Guastella, D.C.; Cantelli, L.; Spadaro, P.; Rundo, F.; Muscato, G.; Giordano, D.; Spampinato, C. Domain adaptation for outdoor robot traversability estimation from RGB data with safety-preserving loss. In Proceedings of the IEEE International Conference on Intelligent Robots and Systems, Las Vegas, NV, USA, 25–29 October 2020; Institute of Electrical and Electronics Engineers Inc.: New York, NY, USA, 2020; pp. 10014–10021.
36. Hosseinpoor, S.; Torresen, J.; Mantelli, M.; Pitto, D.; Kolberg, M.; Maffei, R.; Prestes, E. Traversability Analysis by Semantic Terrain Segmentation for Mobile Robots. In Proceedings of the IEEE International Conference on Automation Science and Engineering, Lyon, France, 23–27 August 2021; IEEE Computer Society: Washington, DC, USA, 2021. [CrossRef]
37. Leung, T.H.Y.; Ignatyev, D.; Zolotas, A. Hybrid Terrain Traversability Analysis in Off-road Environments. In Proceedings of the IEEE International Conference on Robotics and Automation, Philadelphia, PA, USA, 23–27 May 2022; Institute of Electrical and Electronics Engineers Inc.: New York, NY, USA, 2022; pp. 50–56. [CrossRef]
38. Sock, J.; Kim, J.; Min, J.; Kwak, K. Probabilistic traversability map generation using 3D-LIDAR and camera. In Proceedings of the IEEE International Conference on Robotics and Automation, Stockholm, Sweden, 16–21 May 2016; Institute of Electrical and Electronics Engineers Inc.: New York, NY, USA, 2016. [CrossRef]
39. Ahtiainen, J.; Stoyanov, T.; Saarinen, J. Normal Distributions Transform Traversability Maps: LIDAR-Only Approach for Traversability Mapping in Outdoor Environments. *J. Field Robot.* **2017**, *34*, 600–621. [CrossRef]
40. Martínez, J.L.; Morales, J.; Sánchez, M.; Morán, M.; Reina, A.J.; Fernández-Lozano, J.J. Reactive navigation on natural environments by continuous classification of ground traversability. *Sensors* **2020**, *20*, 6423. [CrossRef]
41. An, J. Traversable Region Detection Method in rough terrain using 3D Laser Scanner. *J. Korean Inst. Next Gener. Comput.* **2022**, *18*, 147–158.

42. Lee, H.; Hong, S.; Kim, E. Probabilistic background subtraction in a video-based recognition system. *KSII Trans. Internet Inf. Syst.* **2011**, *5*, 782–804. [CrossRef]
43. Kim, G.; Kim, A. Scan Context: Egocentric Spatial Descriptor for Place Recognition Within 3D Point Cloud Map. In Proceedings of the IEEE/RSJ International Conference on Intelligent Robots and Systems, Madrid, Spain, 1–5 October 2018; IEEE: New York, NY, USA, 2018; pp. 4802–4809. [CrossRef]
44. Thrun, S. The Graph SLAM Algorithm with Applications to Large-Scale Mapping of Urban Structures. *Int. J. Robot. Res.* **2006**, *25*, 403–429. [CrossRef]
45. Douillard, B.; Underwood, J.; Kuntz, N.; Vlaskine, V.; Quadros, A.; Morton, P.; Frenkel, A. On the segmentation of 3D lidar point clouds. In Proceedings of the IEEE International Conference on Robotics and Automation, Shanghai, China, 9–13 May 2011; pp. 2798–2805. [CrossRef]
46. Lai, K.; Fox, D. Object recognition in 3D point clouds using web data and domain adaptation. *Int. J. Robot. Res.* **2010**, *29*, 1019–1037. [CrossRef]
47. Hoover, A.; Jean-Baptiste, G.; Jiang, X.; Flynn, P.J.; Bunke, H.; Goldgof, D.B.; Bowyer, K.; Eggert, D.W.; Fitzgibbon, A.; Fisher, R.B. An Experimental Comparison of Range Image Segmentation Algorithms. *IEEE Trans. Pattern Anal. Mach. Intell.* **1996**, *18*, 673–689. [CrossRef]
48. Leonard, J.; How, J.; Teller, S.; Berger, M.; Campbell, S.; Fiore, G.; Fletcher, L.; Frazzoli, E.; Huang, A.; Karaman, S.; et al. A perception-driven autonomous urban vehicle. *J. Field Robot.* **2009**, *56*, 163–230.
49. Savitzky, A.; Golay, M.J. Smoothing and Differentiation of Data by Simplified Least Squares Procedures. *Anal. Chem.* **1964**, *36*, 1627–1639. [CrossRef]

Disclaimer/Publisher's Note: The statements, opinions and data contained in all publications are solely those of the individual author(s) and contributor(s) and not of MDPI and/or the editor(s). MDPI and/or the editor(s) disclaim responsibility for any injury to people or property resulting from any ideas, methods, instructions or products referred to in the content.

Article

CoSOV1Net: A Cone- and Spatial-Opponent Primary Visual Cortex-Inspired Neural Network for Lightweight Salient Object Detection

Didier Ndayikengurukiye * and Max Mignotte

Département d'Informatique et de Recherche Opérationnelle, Université de Montréal, Montreal, QC H3C 3J7, Canada; mignotte@iro.umontreal.ca
* Correspondence: didier.ndayikengurukiye@umontreal.ca

Abstract: Salient object-detection models attempt to mimic the human visual system's ability to select relevant objects in images. To this end, the development of deep neural networks on high-end computers has recently achieved high performance. However, developing deep neural network models with the same performance for resource-limited vision sensors or mobile devices remains a challenge. In this work, we propose CoSOV1net, a novel lightweight salient object-detection neural network model, inspired by the cone- and spatial-opponent processes of the primary visual cortex (V1), which inextricably link color and shape in human color perception. Our proposed model is trained from scratch, without using backbones from image classification or other tasks. Experiments on the most widely used and challenging datasets for salient object detection show that CoSOV1Net achieves competitive performance (i.e., $F_\beta = 0.931$ on the ECSSD dataset) with state-of-the-art salient object-detection models while having a low number of parameters (1.14 M), low FLOPS (1.4 G) and high FPS (211.2) on GPU (Nvidia GeForce RTX 3090 Ti) compared to the state of the art in lightweight or nonlightweight salient object-detection tasks. Thus, CoSOV1net has turned out to be a lightweight salient object-detection model that can be adapted to mobile environments and resource-constrained devices.

Keywords: lightweight salient object detection; salient object detection; object detection; lightweight neural network; color opponent; cone-opponent; double-opponent; vision sensing

Citation: Ndayikengurukiye, D.; Mignotte, M. CoSOV1Net: A Cone-and Spatial-Opponent Primary Visual Cortex-Inspired Neural Network for Lightweight Salient Object Detection. *Sensors* **2023**, *23*, 6450. https://doi.org/10.3390/s23146450

Academic Editors: Man Qi and Matteo Dunnhofer

Received: 12 June 2023
Revised: 12 July 2023
Accepted: 14 July 2023
Published: 17 July 2023

Copyright: © 2023 by the authors. Licensee MDPI, Basel, Switzerland. This article is an open access article distributed under the terms and conditions of the Creative Commons Attribution (CC BY) license (https://creativecommons.org/licenses/by/4.0/).

1. Introduction

The human visual system (HVS) has the ability to select and process relevant information from among the large amount that is received. This relevant information in an image is called salient objects [1]. Salient object-detection models in computer vision try to mimic this phenomenon by detecting and segmenting salient objects in images. This is an important task, given its many applications in computer vision, such as object tracking, recognition and detection [2], advertisement optimization [3], image/video compression [4], image correction [5], analysis of iconographic illustrations [6], image retrieval [7], aesthetic evaluation [8], image quality evaluation [9], image retargeting [10], image editing [11] and image collages [12], to name a few. Thus, it has been the subject of intensive research in recent years and is still being investigated [13]. Salient object-detection models generally fall into two categories, namely conventional and deep learning-based models, which differ by their feature extraction process. The former use hand-crafted features, while the latter use features learned from a neural network. Thanks to powerful representation learning methods, deep learning-based salient object-detection models have recently shown superior performance over conventional models [13,14]. The high performance of these models is undeniable; however, generally, they are also heavy if we consider their number of parameters and the amount of memory occupied, in addition to their high computational cost and slow detection speed. This makes these models less practical for resource-limited vision sensors or mobile devices that have many constraints on their memory and computational capabilities, as well as for real-time applications [15,16]. Hence, there is a need for lightweight salient object-detection models whose performance is comparable to state-of-the-art models, with the advantages of being deployed on resource-limited vision sensors or mobile devices and having a detection speed that allows them to be used in real-time applications. Existing lightweight salient object-detection models have used different methodologies, such as backbones from nonlightweight classification models [17,18], the imitation of primate hierarchical visual perception [19], human attention mechanisms [16,19], etc.

In this work, we propose an original approach for a new lightweight neural network model, namely CoSOV1Net, for salient object detection, that can therefore be adapted to mobile environments and resource-limited or -constrained devices, with the additional properties of being able to be trained from scratch without having to use backbones developed from image-classification tasks and having few parameters, but with comparable performance with state-of-the-art models.

Given that detecting salient objects is a capability of the human visual system and that a normal human visual system performs this quickly and correctly, we used images or scenes encoding mechanism research advances in neuroscience, especially for the early stage of the human visual system [20–22]. Our strategy in this model is therefore inspired by two neuroscience discoveries in human color perception, namely:

1. The color-opponent encoding in the early stage of the HVS (human visual system) [23–26];
2. The fact that color and pattern are linked inextricably in human color perception [20,27].

Inspired by these neuroscience discoveries, we propose a cone- and spatial-opponent primary visual cortex (CoSOV1) module that extracts features at the spatial level and between color channels at the same time to integrate color in the patterns. This process is applied first on opposing color pair channels two by two and then to grouped feature maps through our deep neural network. Thus, based on the CoSOV1 module, we build a novel lightweight encoder–decoder deep neural network for salient object detection: CoSOV1Net, which has only 1.14 M parameters but comparable performance with state-of-the-art salient object-detection models. CoSOV1Net predicts salient maps at a speed of 4.4 FPS on an Intel CPU, i7-11700F and 211.2 FPS on a Nvidia GeForce RTX 3090 Ti GPU for 384×384 images and it has a low FLOPS = 1.4 G. Therefore, CoSOV1net is a lightweight salient object-detection model that can be adapted for mobile environments and limited-resource devices.

Our contribution is threefold:

- We propose a novel approach to extract features from opposing color pairs in a neural network to exploit the strength of the color-opponent principle from human color perception. This approach permits the acceleration of neural network learning;
- We propose a novel strategy to integrate color in patterns in a neural network by extracting features locally and between color channels at the same time in successively grouped feature maps, which results in a reduction in the number of parameters and the depth of the neural network, while keeping good performance;
- We propose—for the first time, to our knowledge—a novel lightweight salient object-detection neural network architecture based on the proposed approach for learning opposing color pairs along with the strategy of integrating color in patterns. This model has few parameters, but its performance is comparable to state-of-the-art methods.

The rest of this work is organized as follows: Section 2 presents some lightweight models related to this approach; Section 3 presents our proposed lightweight salient object-detection model; Section 4 describes the datasets used, evaluation metrics, our experimental results and the comparison of our model with state-of-the-art models; Section 5 discusses our results; Section 6 concludes this work.

2. Related Work

Many salient object-detection models have been proposed and most of the influential advances in image-based salient object detection have been reviewed by Gupta et al. [13]. Herein, we present some conventional models and lightweight neural network models related to this approach.

2.1. Lightweight Salient Object Detection

In recent years, lightweight salient object-detection models have been proposed with different strategies and architectures. Qin et al. [28] designed U^2net, a lightweight salient object-detection model with a two-level nested Unet [29] neural network able to capture more contextual information from different scales, thanks to the mixture of receptive fields of different sizes. Its advantages are threefold: first, it increases the depth of the whole architecture without increasing the computational cost; second, it is trained from scratch without using pretrained backbones, thus being able to keep feature maps high-resolution; third, it has high accuracy. Its disadvantage is its number of parameters. Other models are based on streamlined architecture to build lightweight deep neural networks. MobileNets [30,31] and ShuffleNets [32,33], along with their variants, are among the latter models. MobileNets [30] uses architecture based on depthwise separable convolution. ShuffleNets [32] uses architecture based on pointwise group convolution and channel shuffle, as well as depthwise convolution, to greatly reduce computational cost while maintaining accuracy. Their advantages are their computational cost, accuracy and speed, while their disadvantages are their number of parameters and their input resolution. Other authors have been inspired by primate or human visual system processes. Thus, Liu et al. [19] designed HVPNet, a lightweight salient object-detection network based on a hierarchical visual perception (HVP) module that mimics the primate visual cortex for hierarchical perception learning, whereas Liu et al. [16] were inspired by human perception attention mechanisms in designing SAMNet, another lightweight salient object-detection network, based on a stereoscopically attentive multiscale (SAM) module that adopts a stereoscopic attention mechanism for effective and efficient multiscale learning. Their advantages are their computational cost and accuracy, while their disadvantages are their number of parameters and their input resolution.

2.2. Color-Opponent Models

Color opponency, which is a human color perception propriety, has inspired many authors who have defined channels or feature maps to tackle their image-processing tasks. Frintrop et al. [34] used three opponent channels—*RG*, *BY* and *I*—to extract features for their salient object-detection model.

To extract features for salient object detection, Ndayikengurukiye and Mignotte [1] used nine (9) opponent channels for RGB color space (RR: red–red; RG: red–green; RB: red–blue; GR: green–red; GG: green–green; GB: green–blue; BR: blue–red; BG: blue–green; BB: blue–blue) with a nonlinear combination, thanks to the OCLTP (opponent color local ternary pattern) texture descriptor, which is an extension of the OCLBP (opponent color local binary pattern) [35,36] and Fastmap [37], which is a fast version of MDS (multidimensional scaling).

Most authors apply the opponent color mechanism to the input image color space channels and not on the resulting feature maps. However, Jain and Healey [38] used opponent features computed from Gabor filter outputs. They computed opponent features by combining information across different spectral bands at different scales obtained via Gabor filters for color texture recognition [38]. Yang et al. [39] proposed a framework based on the color-opponent mechanisms of color-sensitive double-opponent (DO) cells in the human visual system's primary visual cortex (V1) in order to combine brightness and color to maximize the boundary-detection reliability in natural scenes. The advantages of hand-crafted models are their computational cost, number of parameters, speed and input resolution, while their disadvantage is accuracy.

In this work, we propose a model inspired by the human visual system but different from other models, because our model uses the primary visual cortex (V1) cone- and spatial-opponent principle to extract features at channels' spatial levels and between color channels at the same time to integrate color into patterns in a manner allowing for a lightweight deep neural network design with performance comparable with state-of-the-art lightweight salient object-detection models.

3. Materials and Methods

3.1. Introduction

Our model for tackling the challenge of lightweight salient object detection is inspired by the human visual system (HVS)'s early visual color process, especially its cone opponency and spatial opponency in the primary visual cortex (V1). The human retina (located in the inner surface of the eye) has two types of photoreceptors, namely rods and cones. Rods are responsible for monochromatic vision under low levels of illumination, while cones are responsible for color vision at normal levels of illumination. There are three classes of cones: L, M and S. When light is absorbed by cone photoreceptors, the L, M and S cones absorb long-, middle- and short-wavelength visible light, respectively [24,25,27].

The cone signals are then processed by single-opponent retina ganglion cells. The single opponent operates an antagonistic comparison of the cone signals [23,25,26,40]:

- L − M opponent for red–green;
- S − (L + M) opponent for blue–yellow.

The red–green and blue–yellow signals are carried by specific cells (different cells each for red–green and blue–yellow) through the lateral geniculate nucleus (LGN) to the primary visual cortex (V1).

Shapley [27] and Shapley and Hawken [20] showed that the primary visual cortex (V1) plays an important role in color perception through the combined activity of two kinds of color-sensitive cortical neurons, namely single-opponent and double-opponent cells. Single-opponent cells in V1 operate in the same manner as those of retina ganglion cells and provide neuronal signals that can be used for estimating the color of the illumination [27]. Double-opponent cells in V1 compare cone signals across space as well as between cones [21,22,24,27]. Double-opponent cells thus have two opponencies: spatial opponency and cone opponency. These properties permit them to be sensitive to color edges and spatial patterns. They are thus able to inextricably link color and pattern in human color perception [20,27].

As the primary visual cortex (V1) is known to play a major role in visual color perception, as highlighted above, in this work, we propose a deep neural network based on the primary visual cortex (V1) to tackle the challenge of lightweight salient object detection. In particular, we use two neuroscience discoveries in human color perception, namely:

1. The color-opponent encoding in the early stage of the HVS;
2. The fact that color and pattern are inextricably linked in human color perception

These two discoveries in neuroscience inspired us to design a neural network architecture for lightweight salient object detection, which hinges on two main ideas. First, at the beginning of the neural network, our model opposes color channels two by two by grouping them (R-R, R-G, R-B, G-G, G-B, B-B) then extracting the features at the channels' spatial levels and between the color channels from each channel pair at the same time, to integrate color into patterns. Therefore, instead of performing a subtractive comparison or an OCLTP (opponent color linear ternary pattern) like Ndayikengurukiye and Mignotte [1], we let the neural network learn the features that represent the comparison of the two color pairs. Second, this idea of grouping and then extracting the features at the channels' spatial levels and between the color channels at the same time is applied on feature maps at each neural network level until the saliency maps are obtained. This process allows the proposed model to mimic the human visual system's capability of inextricably linking color and pattern in color perception [20,27].

It is this idea that differentiates our model from other models that use depthwise convolution followed by pointwise convolution [30,31] to extract features at each individual color channel level (or feature map) first, not through a group of color channels (or feature maps) at the same time, as our model does. This idea also differentiates our model from models that combine a group of color channels (or feature maps) pixel by pixel first and apply depthwise convolution afterwards [32,33]. The idea of grouping color channels in pairs (or feature map groups) differentiates our model from models that consider all color channels (or feature maps) as a single group while extracting features at color channels' spatial levels and between color channels at the same time.

Our model takes into account nonlinearities in the image at the beginning as well as through our neural network. For this purpose, we use an encoder–decoder neural network type whose core is a module that we call CoSOV1 (cone- and spatial-opponent primary visual cortex).

3.2. CoSOV1 : Cone- and Spatial-Opponent Primary Visual Cortex Module

The CoSOV1 (cone- and spatial-opponent primary visual cortex) module is composed of two parts (see Figure 1).

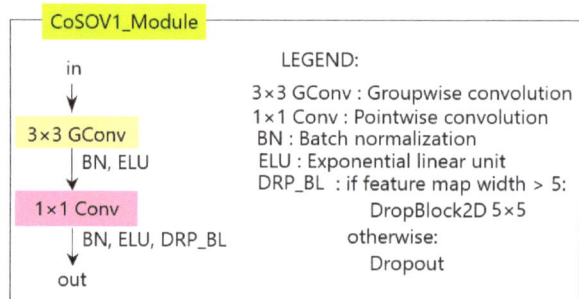

Figure 1. The CoSOV1 (cone- and spatial-opponent primary visual cortex) module is the core of our neural network model.

In the first part, input color channels (or input feature maps) are split into groups of equal depth. Convolution (3×3) operations are then applied to each group of channels (or feature maps) in order to extract features from each group as opposing color channels (or opposing feature maps). This is performed thanks to a set of filters that convolve the group of color channels (or feature maps). Each filter is applied to the color channels (or input feature maps) through a convolution operation that detects local features at all locations on the input. Let $\mathcal{I}^g \in \mathbb{R}^{\mathcal{W} \times \mathcal{H} \times S}$ be an input group of feature maps, where $\mathcal{W}$ and $\mathcal{H}$ are the width and the height of each group's feature map, respectively, and $W \in \mathbb{R}^{3 \times 3 \times S}$, a filter with learned weights, with S being the depth of each group or the number of the channels in each group g, with $g \in \{1, \ldots, \mathcal{G}\}$ (where $\mathcal{G}$ is the number of groups). The output feature map $\mathcal{O}^g \in \mathbb{R}^{\mathcal{W} \times \mathcal{H}}$ for this group g has a pixel value in the (k,l) position, defined as follows:

$$\mathcal{O}^g_{k,l} = \sum_{s=1}^{S} \sum_{i=0}^{2} \sum_{j=0}^{2} W_{i,j,s} \mathcal{I}^g_{k+i-1, l+j-1, s} \quad (1)$$

The weight matrix $W \in \mathbb{R}^{3 \times 3 \times S}$ is the same across the whole group of channels or feature maps. Therefore, each resulting output feature map represents a particular feature at all locations in the input color channels (or input feature maps) [41]. We call the 3×3 convolution on grouped channels (or grouped feature maps) groupwise convolution. The zero padding is applied during the convolution process to keep the input channel size for the output feature maps. After groupwise convolution, we apply the batch normalization transform, which is known to enable faster and more stable training of deep neural networks [42,43]. Let $\mathfrak{B} = \{X_1, \ldots, X_K\}$ be a minibatch that contains K examples from a dataset. The minibatch mean is

$$\mu_\mathfrak{B} = \frac{1}{K} \sum_{k=1}^{K} X_k \qquad (2)$$

and the minibatch variance is

$$\sigma_\mathfrak{B}^2 = \frac{1}{K} \sum_{k=1}^{K} (X_k - \mu_\mathfrak{B})^2 \qquad (3)$$

The batch normalization transform $BN_{\gamma,\beta} : \{X_1, \ldots, X_K\} \longrightarrow \{Y_1, \ldots, Y_K\}$ (γ and β are parameters to be learned):

$$Y_k = \gamma \widehat{X_k} + \beta \qquad (4)$$

where $k \in \{1, \ldots, K\}$ and

$$\widehat{X_k} = \frac{X_k - \mu_\mathfrak{B}}{\sqrt{\sigma_\mathfrak{B}^2 + \epsilon}} \qquad (5)$$

and ϵ is a very small constant to avoid division by zero.

In order to take into account the nonlinearities present in the color channel input (or feature map input), given that groupwise convolution is a linear transformation, batch normalization is followed by a nonlinear function, exponential linear unit (ELU), defined as follows:

$$\text{ELU}(x) = \begin{cases} x & \text{if } x \geq 0, \\ \alpha \times (\exp(x) - 1) & \text{otherwise} \end{cases} \qquad (6)$$

where $\alpha = 1$ by default.

The nonlinear function, which is the activation function, is placed after batch normalization, as recommended by Chollet [44].

The second part of the module searches for the best representation of the obtained feature maps. It is similar to the first part of the module, except for the groupwise convolution, which is replaced by point-wise convolution, but the input feature maps for pointwise convolution in this model are not grouped. Pointwise convolution allows us to learn the filters' weights and thus obtain feature maps that best represent the input channels (or input feature maps) for the salient object-detection task, while having few parameters.

Let $\mathcal{O} \in \mathbb{R}^{\mathcal{W} \times \mathcal{H} \times M}$ be the output of the first part of the module, with M being the number of feature maps in this output and $\mathcal{W}$ and $\mathcal{H}$ being the width and the height, respectively. Let a filter of the learned weights $V \in \mathbb{R}^M$ and $\mathcal{FM} \in \mathbb{R}^{\mathcal{W} \times \mathcal{H}}$ be its output feature map by pointwise convolution. Its pixel value $\mathcal{FM}_{k,l}$ in (k,l) position is:

$$\mathcal{FM}_{k,l} = \sum_{m=1}^{M} V_m \mathcal{O}_{k,l,m} \qquad (7)$$

Thus, $V \in \mathbb{R}^M$ is a vector of learned weights that associates the input feature maps $\mathcal{O} \in \mathbb{R}^{\mathcal{W} \times \mathcal{H} \times M}$ to the feature map $\mathcal{FM} \in \mathbb{R}^{\mathcal{W} \times \mathcal{H}}$, which is the best representation of the latter-mentioned input feature maps. The pointwise convolution in this module uses many filters and thus it outputs many feature maps that are the best representation of the input feature map $\mathcal{O}$. As pointwise convolution is a linear combination, we again apply batch normalization followed by a exponential linear unit function (ELU) on the feature map $\mathcal{FM}$ to obtain the best representation of the input feature maps for the learned weights $V \in \mathbb{R}^M$, which takes into account nonlinearities in the feature maps $\mathcal{O} \in \mathbb{R}^{\mathcal{W} \times \mathcal{H} \times M}$.

Our scheme is different from depthwise separable convolution in that it uses the convolution of a group of channels instead of each channel individually [30,45].

In addition, after the nonlinear function, noise is injected in the resulting feature maps during the neural network learning stage thanks to the dropout process (but not in the prediction stage) to facilitate the learning process. In this model, we use DropBlock [46] if the width of the feature map is greater than 5; otherwise, we use the common dropout [47].

The CoSOV1 module allows our neural network to have few parameters but good performance.

3.3. CoSOV1Net Neural Network Model Architecture

Our proposed model is built on the CoSOV1 module (see Figure 1). It is a neural network of the U-net encoder–decoder type [29] and is illustrated in Figure 2. Thus, our model consists of three main blocks:

1. The input RGB color channel pairing;
2. The encoder;
3. The decoder.

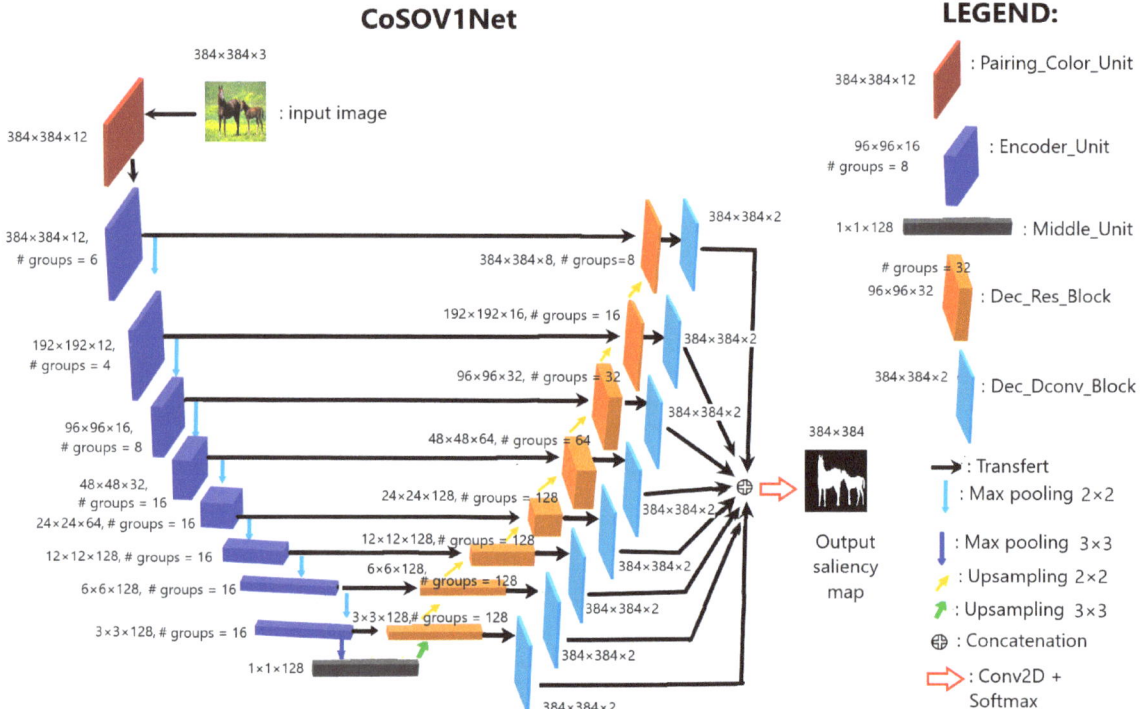

Figure 2. Our model CoSOV1 neural network architecture consisting of 5 blocks : Pairing_Color_Unit, Encoder_Unit, Middle_Unit, Dec_Res_Block and Dec_Dconv_Block.

3.3.1. Input RGB Color Channel Pairing

At this stage, through Pairing_Color_Unit, the input RGB image is paired in six opposing color channel pairs: R-R, R-G, R-B, G-G, G-B and B-B [1,35,48]. These pairs are then concatenated, which gives 12 channels, R, R, R, G, R, B, G, G, G, B, B, B, as illustrated in Figure 3. This is the step for choosing the color channels to oppose. The set of concatenated color channels is then fed to the encoder.

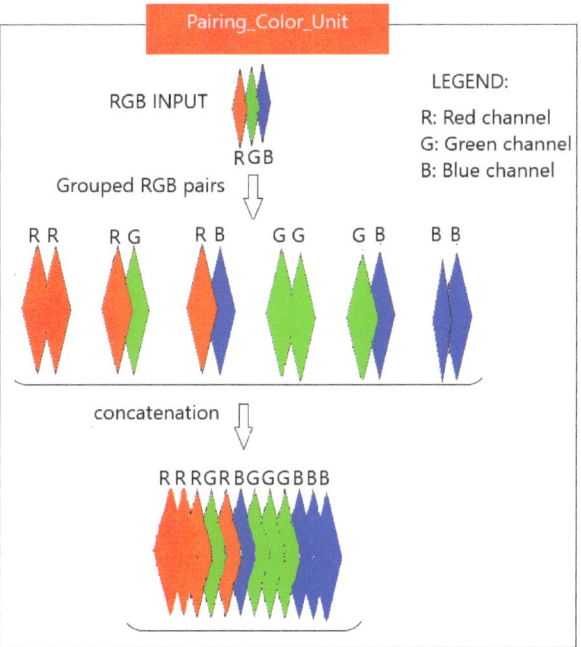

Figure 3. Pairing_Color_Unit: input RGB color image is transformed in 6 opposing color channel pairs; these are then concatenated to obtain 12 color channels.

3.3.2. Encoder

The encoder, in our proposed neural network model, is a convolutional neural network (CNN) [49] where an encoder unit (see Figure 2) is repeated eight times. Each encoder unit is followed by a max pooling (2×2) with strides = 2, except for the eighth neural network level, where the max pooling is 3×3 with strides = 3 (the max pooling is a downsampling operation, like a filtering with a maximum filter). While the size of each feature map is reduced by half, the depth of the feature maps is doubled, except for the first level, where it is kept at 12 and the last two levels, where it is kept at 128 to have few parameters.

The encoder unit (see Figure 4a) is composed of a residual block (Figure 4b) repeated three (3) times.

We used the residual block because this kind of block is known to improve the training of deeper neural networks [50]. The residual block consists of two CoSOV1 modules with a residual link. The reason for all these repetitions is to encode more information and thus allow our network performance to increase.

In the encoder, schematically, as explained above (Section 3.2), the CoSOV1 module (Figure 4c) splits the input channels into groups and applies groupwise convolution (3×3 convolution). Then, pointwise convolution is applied to the outputs of the concatenated groups (see Figure 5 for the first-level input illustration). Each of these convolutions is followed by batch normalization and a nonlinear function (ELU: exponential linear unit activation). After these layers, during the model training, regularization is performed in the CoSOV1 module using the dropout [47] method for small feature maps (dimensions smaller than 5×5) and DropBlock [46]—which is a variant of dropout that zeroes a block instead of pixels individually as dropout does—for feature maps with dimensions greater than 5×5.

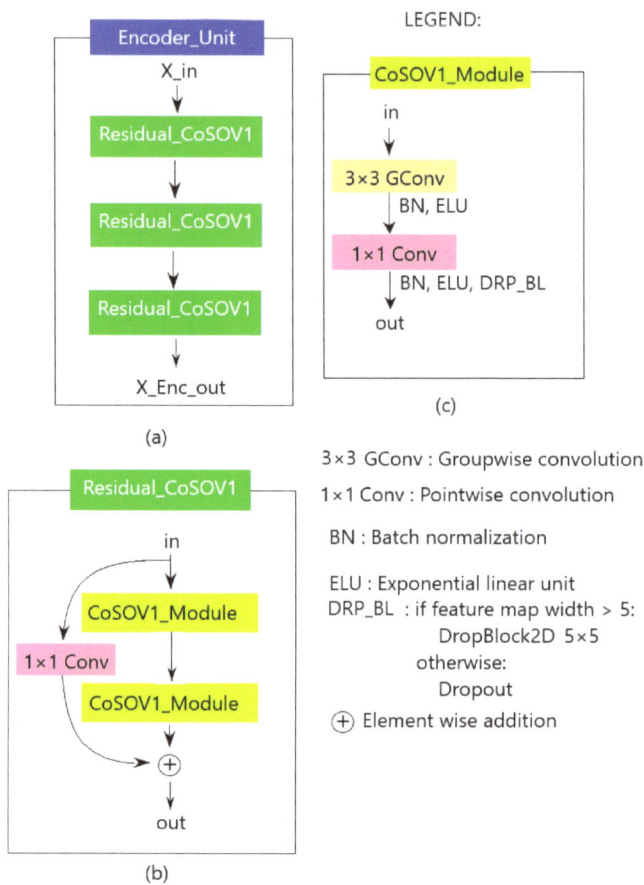

Figure 4. Encoder unit: (**a**) encoder unit; (**b**) the residual block; (**c**) CoSOV1 module.

At its end, the encoder is followed by the middle unit (see Figure 6a), which is the CoSOV1 module (see Figure 6b), where we remove the groupwise convolution—since at this stage, the feature maps are $1 \times 1 \times 128$ in size—and add a residual link.

3.3.3. Decoder

The decoder transforms the features from the encoder to obtain the estimate of the salient object(s) present in the input image. This transformation is achieved through a repeating block, namely the decoder unit (see Figure 7a). The decoder unit consists of two parts: the decoder residual block (see Figure 7b) and the decoder deconvolution block (see Figure 7c). The decoder residual block is a modified CoSOV1 module that allows the model to take into account the output of the corresponding level in the encoder. The output of the decoder residual block takes two directions. On the one hand, it is passed to the next level of the decoder; and on the other, to the second part of the decoder unit, which is the decoder deconvolution block. The latter deconvolves this output, obtaining two feature maps having the size of the input image ($384 \times 384 \times 2$ in our case). At the last level of the decoder, all the outputs from the deconvolution blocks are concatenated and fed to a convolution layer followed by a softmax activation layer, which gives the estimation of the salient object-detection map.

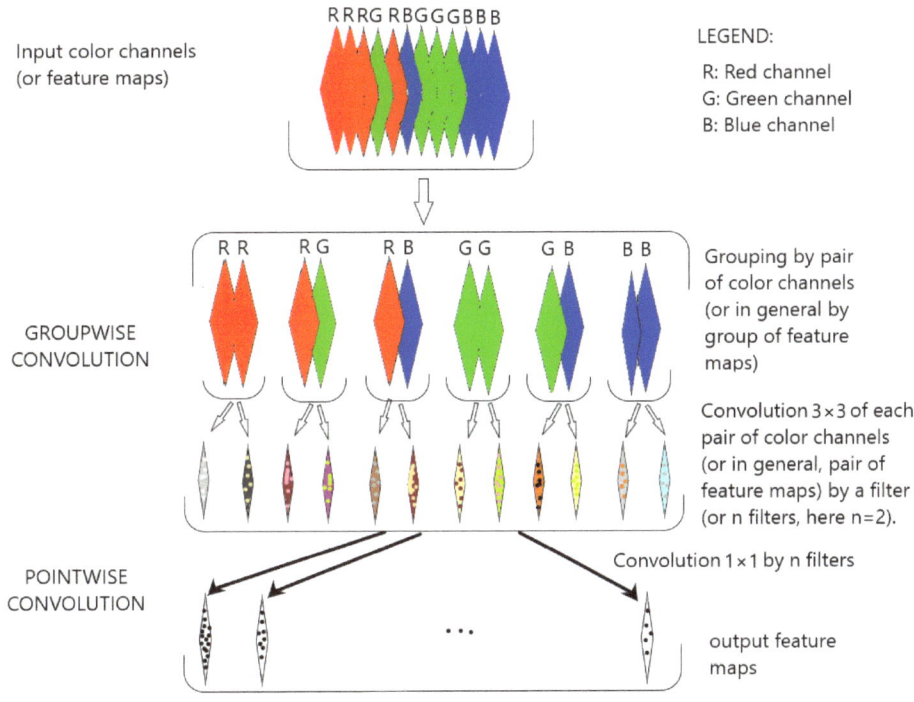

Figure 5. Simplified flowchart in CoSOV1 module for processing pairs of opposing color pairs (or group of feature maps).

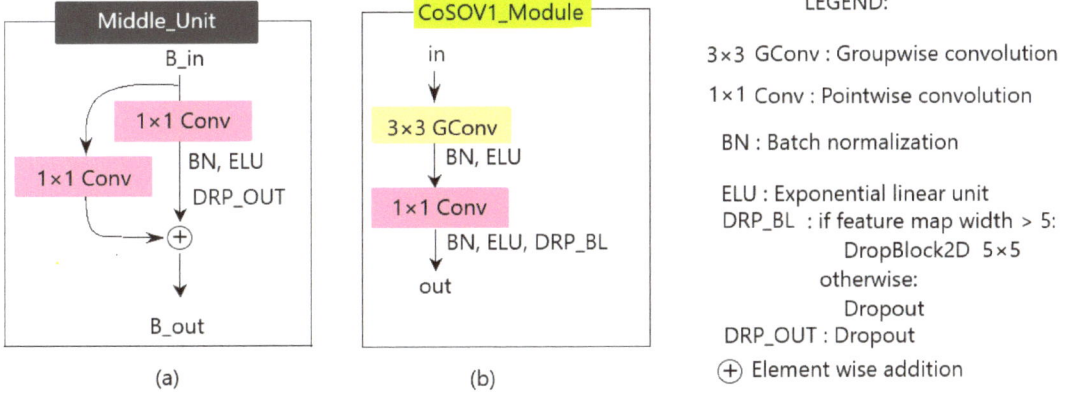

Figure 6. (**a**) The middle unit, (**b**) the CoSOV1 module.

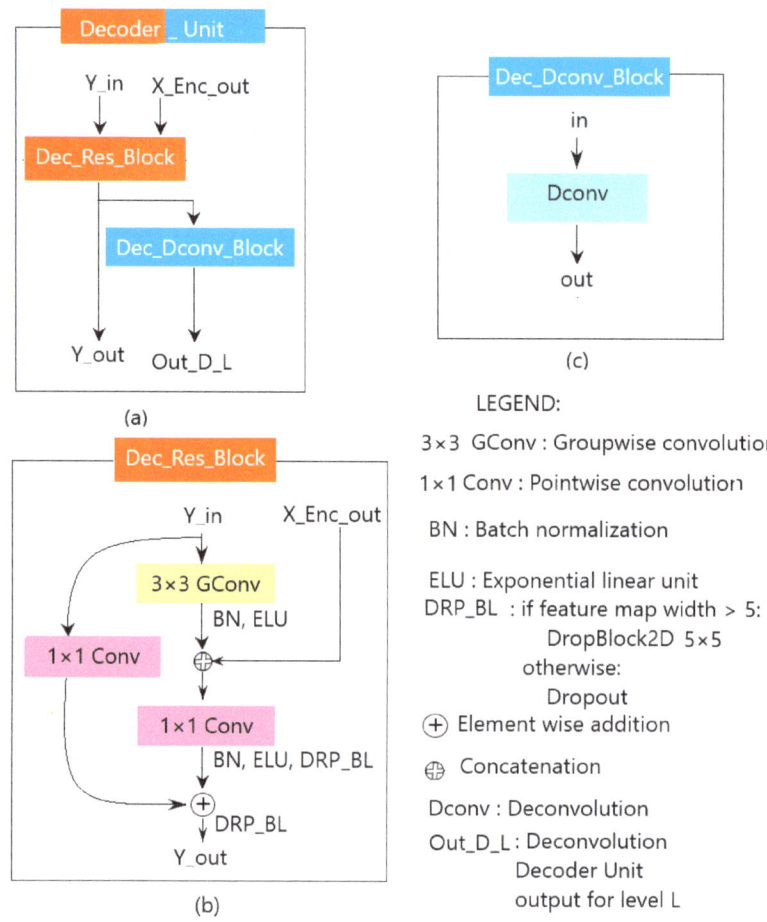

Figure 7. (a) The decoder unit; (b) the decoder residual block; (c) the decoder deconvolution block.

4. Experimental Results

4.1. Implementation Details

For our proposed model implementation, we used the deep learning platform TensorFlow with the Keras deep learning application programming interface (API) [51]. All input images were resized to 384 × 384 and pixel values were normalized (each pixel channel value $\in [0.0, \ldots, 1.0]$ and ground truth pixels $\in \{0, 1\}$). Experiments were conducted on a single GPU, Nvidia GeForce RTX 3090 Ti (24 GB) and an Intel CPU, i7-11700F.

4.2. Datasets

Our proposed model's experiments were conducted on public datasets, which are the most widely used in the field of salient object detection [52]. Thus, we used the Extended Complex Scene Saliency dataset (ECSSD) [53] and the DUT-OMRON (Dalian University of Technology—OMRON Corporation) [54], DUTS [55], HKU-IS [56] and THUR15K [57] datasets.

ECSSD [53] contains 1000 natural images and their ground truths. Many of its images are semantically meaningful but structurally complex for saliency detection [53].

DUT-OMRON [54] contains 5168 images and their binary masks, with diverse variations and complex backgrounds.

The DUTS dataset [55] is divided into DUTS-TR (10,553 training images) and DUTS-TE (5019 test images). We trained and validated our proposed model on the DUTS-TR and DUTS-TE was used for tests.

HKU-IS [56] is composed of 4447 complex images, which contain many disconnected objects with different spatial distributions. Furthermore, it is very challenging for similar foreground/background appearances [58].

THUR15K is a dataset of images taken from the "Flickr" website, divided into five categories (butterfly, coffee mug, dog jump, giraffe, plane), which contains 3000 images. The images of this dataset represent real-world scenes and are considered complex for obtaining salient objects [57] (6232 images with ground truths).

4.3. Model Training Settings

For the reproducibility of the experiments, we set the seed = 123. We trained our proposed model on DUTS-TR (10,553 training images). We split the DUTS-TR dataset into a train set (9472 images) and a validation set (1056 images); that is, approximately 90% of the dataset for the training set and 10% for the validation set. We did not use 25 images because we wanted the training set and the validation set to be divisible by batch size, which is 32.

Our proposed model was trained on scratch without pretrained backbones from image classification (i.e., VGG [59], etc.) or lightweight backbones (i.e., MobileNets [30,31] or ShuffleNets [32,33]). As DUTS-TR is not a big dataset, we used data augmentation during training and many epochs in order to overcome this problem. Indeed, the more epochs, the more the data-augmentation process transforms data. Thus, our proposed model training has two successive stages:

- The first stage is with data augmentation, which is applied to each batch with random transformation (40% zoom in or horizontal flip or vertical flip). This stage has 480 epochs: 240 epochs with learning rate = 0.001 and 240 epochs with learning rate = 0.0001;
- The second stage is without data augmentation. It has 620 epochs: 240 epochs with learning rate = 0.001, followed by 140 epochs with learning rate = 0.0001 and 240 epochs with learning rate = 0.00005.

We also used the same initializer for all layers in the neural network: the HeUniform Keras initializer [60], which draws samples from a uniform distribution within $[-\text{limit}, \text{limit}]$, where limit = $\sqrt{\frac{6}{fan\_in}}$ (fan_in is the number of input units in the weight tensor). The dropout rate was set to 0.2. We used the RMSprop [61] Keras optimizer with default values except for the learning rate; the centered, which was set to true; and the clipnorm = 1. The loss function used was the "sparse_categorical_crossentropy" Keras function; the Keras metric was "SparseCategoricalAccuracy; the Keras check point monitor was "val_sparse_categorical_accuracy".

4.4. Hyperparameters

Hyperparameters such as the ELU activation function, the optimizer, the batch size, the filter size and the learning rates were chosen experimentally by observing the results.

The other hyperparameters were chosen as follows:

- Image size: The best image size was 384×384. We did not choose a small size because we expected to have a small salient object. As we also wanted to have a low computational cost, we did not go beyond this size.
- Number of levels for the encoder: We empirically obtained eight levels as the best number. The choice of image size permitted us to have a maximum of eight levels for the encoder part, given that $384 = 2^7 \times 3$. The size of the feature maps of each level corresponds to the size of those of the previous level divided by 2, except the last level, where the division is by 3.

- Number of levels for the decoder: Eight levels. The number of levels is the same for the encoder part and the decoder part.
- Number of layers: At each level, we chose to use an encoder unit that has an equal number of layers for all levels and a decoder unit that has an equal number of layers for all levels. The number of layers was obtained experimentally.
- Number of filters: We also experimentally chose the number of filters keeping in mind the minimum parameters; the encoder's number of filters was 12, 16, 32, 64, 128, 128, 128 and 128, respectively, for the first, second, ..., seventh and eighth levels; the decoder residual bloc number of filters was 128, 128, 128, 128, 64, 32, 16 and 8, respectively, for the eighth, seventh, sixth, ..., second and first levels. For the decoder deconvolution blocs, at each level, the number of filters was 2.
- The use of batch normalization: Batch normalization is known to enable faster and more stable training for deep neural networks [42,43]. So, we decided to use it.
- Use of dropout: The dropout process injects noise in the resulting feature maps during the neural network learning stage (but not in the prediction stage) to facilitate the learning process. In this model, we used DropBlock [46] if the width of the feature map was greater than 5; otherwise, we used the common dropout [47]. The best results were obtained for DropBlock size = 5 × 5 and rate = 0.1 (the authors' paper suggested a value between 0.05 and 0.25). For the common dropout, the best rate was 0.2, obtained experimentally.

As our proposed model, CoSOV1Net does not use pretrained backbones and the input image is resized to 384 × 384; it has the advantage of good resolution.

4.5. Evaluation Metrics

4.5.1. Accuracy

The metrics used to evaluate our proposed model accuracy were F_β measure, MAE (mean absolute error) and weighted F_β^w measure [62]. We also used precision, precision–recall and F_β measure curves.

Let M be the binary mask obtained for the predicted saliency probability map, given a threshold in the range of $[0, 1)$ and with G being the corresponding ground truth:

$$\text{Precision} = \frac{|M \cap G|}{|M|} \quad (8)$$

$$\text{Recall} = \frac{|M \cap G|}{|G|} \quad (9)$$

$\cap$: set intersection symbol; $|.|$: the number of pixels whose values are not zeros.

The F_β-measure (F_β) is the weighted harmonic mean of precision and recall:

$$F_\beta = \frac{(1+\beta^2) \times \text{Precision} \times \text{Recall}}{\beta^2 \times \text{Precision} + \text{Recall}} \quad (10)$$

During evaluation, $\beta^2 = 0.3$, as it is often suggested [16,58].

Let $\overline{S}$ be the saliency map estimation with pixel values normalized in order to be in $[0.0, \ldots, 1.0]$ and $\overline{G}$; its ground truth also normalized in $\{0;1\}$. The MAE (mean absolute error) is:

$$MAE = \frac{1}{W \times H} \sum_{x=1}^{W} \sum_{y=1}^{H} |\overline{S}(x,y) - \overline{G}(x,y)| \quad (11)$$

where W and H are the width and the height, respectively, of the above maps ($\overline{S}$ and $\overline{G}$).

The F_β^w measure [62] fixes the interpolation flaw, dependence flaw and equal importance flaw in traditional evaluation metrics and its value is:

$$F_\beta^w = (1 + \beta^2) \frac{\text{Precision}^w \times \text{Recall}^w}{\beta^2 \times \text{Precision}^w + \text{Recall}^w} \quad (12)$$

Precision^w and Recall^w are the weighted precision and the weighted recall, respectively.

4.5.2. Lightweight Measures

Since we propose a lightweight salient object-detection model in this work, we therefore also evaluate the model with lightweight measures: the number of parameters, the saliency map estimation speed (FPS: frames per second) and the computational cost by measuring the FLOPS (the number of floating-point operations). The FLOPS is related to the device's energy consumption (the higher the FLOPS, the higher the energy consumption). The floating-point operation numbers are computed as follows [63]:

- For a convolution layer with n filters of size $k \times k$ applied to $W \times H \times C$ feature maps (W: width; H: height; C: channels), with P: number of parameters:

$$\text{FLOPS} = W \times H \times P \quad (13)$$

- For a max-pooling layer or an upsampling layer with a window of size $sz \times sz$ on $W \times H \times C$ feature maps (W: width; H: height; C: channels):

$$\text{FLOPS} = W \times H \times C \times sz \times sz \quad (14)$$

4.6. Comparison with State of the Art

We compare our proposed model with 20 state-of-the-art salient object detection and 10 state-of-the-art lightweight salient object-detection models. We divided these methods because the lightweight methods outperform others with respect to lightweight measures. However, the lightweight methods' accuracy is lower than the accuracy of those with huge parameters. We mainly used the salient object-detection results provided by Liu et al. [16], except for the F_β measure and precision–recall curves, where we used saliency maps provided by these authors. We also used saliency maps provided by the HVPNet authors [19] to compute HVPNet F_β^ω measures.

In this section, we describe the comparison with the 20 salient object-detection models, namely DRFI [64], DCL [65], DHSNet [66], RFCN [67], NLDF [68], DSS [69], Amulet [18], UCF [70], SRM [71], PiCANet [17], BRN [72], C2S [73], RAS [74], DNA [75], CPD [76], BASNet [77], AFNet [78], PoolNet [79], EGNet [80] and BANet [81].

Table 1 shows that our proposed model CoSOV1Net outperforms all 20 state-of-the-art salient object-detection models for lightweight measures (#parameters, FLOPS and FPS) by a large margin (i.e., the best among them for FLOPS is DHSNet [66], with FLOPS = 15.8 G and $F_\beta = 0.903$ for ECSSD; the worst is EGNet [80], with FLOPS = 270.8 G and $F_\beta = 0.938$ for ECSSD; meanwhile, our proposed model, CoSOV1Net, has FLOPS = 1.4 G, and its $F_\beta = 0.931$ for ECSSD) (see Table 1).

Table 1 also shows that CoSOV1Net is among the top 6 models for ECSSD, among the top 7 for DUT-OMRON and around the top 10 for the other three datasets for the F-measure. Tables 2 and 3 compare our model with the state-of-the-art models for the MAE and F_β^ω measures, respectively. From this comparison, we see that our model is ranked around the top 10 for all four datasets and is ranked 15th for the HKU-IS dataset. This demonstrates that our model is also competitive with respect to the performance of state-of-the-art models.

Tables 1–3 show that our proposed model, CoSOV1Net, clearly has the advantage of the number of parameters, computational cost and speed over salient object detection. They also show that its performance is closer to the best among them.

Table 1. Our proposed model F-measure ($F_\beta \uparrow$, $\beta^2 = 0.3$) compared with 20 state-of-the-art models (best value in bold) [# Param: number of parameters, ↑: great is best, ↓: small is the best].

Methods	# Param (M) ↓	FLOPS (G) ↓	Speed (FPS) ↑	ECSSD	DUT-OMRON	DUTS-TE	HKU-IS	THUR15K
DRFI [64]	-	-	0.1	0.777	0.652	0.649	0.774	0.670
DCL [65]	66.24	224.9	1.4	0.895	0.733	0.785	0.892	0.747
DHSNet [66]	94.04	15.8	10.0	0.903	-	0.807	0.889	0.752
RFCN [67]	134.69	102.8	0.4	0.896	0.738	0.782	0.892	0.754
NLDF [68]	35.49	263.9	18.5	0.902	0.753	0.806	0.902	0.762
DSS [69]	62.23	114.6	7.0	0.915	0.774	0.827	0.913	0.770
Amulet [18]	33.15	45.3	9.7	0.913	0.743	0.778	0.897	0.755
UCF [70]	23.98	61.4	12.0	0.901	0.730	0.772	0.888	0.758
SRM [71]	43.74	20.3	12.3	0.914	0.769	0.826	0.906	0.778
PiCANet [17]	32.85	37.1	5.6	0.923	0.766	0.837	0.916	0.783
BRN [72]	126.35	24.1	3.6	0.919	0.774	0.827	0.910	0.769
C2S [73]	137.03	20.5	16.7	0.907	0.759	0.811	0.898	0.775
RAS [74]	20.13	35.6	20.4	0.916	0.785	0.831	0.913	0.772
DNA [75]	20.06	82.5	25.0	0.935	0.799	0.865	0.930	0.793
CPD [76]	29.23	59.5	68.0	0.930	0.794	0.861	0.924	0.795
BASNet [77]	87.06	127.3	36.2	0.938	**0.805**	0.859	0.928	0.783
AFNet [78]	37.11	38.4	21.6	0.930	0.784	0.857	0.921	0.791
PoolNet [79]	53.63	123.4	39.7	0.934	0.791	0.866	0.925	**0.800**
EGNet [80]	108.07	270.8	12.7	0.938	0.794	0.870	0.928	**0.800**
BANet [81]	55.90	121.6	12.5	**0.940**	0.803	**0.872**	**0.932**	0.796
CoSOV1Net (OURS)	**1.14**	**1.4**	**211.2**	0.931	0.789	0.833	0.912	0.773

Table 2. Our proposed model MAE (↓) compared with 20 state-of-the-art models (best performance in bold) [# Param: number of parameters, ↑: great is the best, ↓: small is the best].

Methods	# Param (M) ↓	FLOPS (G) ↓	Speed (FPS) ↑	ECSSD	DUT-OMRON	DUTS-TE	HKU-IS	THUR15K
DRFI [64]	-	-	0.1	0.161	0.138	0.154	0.146	0.150
DCL [65]	66.24	224.9	1.4	0.080	0.095	0.082	0.063	0.096
DHSNet [66]	94.04	15.8	10.0	0.062	-	0.066	0.053	0.082
RFCN [67]	134.69	102.8	0.4	0.097	0.095	0.089	0.080	0.100
NLDF [68]	35.49	263.9	18.5	0.066	0.080	0.065	0.048	0.080
DSS [69]	62.23	114.6	7.0	0.056	0.066	0.056	0.041	0.074
Amulet [18]	33.15	45.3	9.7	0.061	0.098	0.085	0.051	0.094
UCF [70]	23.98	61.4	12.0	0.071	0.120	0.112	0.062	0.112
SRM [71]	43.74	20.3	12.3	0.056	0.069	0.059	0.046	0.077
PiCANet [17]	32.85	37.1	5.6	0.049	0.068	0.054	0.042	0.083
BRN [72]	126.35	24.1	3.6	0.043	0.062	0.050	0.036	0.076
C2S [73]	137.03	20.5	16.7	0.057	0.072	0.062	0.046	0.083
RAS [74]	20.13	35.6	20.4	0.058	0.063	0.059	0.045	0.075
DNA [75]	20.06	82.5	25.0	0.041	**0.056**	0.044	**0.031**	0.069
CPD [76]	29.23	59.5	68.0	0.044	0.057	0.043	0.033	**0.068**
BASNet [77]	87.06	127.3	36.2	0.040	**0.056**	0.048	0.032	0.073
AFNet [78]	37.11	38.4	21.6	0.045	0.057	0.046	0.036	0.072
PoolNet [79]	53.63	123.4	39.7	0.048	0.057	0.043	0.037	**0.068**
EGNet [80]	108.07	270.8	12.7	0.044	**0.056**	0.044	0.034	0.070
BANet [81]	55.90	121.6	12.5	**0.038**	0.059	**0.040**	**0.031**	**0.068**
CoSOV1Net (OURS)	**1.14**	**1.4**	**211.2**	0.051	0.064	0.057	0.045	0.076

Table 3. Our proposed model weighted F-measure (F_β^ω ↑, $\beta^2 = 1$) compared with 20 state-of-the-art models (best value in bold) [# Param: number of parameters, ↑: great is the best, ↓: small is the best].

Methods	# Param (M) ↓	FLOPS (G) ↓	Speed (FPS) ↑	ECSSD	DUT-OMRON	DUTS-TE	HKU-IS	THUR15K
DRFI [64]	-	-	0.1	0.548	0.424	0.378	0.504	0.444
DCL [65]	66.24	224.9	1.4	0.782	0.584	0.632	0.770	0.624
DHSNet [66]	94.04	15.8	10.0	0.837	-	0.705	0.816	0.666
RFCN [67]	134.69	102.8	0.4	0.725	0.562	0.586	0.707	0.591
NLDF [68]	35.49	263.9	18.5	0.835	0.634	0.710	0.838	0.676
DSS [69]	62.23	114.6	7.0	0.864	0.688	0.752	0.862	0.702
Amulet [18]	33.15	45.3	9.7	0.839	0.626	0.657	0.817	0.650
UCF [70]	23.98	61.4	12.0	0.805	0.573	0.595	0.779	0.613
SRM [71]	43.74	20.3	12.3	0.849	0.658	0.721	0.835	0.684
PiCANet [17]	32.85	37.1	5.6	0.862	0.691	0.745	0.847	0.687
BRN [72]	126.35	24.1	3.6	0.887	0.709	0.774	0.875	0.712
C2S [73]	137.03	20.5	16.7	0.849	0.663	0.717	0.835	0.685
RAS [74]	20.13	35.6	20.4	0.855	0.695	0.739	0.849	0.691
DNA [75]	20.06	82.5	25.0	0.897	0.729	0.797	**0.889**	0.723
CPD [76]	29.23	59.5	68.0	0.889	0.715	0.799	0.879	**0.731**
BASNet [77]	87.06	127.3	36.2	0.898	**0.751**	0.802	**0.889**	0.721
AFNet [78]	37.11	38.4	21.6	0.880	0.717	0.784	0.869	0.719
PoolNet [79]	53.63	123.4	39.7	0.875	0.710	0.783	0.864	0.724
EGNet [80]	108.07	270.8	12.7	0.886	0.727	0.796	0.876	0.727
BANet [81]	55.90	121.6	12.5	**0.901**	0.736	**0.810**	**0.889**	0.730
CoSOV1Net (OURS)	**1.14**	**1.4**	**211.2**	0.861	0.696	0.731	0.834	0.688

We also compared CoSOV1Net with the state-of-the-art lightweight salient object-detection models MobileNet [30], MobileNetV2 [31], ShuffleNet [32], ShuffleNetV2 [33], ICNet [82], BiSeNet R18 [83], BiSeNet X39 [83], DFANet [84], HVPNet [19] and SAMNet [16].

For the comparison with state-of-the-art lightweight models, Table 4 shows that our proposed model outperforms these state-of-the-art lightweight models in parameter numbers and the F_β measure for the ECSSD dataset and is competitive for other measures and datasets. Table 5 shows that our model outperforms these state-of-the-art lightweight models for the MAE measure for the ECSSD and DUTS-TE datasets and is ranked first ex aequo with HVPNet for DUT-OMRON, first ex aequo with HVPNet and SAMNet for the HKU-IS dataset and second for the THUR15K dataset. Our model also outperforms these state-of-the-art lightweight models for the F_β^ω measure for ECSSD and DUTS-TE and is competitive for the three other datasets (see Table 6).

Tables 4–6 show that CoSOV1Net clearly has the advantage of the number of parameters over the lightweight salient object detection. They also show that its performance is closer to the best among them. Thus, CoSOV1Net has the advantage of performance.

Regarding computational cost, CoSOV1Net has an advantage over half of the state-of-the-art lightweight salient object-detection models. Overall, we can conclude that it has an advantage in terms of computational cost.

4.7. Comparison with SAMNet and HVPNet State of the Art

We chose to compare our CoSOV1Net model specifically with SAMNet [16] and HVPNet [19] because they are among the best state-of-the-art models.

Figure 8 shows that precision curves for ECSSD and HKU-IS datasets highlight that CoSOV1Net slightly dominates the SAMNet and HVPNet state-of-the-art lightweight salient object-detection models and that there is no clear domination for the DUT-OMRON, DUTS-TE and THUR15K precision curves between the three models. Therefore, the proposed model CoSOV1Net is competitive with these two state-of-the-art lightweight salient object-detection models with respect to precision.

Table 4. Our proposed model's F-measure (F_β ↑, $\beta^2 = 0.3$) compared with state-of-the-art lightweight salient object-detection models (best value in bold) [# Param: number of parameters, ↑: great is the best, ↓: small is the best].

Methods	# Param (M) ↓	FLOPS (G) ↓	Speed (FPS) ↑	ECSSD	DUT-OMRON	DUTS-TE	HKU-IS	THUR15K
MobileNet * [30]	4.27	2.2	295.8	0.906	0.753	0.804	0.895	0.767
MobileNetV2 * [31]	2.37	0.8	446.2	0.905	0.758	0.798	0.890	0.766
ShuffleNet * [32]	1.80	0.7	406.9	0.907	0.757	0.811	0.898	0.771
ShuffleNetV2 * [33]	1.60	**0.5**	**452.5**	0.901	0.746	0.789	0.884	0.755
ICNet [82]	6.70	6.3	75.1	0.918	0.773	0.810	0.898	0.768
BiSeNet R18 [83]	13.48	25.0	120.5	0.909	0.757	0.815	0.902	0.776
BiSeNet X39 [83]	1.84	7.3	165.8	0.901	0.755	0.787	0.888	0.756
DFANet [84]	1.83	1.7	91.4	0.896	0.750	0.791	0.884	0.757
HVPNet [19]	1.23	1.1	333.2	0.925	**0.799**	**0.839**	**0.915**	**0.787**
SAMNet [16]	1.33	**0.5**	343.2	0.925	0.797	0.835	**0.915**	0.785
CoSOV1Net (OURS)	**1.14**	1.4	211.2	**0.931**	0.789	0.833	0.912	0.773

* SAMNet, where the encoder is replaced by this backbone.

Table 5. Our proposed model MAE (↓) compared with state-of-the art lightweight salient object-detection models (best value in bold) [# Param: number of parameters, ↑: great is the best, ↓: small is the best].

Methods	# Param (M) ↓	FLOPS (G) ↓	Speed (FPS) ↑	ECSSD	DUT-OMRON	DUTS-TE	HKU-IS	THUR15K
MobileNet * [30]	4.27	2.2	295.8	0.064	0.073	0.066	0.052	0.081
MobileNetV2 * [31]	2.37	0.8	446.2	0.066	0.075	0.070	0.056	0.085
ShuffleNet * [32]	1.80	0.7	406.9	0.062	0.069	0.062	0.050	0.078
ShuffleNetV2 * [33]	1.60	**0.5**	**452.5**	0.069	0.076	0.071	0.059	0.086
ICNet [82]	6.70	6.3	75.1	0.059	0.072	0.067	0.052	0.084
BiSeNet R18 [83]	13.48	25.0	120.5	0.062	0.072	0.062	0.049	0.080
BiSeNet X39 [83]	1.84	7.3	165.8	0.070	0.078	0.074	0.059	0.090
DFANet [84]	1.83	1.7	91.4	0.073	0.078	0.075	0.061	0.089
HVPNet [19]	1.23	1.1	333.2	0.055	**0.064**	0.058	**0.045**	0.076
SAMNet [16]	1.33	**0.5**	343.2	0.053	0.065	0.058	**0.045**	**0.077**
CoSOV1Net (OURS)	**1.14**	1.4	211.2	**0.051**	**0.064**	**0.057**	**0.045**	0.076

* SAMNet, where the encoder is replaced by this backbone.

Table 6. Our proposed model's weighted F-measure (F_β^ω ↑, $\beta^2 = 1$) compared with lightweight salient object-detection models (best value in bold) [# Param: number of parameters, ↑: great is the best, ↓: small is the best].

Methods	# Param (M) ↓	FLOPS (G) ↓	Speed (FPS) ↑	ECSSD	DUT-OMRON	DUTS-TE	HKU-IS	THUR15K
MobileNet * [30]	4.27	2.2	295.8	0.829	0.656	0.696	0.816	0.675
MobileNetV2 * [31]	2.37	0.8	446.2	0.820	0.651	0.676	0.799	0.660
ShuffleNet * [32]	1.80	0.7	406.9	0.831	0.667	0.709	0.820	0.683
ShuffleNetV2 * [33]	1.60	**0.5**	**452.5**	0.812	0.637	0.665	0.788	0.652
ICNet [82]	6.70	6.3	75.1	0.838	0.669	0.694	0.812	0.668
BiSeNet R18 [83]	13.48	25.0	120.5	0.829	0.648	0.699	0.819	0.675
BiSeNet X39 [83]	1.84	7.3	165.8	0.802	0.632	0.652	0.784	0.641
DFANet [84]	1.83	1.7	91.4	0.799	0.627	0.652	0.778	0.639
HVPNet [19]	1.23	1.1	333.2	0.854	**0.699**	0.730	**0.839**	**0.696**
SAMNet [16]	1.33	**0.5**	343.2	0.855	**0.699**	0.729	0.837	0.693
CoSOV1Net (OURS)	**1.14**	1.4	211.2	**0.861**	0.696	**0.731**	0.834	0.688

* SAMNet, where the encoder is replaced by this backbone.

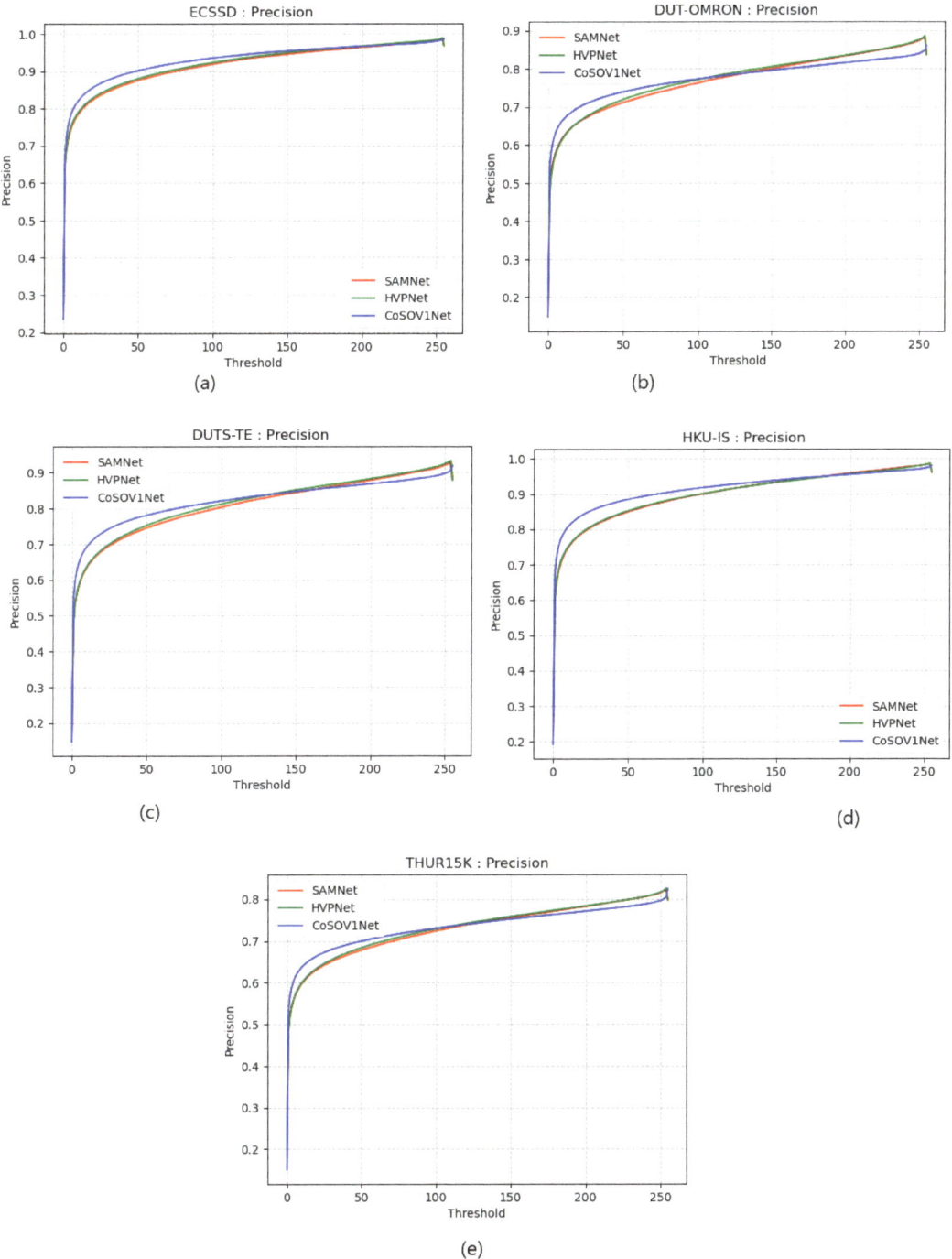

Figure 8. Precision curves for (**a**) ECSSD, (**b**) DUT-OMRON, (**c**) DUTS-TE, (**d**) HKU-IS and (**e**) THUR15K datasets.

Figure 9 shows that the three models' precision–recall curves (for the five datasets used: ECSSD, DUT-OMRON, DUTS-TE, HKU-IS and THUR15K) are very close to each other. Therefore, the proposed model is competitive with these two state-of-the-art lightweight salient object-detection models with respect to precision–recall.

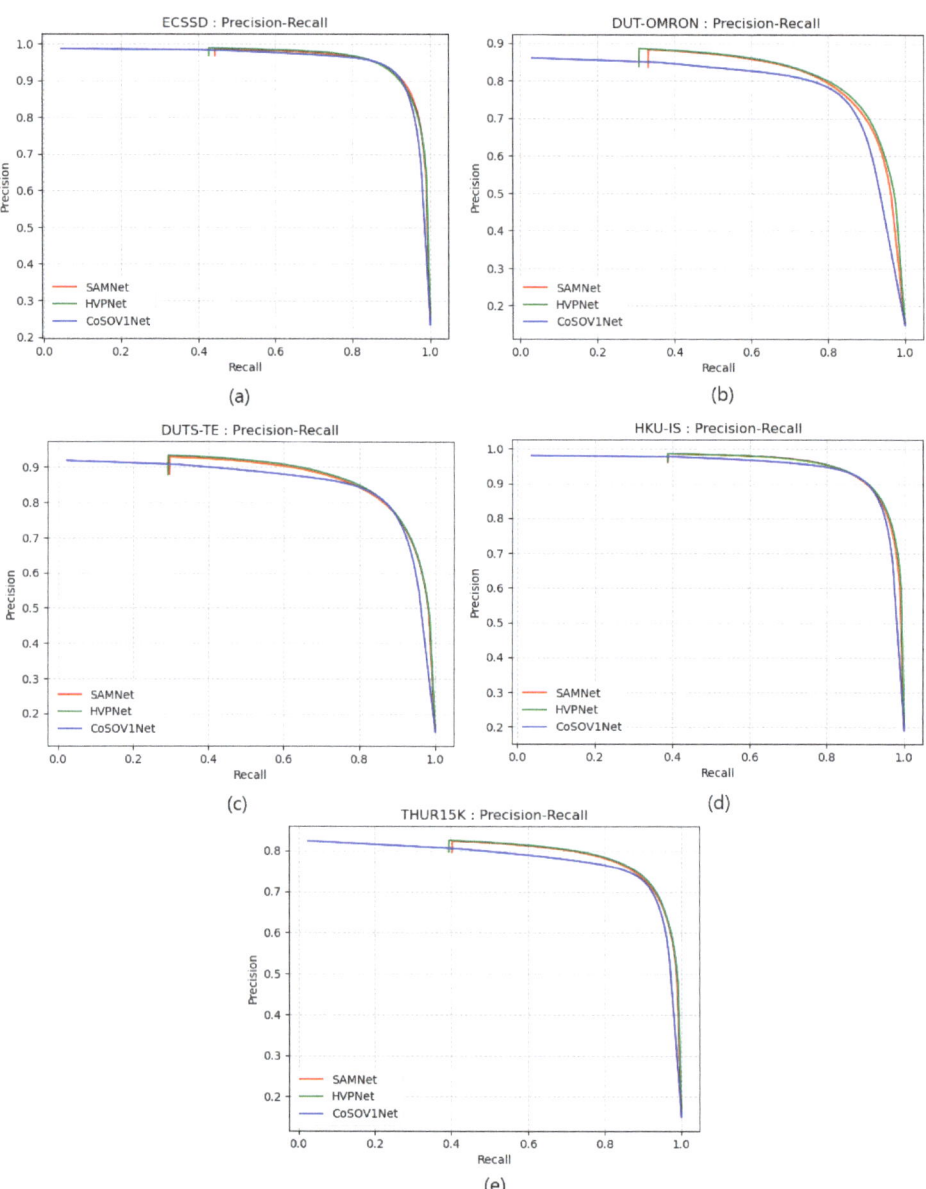

Figure 9. Precision–recall curves for (**a**) ECSSD, (**b**) DUT-OMRON, (**c**) DUTS-TE, (**d**) HKU-IS and (**e**) THUR15K datasets.

Figure 10 shows that the three models' F_β measure curves (for the five datasets used: ECSSD, DUT-OMRON, DUTS-TE, HKU-IS and THUR15K) are very close to each other. The CoSOV1Net model slightly dominates the two state-of-the-art lightweight salient object-detection models for thresholds ≤ 150 and the two state-of-the-art models slightly dominate for thresholds ≥ 150. Thus, there is no clear dominance for one model among the three. This proves that our CoSOV1Net model is comparable to these state-of-the-art lightweight salient object-detection models while having the advantage of a low number of parameters compared to them.

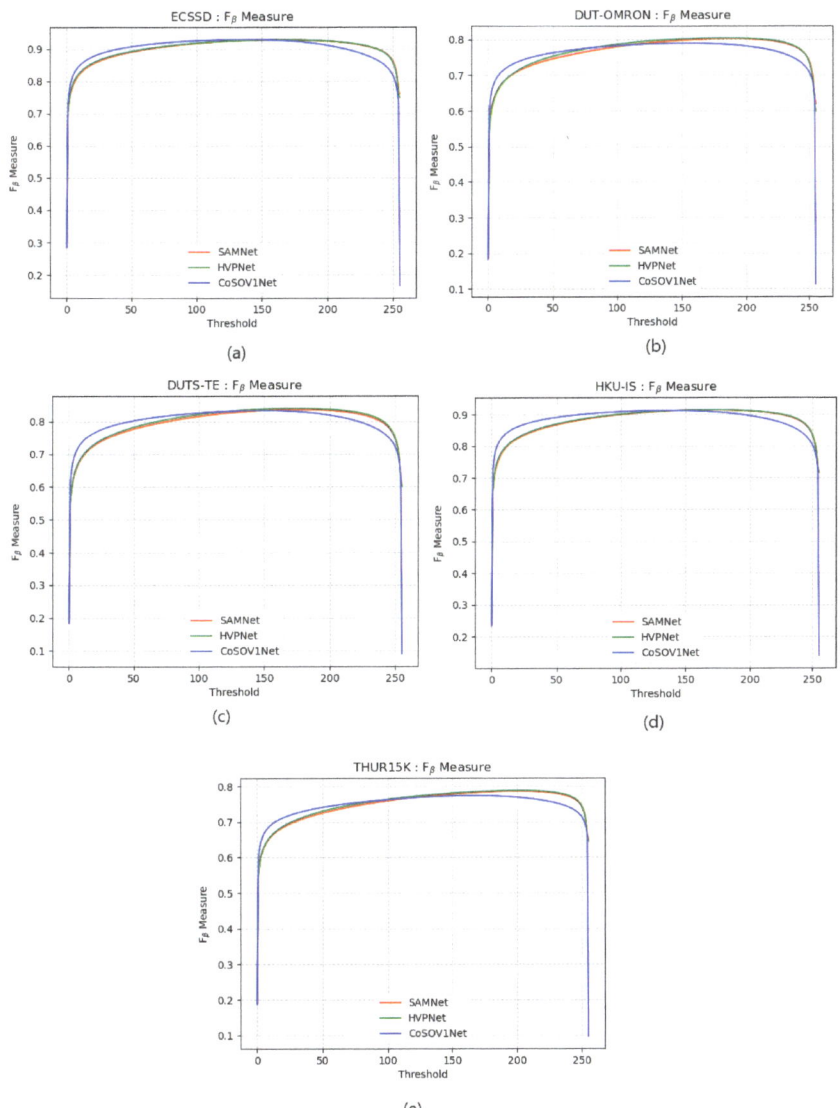

Figure 10. F_β measure curves for (**a**) ECSSD, (**b**) DUT-OMRON, (**c**) DUTS-TE, (**d**) HKU-IS and (**e**) THUR15K datasets.

For qualitative comparison, Figure 11 shows some images highlighting that our proposed model (CoSOV1Net) is competitive with regard to the state-of-the-art SAMNet [16] and HVPNet [19] models, which are among the best ones.

Images from rows 1 and 2 show a big salient object on a cloudy background and a big object on a complex background, respectively: CoSOV1Net (ours) performs better than HVPNet on these saliency maps. Row 3 shows salient objects with the same colors and row 4 shows salient objects with multiple colors: the SAMNet and CoSOV1Net saliency maps are slightly identical and the HVPNet saliency map is slightly better. Row 5 shows n image with three salient objects with different sizes and colors: two are big and one is very small; the CoSOV1Net saliency map is better than SAMNet's and HVPNet's. Row 6 shows red salient objects on a black and yellow background; SAMNet's saliency map is the worst, while CoSOV1Net and HVPNet perform well on that image. Row 7 shows a complex background and multiple salient objects with different colors: CoSOV1Net performs better than SAMNet and HVPNet. Row 8 shows tiny salient objects: the three models perform well. On row 9, SAMNet has the worst performance, while CoSOV1Net is the best. Row 10 shows colored glasses as salient objects: the CoSOV1Net performance is better than SAMNet's and HVPNet's. On row 11, SAMNet has the worst performance. On row 12 and 13, CoSOV1Net has the best performance. Row 18 shows a submarine image: CoSOV1Net is better than SAMNet.

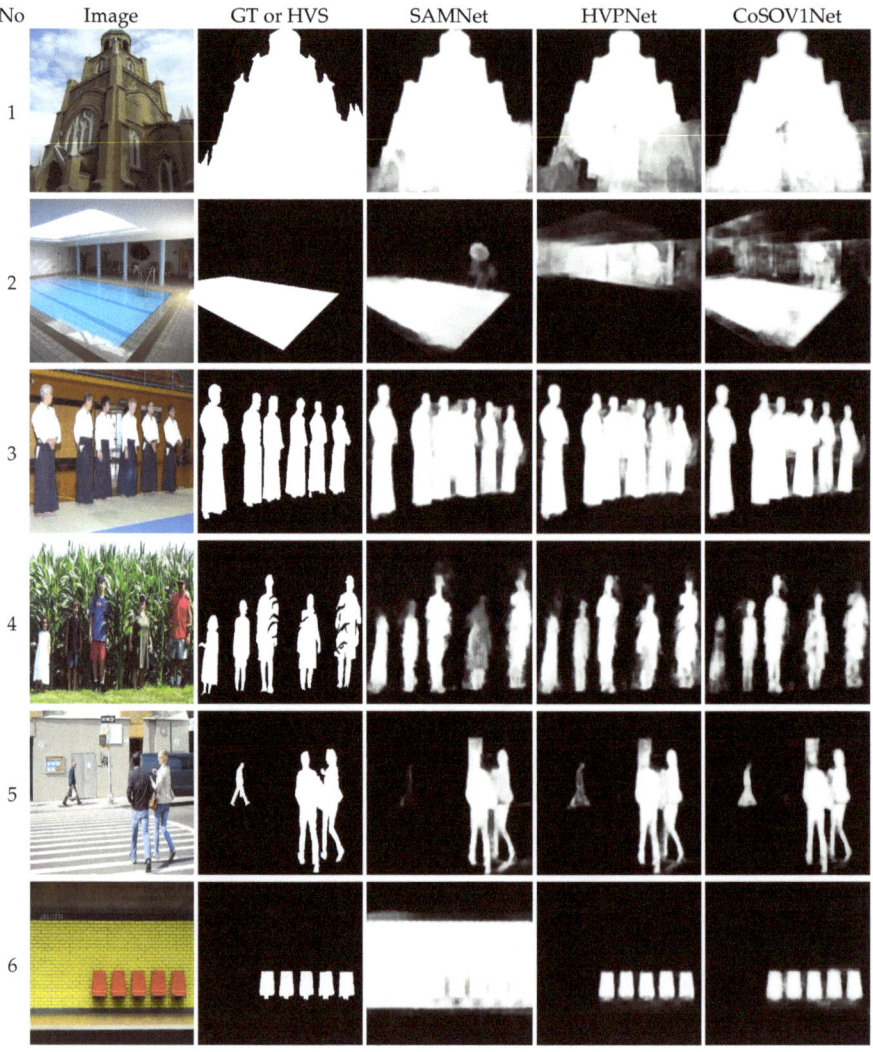

Figure 11. *Cont.*

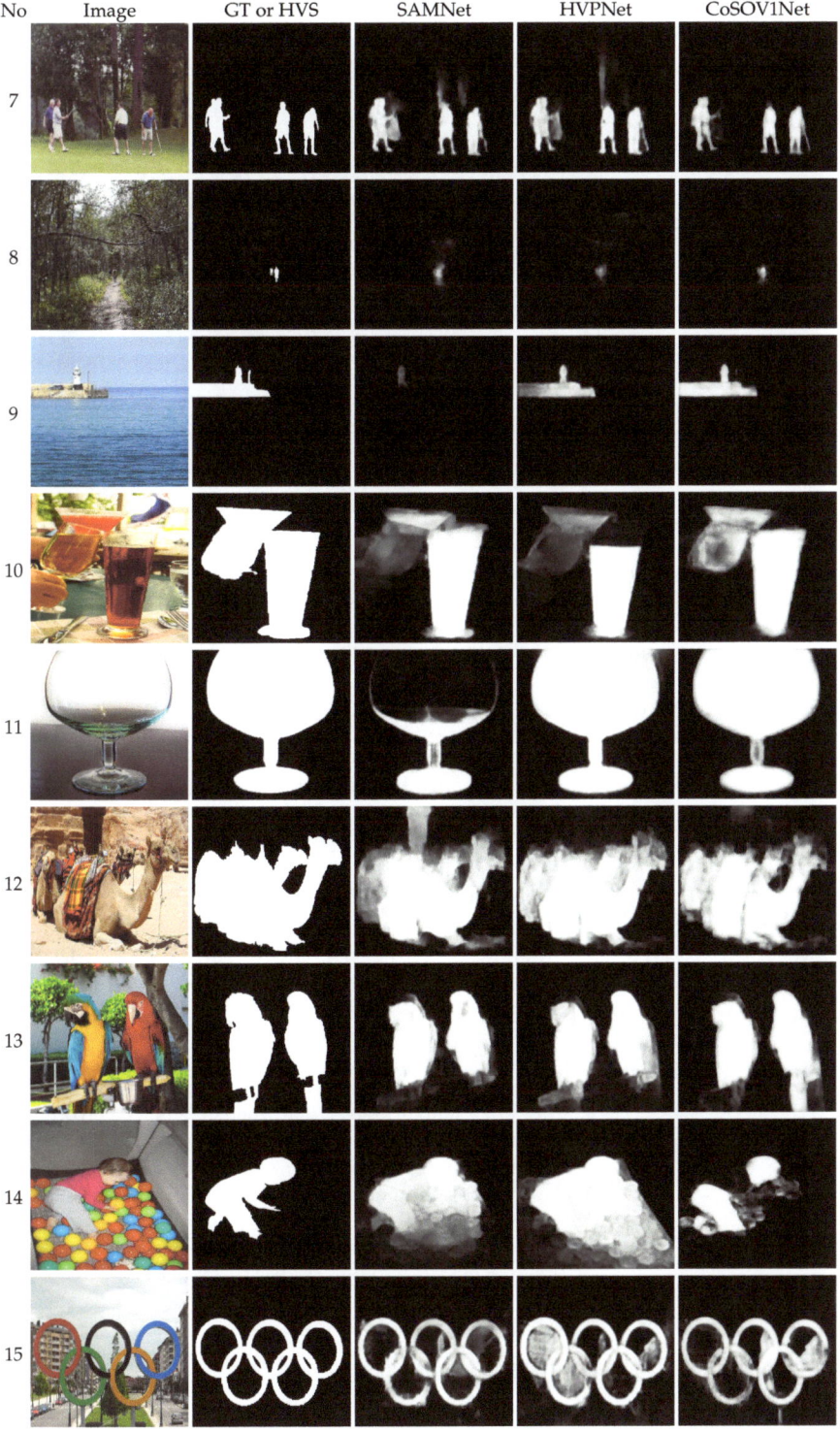

Figure 11. *Cont.*

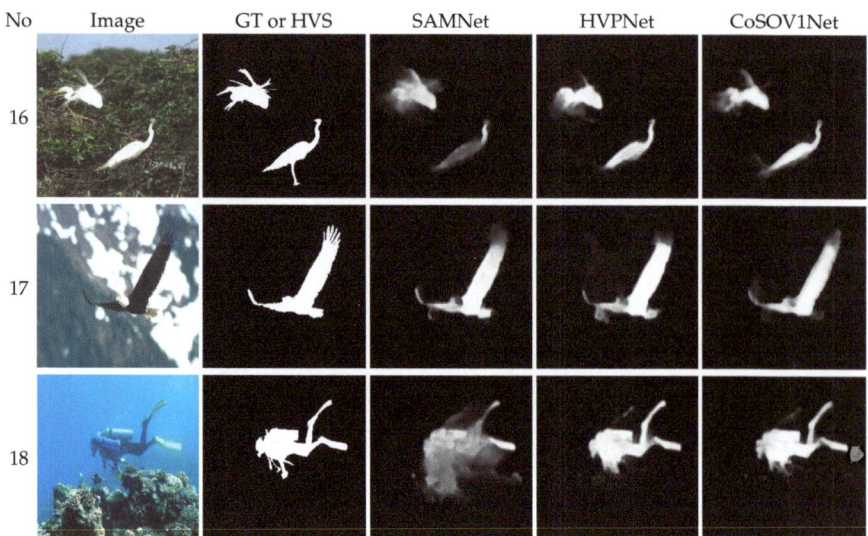

Figure 11. Comparison between SAMNet [16], HVPNet [19] and our proposed model, CoSOV1Net, on some image saliency maps: 1st column: images; 2nd column: ground truth or human visual system saliency map; 3rd column: SAMNet; 4th column: HVPNet; 5th column: CoSOV1Net (ours).

Figures 8–11 confirm that CoSOV1Net has an advantage on performance.

5. Discussion

The results show the performance of our model, CoSOV1Net, for accuracy measures and lightweight measures. CoSOV1Net's rank, when compared to state-of-the-art models, shows that it behaves as a lightweight salient object-detection model by dominating lightweight measures and having good performance for accuracy measures (see Table 7).

Table 7. Our proposed model (CoSOV1Net)'s ranking with respect to existing salient object detection [# Param: number of parameters, ↑: great is the best, ↓: small is the best].

Measure	# Param (M) ↓	FLOPS (G) ↓	Speed (FPS) ↑	ECSSD	DUT-OMRON	DUTS-TE	HKU-IS	THUR15K
F_β	1st	1st	1st	6th	7th	9th	11th	11th
MAE	1st	1st	1st	10th	10th	11th	11th	10th
F_β^ω	1st	1st	1st	11th	9th	11th	15th	11th

The results also show that when CoSOV1Net is compared to state-of-the-art lightweight salient object-detection models, its measure results are generally ranked among the best for the datasets and measures used (see Table 8). Thus, we can conclude that CoSOV1Net behaves as a competitive lightweight salient object-detection model.

Table 8. Our proposed model (CoSOV1Net)'s ranking with respect to lightweight salient object-detection models [# Param: number of parameters, ↑: great is the best, ↓: small is the best].

Measure	# Param (M) ↓	FLOPS (G) ↓	Speed (FPS) ↑	ECSSD	DUT-OMRON	DUTS-TE	HKU-IS	THUR15K
F_β	1st	6th	7th	1st	3rd	3rd	3rd	4th
MAE	1st	6th	7th	1st	1st	1st	1st	2nd
F_β^ω	1st	6th	7th	1st	3rd	1st	3rd	3rd

As we did not use backbones from image classification (i.e., VGG [59], ...) or lightweight backbones (i.e., MobileNets [30,31] or ShuffleNets [32,33]), we conclude that CoSOV1Net's performance is intrinsic to this model itself.

Finally, putting together the measures for salient object-detection models and lightweight salient object-detection models in a graphic, we noticed that the CoSOV1Net model is located for F_β measures with respect to FLOPS and for the number of parameters in the top left, while for the FPS measure, it is located in the top right, thus demonstrating its performance as a lightweight salient object-detection model (see Figure 12). This shows that CoSOV1Net is competitive with the best state-of-the-art models used.

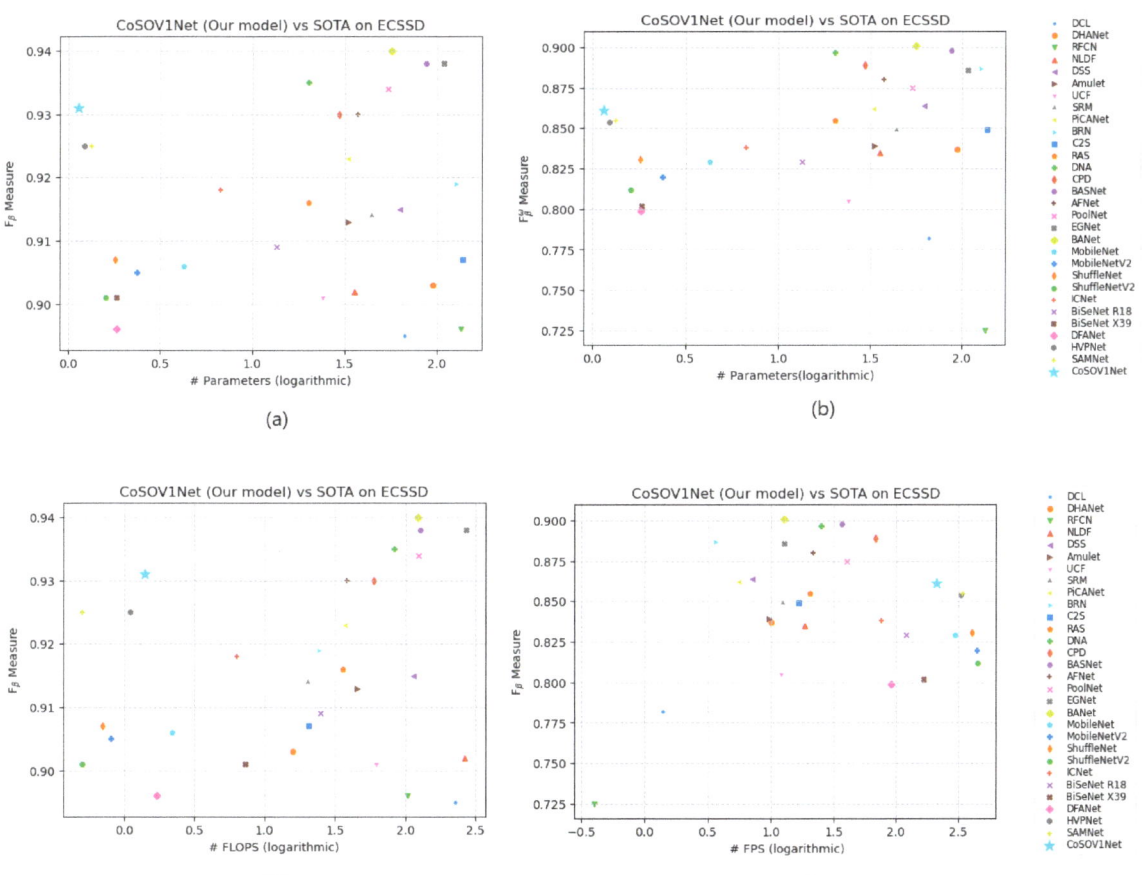

Figure 12. Example of trade-off between (**a**) F_β measure and #parameters; (**b**) F_β^ω measure and #parameters; (**c**) F_β measure and FLOPS; (**d**) F_β measure and FPS, for ECSSD.

The quantitative and the qualitative comparisons with SAMNet [16] and HVPNet [19] showed that our proposed model has good performance, given that these state-of-the-art models are among the best ones.

6. Conclusions

In this work, we present a lightweight salient object-detection deep neural network, CoSOV1Net, with a very low number of parameters (1.14 M), a low floating-point operations number (FLOPS = 1.4 G) and thus low computational cost and respectable speed (FPS = 211.2 on GPU: Nvidia GeForce RTX 3090 Ti), yet with comparable performance with state-of-the-art salient object-detection models that use significantly more parameters, and other lightweight salient object-detection models such as SAMNet [16] and HVPNet [19].

The novelty of our proposed model (CoSOV1Net) is that it uses the principle of integrating color in pattern in a salient object-detection deep neural network, since according to Shapley [27] and Shapley and Hawken [20], color and pattern are inextricably linked in color human perception. This is implemented by taking inspiration from the primary visual cortex (V1) cells, especially cone- and spatial-opponent cells. Thus, our method extracts features at the color channels' spatial level and between the color channels at the same time on a pair of opposing color channels. The idea of grouping color pushed us to group feature maps through the neural network and extract features at the spatial level and between feature maps, as carried out for color channels.

Our results showed that this strategy generates a model that is very promising, competitive with most state-of-the-art salient object-detection and lightweight salient object-detection models and practical for mobile environments and limited-resource devices.

In future work, our proposed CoSOV1Net model, based on integrating color into patterns, can be improved by coupling it with the human visual system attention mechanism, which is the basis of many lightweight models, to tackle its speed limitation and thus produce a more efficient lightweight salient object-detection model.

Author Contributions: Conceptualization, D.N. and M.M.; methodology, D.N.; software, D.N.; validation, D.N.; formal analysis, D.N.; investigation, D.N.; resources, D.N. and M.M.; data curation, D.N.; writing—original draft preparation, D.N.; writing—review and editing, D.N. and M.M.; visualization, D.N.; supervision, M.M.; project administration, M.M.; funding acquisition, M.M. All authors have read and agreed to the published version of the manuscript.

Funding: This research was funded by individual discovery grant number RGPIN-2022-03654.

Institutional Review Board Statement: Not applicable.

Informed Consent Statement: Not applicable.

Data Availability Statement: The ECSSD dataset is available at 09 June 2023. https://www.cse.cuhk.edu.hk/leojia/projects/hsaliency/dataset.html. The DUT-OMRON dataset is available at 09 June 2023. http://saliencydetection.net/dut-omron/. The DUTS-TR and DUTS-TE datasets are available at 09 June 2023 http://saliencydetection.net/duts/. The HKU-IS dataset is available at 09 June 2023. https://i.cs.hku.hk/~yzyu/research/deep_saliency.html. The THUR15K dataset is available at 09 June 2023. https://mmcheng.net/code-data/. The SAMNet [16] model's datasets are available at 09 June 2023. https://github.com/yun-liu/FastSaliency/tree/master/SaliencyMaps/SAMNet. The HVPNet [19] model's datasets are available at 09 June 2023. https://github.com/yun-liu/FastSaliency/tree/master/SaliencyMaps/HVPNet. Our data results (CoSOV1Net saliency maps) are available at 16 July 2023. http://www.iro.umontreal.ca/~mignotte/ResearchMaterial/COSOV1NET-Data/.

Acknowledgments: The authors would like to thank the NSERC (Natural Sciences and Engineering Research Council of Canada) for having supported this research work via the individual discovery grant program (RGPIN-2022-03654).

Conflicts of Interest: The authors declare no conflict of interest.

References

1. Ndayikengurukiye, D.; Mignotte, M. Salient Object Detection by LTP Texture Characterization on Opposing Color Pairs under SLICO Superpixel Constraint. *J. Imaging* **2022**, *8*, 110. [CrossRef] [PubMed]
2. Smeulders, A.W.; Chu, D.M.; Cucchiara, R.; Calderara, S.; Dehghan, A.; Shah, M. Visual tracking: An experimental survey. *IEEE Trans. Pattern Anal. Mach. Intell.* **2013**, *36*, 1442–1468.
3. Pieters, R.; Wedel, M. Attention capture and transfer in advertising: Brand, pictorial and text-size effects. *J. Mark.* **2004**, *68*, 36–50. [CrossRef]
4. Itti, L. Automatic foveation for video compression using a neurobiological model of visual attention. *IEEE Trans. Image Process.* **2004**, *13*, 1304–1318. [CrossRef]
5. Li, J.; Feng, X.; Fan, H. Saliency-based image correction for colorblind patients. *Comput. Vis. Media* **2020**, *6*, 169–189. [CrossRef]

6. Pinciroli Vago, N.O.; Milani, F.; Fraternali, P.; da Silva Torres, R. Comparing CAM algorithms for the identification of salient image features in iconography artwork analysis. *J. Imaging* **2021**, *7*, 106. [CrossRef]
7. Gao, Y.; Shi, M.; Tao, D.; Xu, C. Database saliency for fast image retrieval. *IEEE Trans. Multimed.* **2015**, *17*, 359–369. [CrossRef]
8. Wong, L.K.; Low, K.L. Saliency-enhanced image aesthetics class prediction. In Proceedings of the 2009 16th IEEE International Conference on Image Processing (ICIP), Cairo, Egypt, 7–10 November 2009; pp. 997–1000.
9. Liu, H.; Heynderickx, I. Studying the added value of visual attention in objective image quality metrics based on eye movement data. In Proceedings of the 2009 16th IEEE International Conference on Image Processing (ICIP), Cairo, Egypt, 7–10 November 2009; pp. 3097–3100.
10. Chen, L.Q.; Xie, X.; Fan, X.; Ma, W.Y.; Zhang, H.J.; Zhou, H.Q. A visual attention model for adapting images on small displays. *Multimed. Syst.* **2003**, *9*, 353–364. [CrossRef]
11. Chen, T.; Cheng, M.M.; Tan, P.; Shamir, A.; Hu, S.M. Sketch2photo: Internet image montage. *ACM Trans. Graph. (TOG)* **2009**, *28*, 1–10.
12. Huang, H.; Zhang, L.; Zhang, H.C. Arcimboldo-like collage using internet images. In Proceedings of the 2011 SIGGRAPH Asia Conference, Hong Kong, China, 12–15 December 2011; pp. 1–8.
13. Gupta, A.K.; Seal, A.; Prasad, M.; Khanna, P. Salient object detection techniques in computer vision—A survey. *Entropy* **2020**, *22*, 1174. [CrossRef]
14. Wang, W.; Lai, Q.; Fu, H.; Shen, J.; Ling, H.; Yang, R. Salient object detection in the deep learning era: An in-depth survey. *IEEE Trans. Pattern Anal. Mach. Intell.* **2021**, *44*, 3239–3259. [CrossRef]
15. Gao, S.H.; Tan, Y.Q.; Cheng, M.M.; Lu, C.; Chen, Y.; Yan, S. Highly efficient salient object detection with 100k parameters. In Proceedings of the Computer Vision–ECCV 2020: 16th European Conference, Glasgow, UK, 23–28 August 2020; Proceedings, Part VI; Springer: Berlin/Heidelberg, Germany, 2020; pp. 702–721.
16. Liu, Y.; Zhang, X.Y.; Bian, J.W.; Zhang, L.; Cheng, M.M. SAMNet: Stereoscopically attentive multi-scale network for lightweight salient object detection. *IEEE Trans. Image Process.* **2021**, *30*, 3804–3814. [CrossRef] [PubMed]
17. Liu, N.; Han, J.; Yang, M.H. Picanet: Learning pixel-wise contextual attention for saliency detection. In Proceedings of the IEEE Conference on Computer Vision and Pattern Recognition, Salt Lake City, UT, USA, 18–23 June 2018; pp. 3089–3098.
18. Zhang, P.; Wang, D.; Lu, H.; Wang, H.; Ruan, X. Amulet: Aggregating multi-level convolutional features for salient object detection. In Proceedings of the IEEE International Conference on Computer Vision, Venice, Italy, 22–29 October 2017; pp. 202–211.
19. Liu, Y.; Gu, Y.C.; Zhang, X.Y.; Wang, W.; Cheng, M.M. Lightweight salient object detection via hierarchical visual perception learning. *IEEE Trans. Cybern.* **2020**, *51*, 4439–4449. [CrossRef] [PubMed]
20. Shapley, R.; Hawken, M.J. Color in the cortex: Single-and double-opponent cells. *Vis. Res.* **2011**, *51*, 701–717. [CrossRef] [PubMed]
21. Kruger, N.; Janssen, P.; Kalkan, S.; Lappe, M.; Leonardis, A.; Piater, J.; Rodriguez-Sanchez, A.J.; Wiskott, L. Deep hierarchies in the primate visual cortex: What can we learn for computer vision? *IEEE Trans. Pattern Anal. Mach. Intell.* **2012**, *35*, 1847–1871. [CrossRef]
22. Nunez, V.; Shapley, R.M.; Gordon, J. Cortical double-opponent cells in color perception: Perceptual scaling and chromatic visual evoked potentials. *i-Perception* **2018**, *9*, 2041669517752715. [CrossRef]
23. Conway, B.R. Color vision, cones and color-coding in the cortex. *Neuroscientist* **2009**, *15*, 274–290. [CrossRef]
24. Conway, B.R. Spatial structure of cone inputs to color cells in alert macaque primary visual cortex (V-1). *J. Neurosci.* **2001**, *21*, 2768–2783. [CrossRef]
25. Hunt, R.W.G.; Pointer, M.R. *Measuring Colour*; John Wiley & Sons: Hoboken, NJ, USA, 2011.
26. Engel, S.; Zhang, X.; Wandell, B. Colour tuning in human visual cortex measured with functional magnetic resonance imaging. *Nature* **1997**, *388*, 68–71. [CrossRef]
27. Shapley, R. Physiology of color vision in primates. In *Oxford Research Encyclopedia of Neuroscience*; Oxford University Press: Oxford, UK, 2019.
28. Qin, X.; Zhang, Z.; Huang, C.; Dehghan, M.; Zaiane, O.R.; Jagersand, M. U2-Net: Going deeper with nested U-structure for salient object detection. *Pattern Recognit.* **2020**, *106*, 107404. [CrossRef]
29. Ronneberger, O.; Fischer, P.; Brox, T. U-net: Convolutional networks for biomedical image segmentation. In Proceedings of the Medical Image Computing and Computer-Assisted Intervention–MICCAI 2015: 18th International Conference, Munich, Germany, 5–9 October 2015; Proceedings, Part III 18; Springer: Berlin/Heidelberg, Germany, 2015; pp. 234–241.
30. Howard, A.G.; Zhu, M.; Chen, B.; Kalenichenko, D.; Wang, W.; Weyand, T.; Andreetto, M.; Adam, H. Mobilenets: Efficient convolutional neural networks for mobile vision applications. *arXiv* **2017**, arXiv:1704.04861.
31. Sandler, M.; Howard, A.; Zhu, M.; Zhmoginov, A.; Chen, L.C. Mobilenetv2: Inverted residuals and linear bottlenecks. In Proceedings of the IEEE Conference on Computer Vision and Pattern Recognition, Salt Lake City, UT, USA, 18–23 June 2018; pp. 4510–4520.
32. Zhang, X.; Zhou, X.; Lin, M.; Sun, J. Shufflenet: An extremely efficient convolutional neural network for mobile devices. In Proceedings of the IEEE Conference on Computer Vision and Pattern Recognition, Salt Lake City, UT, USA, 18–23 June 2018; pp. 6848–6856.
33. Ma, N.; Zhang, X.; Zheng, H.T.; Sun, J. Shufflenet v2: Practical guidelines for efficient cnn architecture design. In Proceedings of the European Conference on Computer Vision (ECCV), Munich, Germany, 8–14 September 2018; pp. 116–131.

34. Frintrop, S.; Werner, T.; Martin Garcia, G. Traditional saliency reloaded: A good old model in new shape. In Proceedings of the IEEE Conference on Computer Vision and Pattern Recognition, Boston, MA, USA, 7–12 June 2015; pp. 82–90.
35. Mäenpää, T.; Pietikäinen, M. Classification with color and texture: Jointly or separately? *Pattern Recognit.* **2004**, *37*, 1629–1640. [CrossRef]
36. Chan, C.H.; Kittler, J.; Messer, K. Multispectral local binary pattern histogram for component-based color face verification. In Proceedings of the 2007 First IEEE International Conference on Biometrics: Theory, Applications and Systems, Crystal City, Virginia, 27–29 September 2007; pp. 1–7.
37. Faloutsos, C.; Lin, K.I. *FastMap: A Fast Algorithm for Indexing, Data-Mining and Visualization of Traditional and Multimedia Datasets*; ACM: Rochester, NY, USA, 1995; Volume 24.
38. Jain, A.; Healey, G. A multiscale representation including opponent color features for texture recognition. *IEEE Trans. Image Process.* **1998**, *7*, 124–128. [CrossRef] [PubMed]
39. Yang, K.F.; Gao, S.B.; Guo, C.F.; Li, C.Y.; Li, Y.J. Boundary detection using double-opponency and spatial sparseness constraint. *IEEE Trans. Image Process.* **2015**, *24*, 2565–2578. [CrossRef] [PubMed]
40. Hurvich, L.M.; Jameson, D. An opponent-process theory of color vision. *Psychol. Rev.* **1957**, *64*, 384. [CrossRef] [PubMed]
41. Farabet, C.; Couprie, C.; Najman, L.; LeCun, Y. Learning hierarchical features for scene labeling. *IEEE Trans. Pattern Anal. Mach. Intell.* **2012**, *35*, 1915–1929. [CrossRef]
42. Ioffe, S.; Szegedy, C. Batch normalization: Accelerating deep network training by reducing internal covariate shift. In Proceedings of the International Conference on Machine Learning. Pmlr, Lille, France, 7–9 July 2015; pp. 448–456.
43. Santurkar, S.; Tsipras, D.; Ilyas, A.; Madry, A. How does batch normalization help optimization? *Adv. Neural Inf. Process. Syst.* **2018**, *31*, 2483–2493 .
44. Chollet, F. *Deep Learning with Python*; Simon and Schuster: Manhattan, NY, USA, 2021.
45. Chollet, F. Xception: Deep learning with depthwise separable convolutions. In Proceedings of the IEEE Conference on Computer Vision and Pattern Recognition, Honolulu, HI, USA, 21–26 July 2017; pp. 1251–1258.
46. Ghiasi, G.; Lin, T.Y.; Le, Q.V. Dropblock: A regularization method for convolutional networks. *Adv. Neural Inf. Process. Syst.* **2018**, *31*, 10727–10737.
47. Srivastava, N.; Hinton, G.; Krizhevsky, A.; Sutskever, I.; Salakhutdinov, R. Dropout: A simple way to prevent neural networks from overfitting. *J. Mach. Learn. Res.* **2014**, *15*, 1929–1958.
48. Pietikäinen, M.; Hadid, A.; Zhao, G.; Ahonen, T. *Computer Vision Using Local Binary Patterns*; Springer Science & Business Media: Berlin/Heidelberg, Germany, 2011; Volume 40.
49. Voulodimos, A.; Doulamis, N.; Doulamis, A.; Protopapadakis, E. Deep learning for computer vision: A brief review. *Comput. Intell. Neurosci.* **2018**, *2018*, 7068349. [CrossRef]
50. He, K.; Zhang, X.; Ren, S.; Sun, J. Deep residual learning for image recognition. In Proceedings of the IEEE Conference on Computer Vision and Pattern Recognition, Las Vegas, NV, USA, 27–30 June 2016; pp. 770–778.
51. Chollet, F. Keras. 2015. Available online: https://keras.io (accessed on 9 June 2023).
52. Borji, A.; Cheng, M.M.; Jiang, H.; Li, J. Salient object detection: A benchmark. *IEEE Trans. Image Process.* **2015**, *24*, 5706–5722. [CrossRef] [PubMed]
53. Shi, J.; Yan, Q.; Xu, L.; Jia, J. Hierarchical image saliency detection on extended CSSD. *IEEE Trans. Pattern Anal. Mach. Intell.* **2016**, *38*, 717–729. [CrossRef] [PubMed]
54. Yang, C.; Zhang, L.; Lu, H.; Ruan, X.; Yang, M.H. Saliency detection via graph-based manifold ranking. In Proceedings of the IEEE Conference on Computer Vision and Pattern Recognition, Washington, DC, USA, 23–28 June 2013; pp. 3166–3173.
55. Wang, L.; Lu, H.; Wang, Y.; Feng, M.; Wang, D.; Yin, B.; Ruan, X. Learning to detect salient objects with image-level supervision. In Proceedings of the IEEE Conference on Computer Vision and Pattern Recognition, Honolulu, HI, USA, 21–26 July 2017; pp. 136–145.
56. Li, G.; Yu, Y. Visual saliency based on multiscale deep features. In Proceedings of the IEEE Conference on Computer Vision and Pattern Recognition, Boston, MA, USA, 7–12 June 2015; pp. 5455–5463.
57. Cheng, M.M.; Mitra, N.J.; Huang, X.; Hu, S.M. Salientshape: Group saliency in image collections. *Vis. Comput.* **2014**, *30*, 443–453. [CrossRef]
58. Feng, W.; Li, X.; Gao, G.; Chen, X.; Liu, Q. Multi-scale global contrast CNN for salient object detection. *Sensors* **2020**, *20*, 2656. [CrossRef] [PubMed]
59. Simonyan, K.; Zisserman, A. Very deep convolutional networks for large-scale image recognition. *arXiv* **2014**, arXiv:1409.1556.
60. He, K.; Zhang, X.; Ren, S.; Sun, J. Delving deep into rectifiers: Surpassing human-level performance on imagenet classification. In Proceedings of the IEEE International Conference on Computer Vision, Santiago, Chile, 7–13 December 2015; pp. 1026–1034.
61. Tieleman, T.; Hinton, G. Lecture 6.5-rmsprop: Divide the gradient by a running average of its recent magnitude. *COURSERA Neural Netw. Mach. Learn.* **2012**, *4*, 26–31.
62. Margolin, R.; Zelnik-Manor, L.; Tal, A. How to evaluate foreground maps? In Proceedings of the IEEE Conference on Computer Vision and Pattern Recognition, Washington, DC, USA, 23–28 June 2014; pp. 248–255.
63. Varadarajan, V.; Garg, D.; Kotecha, K. An efficient deep convolutional neural network approach for object detection and recognition using a multi-scale anchor box in real-time. *Future Internet* **2021**, *13*, 307. [CrossRef]

64. Jiang, H.; Wang, J.; Yuan, Z.; Wu, Y.; Zheng, N.; Li, S. Salient object detection: A discriminative regional feature integration approach. In Proceedings of the IEEE Conference on Computer Vision and Pattern Recognition, Portland, OR, USA, 23–28 June 2013; pp. 2083–2090.
65. Li, G.; Yu, Y. Deep contrast learning for salient object detection. In Proceedings of the IEEE Conference on Computer Vision and Pattern Recognition, Las Vegas, NV, USA, 27–30 June 2016; pp. 478–487.
66. Liu, N.; Han, J. Dhsnet: Deep hierarchical saliency network for salient object detection. In Proceedings of the IEEE Conference on Computer Vision and Pattern Recognition, Las Vegas, NV, USA, 27–30 June 2016; pp. 678–686.
67. Wei, J.; Zhong, B. Saliency detection using fully convolutional network. In Proceedings of the 2018 Chinese Automation Congress (CAC), Xi'an, China, 30 November–2 December 2018; pp. 3902–3907.
68. Luo, Z.; Mishra, A.; Achkar, A.; Eichel, J.; Li, S.; Jodoin, P.M. Non-local deep features for salient object detection. In Proceedings of the IEEE Conference on Computer Vision and Pattern Recognition, Honolulu, HI, USA, 21–26 July 2017; pp. 6609–6617.
69. Hou, Q.; Cheng, M.M.; Hu, X.; Borji, A.; Tu, Z.; Torr, P.H. Deeply supervised salient object detection with short connections. In Proceedings of the IEEE Conference on Computer Vision and Pattern Recognition, Honolulu, HI, USA, 21–26 July 2017; pp. 3203–3212.
70. Zhang, P.; Wang, D.; Lu, H.; Wang, H.; Yin, B. Learning uncertain convolutional features for accurate saliency detection. In Proceedings of the IEEE International Conference on Computer Vision, Venice, Italy, 22–29 October 2017; pp. 212–221.
71. Wang, T.; Borji, A.; Zhang, L.; Zhang, P.; Lu, H. A stagewise refinement model for detecting salient objects in images. In Proceedings of the IEEE International Conference on Computer Vision, Venice, Italy, 22–29 October 2017; pp. 4019–4028.
72. Wang, T.; Zhang, L.; Wang, S.; Lu, H.; Yang, G.; Ruan, X.; Borji, A. Detect globally, refine locally: A novel approach to saliency detection. In Proceedings of the IEEE Conference on Computer Vision and Pattern Recognition, Salt Lake City, UT, USA, 18–23 June 2018; pp. 3127–3135.
73. Li, X.; Yang, F.; Cheng, H.; Liu, W.; Shen, D. Contour knowledge transfer for salient object detection. In Proceedings of the European Conference on Computer Vision (ECCV), Munich, Germany, 8–14 September 2018; pp. 355–370.
74. Chen, S.; Tan, X.; Wang, B.; Hu, X. Reverse attention for salient object detection. In Proceedings of the European Conference on Computer Vision (ECCV), Munich, Germany, 8–14 September 2018; pp. 234–250.
75. Liu, Y.; Cheng, M.M.; Zhang, X.Y.; Nie, G.Y.; Wang, M. DNA: Deeply supervised nonlinear aggregation for salient object detection. *IEEE Trans. Cybern.* **2021**, *52*, 6131–6142. [CrossRef]
76. Wu, Z.; Su, L.; Huang, Q. Cascaded partial decoder for fast and accurate salient object detection. In Proceedings of the IEEE/CVF Conference on Computer Vision and Pattern Recognition, Long Beach, CA, USA, 15–20 June 2019; pp. 3907–3916.
77. Qin, X.; Zhang, Z.; Huang, C.; Gao, C.; Dehghan, M.; Jagersand, M. Basnet: Boundary-aware salient object detection. In Proceedings of the IEEE/CVF Conference on Computer Vision and Pattern Recognition, Long Beach, CA, USA, 15–20 June 2019; pp. 7479–7489.
78. Feng, M.; Lu, H.; Ding, E. Attentive feedback network for boundary-aware salient object detection. In Proceedings of the IEEE/CVF Conference on Computer Vision and Pattern Recognition, Long Beach, CA, USA, 15–20 June 2019; pp. 1623–1632.
79. Liu, J.J.; Hou, Q.; Cheng, M.M.; Feng, J.; Jiang, J. A simple pooling-based design for real-time salient object detection. In Proceedings of the IEEE/CVF Conference on Computer Vision and Pattern Recognition, Long Beach, CA, USA, 15–20 June 2019; pp. 3917–3926.
80. Zhao, J.X.; Liu, J.J.; Fan, D.P.; Cao, Y.; Yang, J.; Cheng, M.M. EGNet: Edge guidance network for salient object detection. In Proceedings of the IEEE/CVF International Conference on Computer Vision, Seoul, Korea, 27 October–2 November 2019; pp. 8779–8788.
81. Su, J.; Li, J.; Zhang, Y.; Xia, C.; Tian, Y. Selectivity or invariance: Boundary-aware salient object detection. In Proceedings of the IEEE/CVF International Conference on Computer Vision, Seoul, Korea, 27 October–2 November 2019; pp. 3799–3808.
82. Zhao, H.; Qi, X.; Shen, X.; Shi, J.; Jia, J. Icnet for real-time semantic segmentation on high-resolution images. In Proceedings of the European Conference on Computer Vision (ECCV), Munich, Germany, 8–14 September 2018; pp. 405–420.
83. Yu, C.; Wang, J.; Peng, C.; Gao, C.; Yu, G.; Sang, N. Bisenet: Bilateral segmentation network for real-time semantic segmentation. In Proceedings of the European conference on computer vision (ECCV), Munich, Germany, 8–14 September 2018; pp. 325–341.
84. Li, H.; Xiong, P.; Fan, H.; Sun, J. Dfanet: Deep feature aggregation for real-time semantic segmentation. In Proceedings of the IEEE/CVF Conference on Computer Vision and Pattern Recognition, Long Beach, CA, USA, 15–20 June 2019; pp. 9522–9531.

Disclaimer/Publisher's Note: The statements, opinions and data contained in all publications are solely those of the individual author(s) and contributor(s) and not of MDPI and/or the editor(s). MDPI and/or the editor(s) disclaim responsibility for any injury to people or property resulting from any ideas, methods, instructions or products referred to in the content.

Article

A Novel Approach for Apple Freshness Prediction Based on Gas Sensor Array and Optimized Neural Network

Wei Wang [1], Weizhen Yang [1], Maozhen Li [1,2,*], Zipeng Zhang [1] and Wenbin Du [1]

1 School of Information and Communication Engineering, North University of China, Taiyuan 030051, China
2 Department of Electronic and Electrical Engineering, Brunel University London, Uxbridge UB8 3PH, UK
* Correspondence: maozhen.li@brunel.ac.uk

Abstract: Apple is an important cash crop in China, and the prediction of its freshness can effectively reduce its storage risk and avoid economic loss. The change in the concentration of odor information such as ethylene, carbon dioxide, and ethanol emitted during apple storage is an important feature to characterize the freshness of apples. In order to accurately predict the freshness level of apples, an electronic nose system based on a gas sensor array and wireless transmission module is designed, and a neural network prediction model using an improved Sparrow Search Algorithm (SSA) based on chaotic sequence (Tent) to optimize Back Propagation (BP) is proposed. The odor information emitted by apples is studied to complete an apple freshness prediction. Furthermore, by fitting the relationship between the prediction coefficient and the input vector, the accuracy benchmark of the prediction model is set, which further improves the prediction accuracy of apple odor information. Compared with the traditional prediction method, the system has the characteristics of simple operation, low cost, reliable results, mobile portability, and it avoids the damage to apples in the process of freshness prediction to realize non-destructive testing.

Keywords: gas sensor array; freshness prediction; chaotic sequence; sparrow search

Citation: Wang, W.; Yang, W.; Li, M.; Zhang, Z.; Du, W. A Novel Approach for Apple Freshness Prediction Based on Gas Sensor Array and Optimized Neural Network. *Sensors* **2023**, *23*, 6476. https://doi.org/10.3390/s23146476

Academic Editor: Maria Strianese

Received: 16 June 2023
Revised: 13 July 2023
Accepted: 15 July 2023
Published: 17 July 2023

Copyright: © 2023 by the authors. Licensee MDPI, Basel, Switzerland. This article is an open access article distributed under the terms and conditions of the Creative Commons Attribution (CC BY) license (https://creativecommons.org/licenses/by/4.0/).

1. Introduction

1.1. Background

Apples are one of the most commonly consumed fruits by people. China's apple production accounts for one-seventh of the world's output, and it is an important cash crop in China. The freshness of apples is the most important indicator to evaluate the quality of apples, which directly affects the sales of apples. If the shelf life of apples can be accurately predicted, it will provide an effective guarantee for quality and output value.

Fruit and vegetable freshness prediction technology has a long history, and its freshness prediction methods [1] mainly include fuzzy sense, dielectric property, mechanical property, acoustic property, near-infrared spectroscopy, and electronic nose detection technology. Fuzzy sense mainly relies on individuals to judge the feel, smell, and experience of objects, which are highly subjective. The dielectric property is detected by using the dielectric constant of the fruit, which can be used for the detection of fruit sugar content and moisture content. Acoustic characteristics are detected using acoustic properties such as fruit reflection, scattering, transmission, and attenuation. Kinetic modeling is a technique that uses the relevant mechanical properties of fruits for testing. Near-infrared spectroscopy is the use of fruit to detect the absorption, reflection, scattering, transmission, and other characteristics of light. The above four detection methods are generally for a single detection object, which needs to be judged one by one; the detection efficiency is relatively low; and the requirements for equipment are high.

The smell of the same fruit at different growth stages is different. The odor between different varieties is also different, and the electronic nose is used to simulate the biological olfactory function to analyze and identify the odor for detection. In order to meet the requirements of rapid and non-destructive real-time monitoring of food freshness, electronic nose technology based on gas sensors has developed rapidly, and in recent years, more and more research on the freshness prediction of food has been applied by electronic nose technology.

In 2008, Antihus [2] used the PEN2 electronic nose, principal component analysis method, and LDA (Linear Discriminant Analysis) algorithm to monitor the shelf life of tomato storage, which realized the monitoring and differentiation of tomatoes with different storage times but did not realize the freshness prediction of tomatoes.

In 2013, Hui Guohua [3] proposed a storage time prediction method for Fuji apples based on electronic nose, which used the random resonance method to calculate the gas concentration data collected by electronic nose and established a prediction relationship with the storage time of apples, which detected a single apple sampling time of 3 h and a long detection time.

In 2016, Alireza Sanaeifar [4] used a low cost electronic nose to detect bananas with different shelf lives and used SVM (support vector machine) technology to predict various quality indicators of bananas, which had a good prediction effect on soluble solids and hardness but a poor prediction effect on PH and titratable acid, so that the overall freshness prediction effect of bananas did not reach the expected level.

In 2019, Wojciech Wojnowski [5] et al. proposed a prediction method for the bioamine index of refrigerated chicken based on an electronic nose. Using a modular electronic nose and a special sample chamber to analyze the volatile components of chicken and a BP neural network for data modeling, the results show that it can accurately predict the biogenic amine index of chicken. In this experiment, the biogenic amine index of chicken was predicted, but the quality of the chicken was not evaluated.

In 2020, Parthasarathy Srinivasan [6] used a self-designed electronic nose to predict the quality changes of Pacific white shrimp during storage and determined its quality by measuring PH value, determining microbial content, texture analysis, and sensory evaluation, and the identification rate of white shrimp stored at low temperatures was as high as 96.29% through the Soft-max algorithm.

In 2019, Feng Lei [7] from Jiangnan University applied electronic nose and low-field nuclear magnetic resonance technology to study the freshness of cucumbers, cherries, and tomatoes, and by monitoring their flavor characteristics and the change of moisture status, the PLS (Partial Least-Squares) algorithm model was used to predict the hardness, soluble solids, and color difference of cucumbers and tomatoes and the quality changes of cucumbers during storage. In the model, the detection cost of low-field NMR (Nuclear Magnetic Resonance) technology is high, and it is difficult to popularize and practice.

In 2020, Chen Shaoxia [8] from Nanjing Agricultural University used an electronic nose and near-infrared spectroscopy to predict the quality loss rate, VC (vitamin C) content, hardness, and other quality indicators of baby vegetables during storage. The results show that the combination of the two technologies has a good predictive effect on quality indicators. However, the combination of the two technologies increases the detection cost and makes the experimental process too complex and cumbersome. The infrared spectroscopy technology has high requirements for light source selection, which is not suitable for large-scale promotion.

In 2022, Zhang Man [9] proposed a method for predicting the freshness of cold fresh mutton based on a BP neural network using gas information and established a prediction model for the physical and chemical indicators (hardness, pH value, color, TVB-N (total volatile basic nitrogen) content) that characterize the freshness of mutton by detecting the environmental gas content. The results showed that the coefficient of prediction of physicochemical properties was above 0.9, indicating that the BP neural network has a good prediction effect. In the implementation method, the amount of data is large, and the training time is long.

In summary, through the analysis of the research status at home and abroad, the use of an electronic nose to predict the freshness of vegetables and fruits is achievable, and the detection process has the advantages of non-contact and batch detection. At present, among the relevant detection methods, there are problems such as a complex detection process, a high detection cost, an excessive amount of collected data, and a long detection time.

1.2. Related Work

In recent years, our research group has completed a number of studies on electronic nose freshness detection technology for apples. Guo et al. [10] built an apple freshness test platform using odor recognition technology combined with fuzzy sensory algorithms and dielectric properties to measure apple quality, and the accuracy of apple freshness determination was 93.75%. Liu et al. [11] used a self-made odor recognition system to complete a rapid evaluation of the classification of the freshness of Fuji apples. The principal component analysis algorithm was used to detect the freshness features of apples within 1 min, with an accuracy of 95.33%. Yan et al. [12] connected the smell of apples with their sweetness through the gas sensor array and realized the classification of the sweetness of apples. The CPSO-BP (BP optimization by Chaotic Particle Swarm Optimization) neural network algorithm was adopted with an accuracy of 83.33%, which was comparable to the detection accuracy of commercial near-infrared spectroscopy analyzers and realized the non-destructive testing of the sweetness of apples.

Among the existing research results [13], an odor recognition system for evaluating the freshness of Fuji apples was designed. By collecting and detecting the aroma emitted by apples, cluster analysis and a classification model are established using stable system response values. The continuous projection algorithm is used to optimize the sensor array, solve the collinearity and overlap problems, and eliminate abnormal and redundant sensors. It uses a ZigBee wireless sensor network to send data to the upward computer and uses a BP neural network algorithm optimized by the hybrid leapfrog algorithm to recognize the gas data, which improves the training speed and accuracy of the neural network. The experimental results showed that the accuracy of the method was 98.67%, and it could identify the freshness of Fuji apples quickly and comprehensively.

Apple odor information is feasible and reliable for apple quality detection, and on this basis, apple freshness prediction is further realized. Combined with the existing research results, The main work of this study is as follows: (1) This study detects the gas concentration released by apples during storage by the self-designed electronic nose system, accurately characterizes the freshness of apples by using a sensor array composed of ethylene, ethanol, oxygen, and carbon dioxide, and uses a WSN (wireless sensor network) as a means of information transmission. (2) Taking the prediction of the future freshness of apples as the starting point, the SSA optimization BP neural network added to Tent is used to further optimize the network to complete the prediction of apple odor characteristics. The Tent-SSA-BP model for apple freshness prediction is established, and finally, low cost, lossless, and efficient apple freshness prediction is realized.

2. Technical Principles

2.1. Freshness Prediction System

Using a self-made dual-chip wireless acquisition and processing system, the overall function is completed by two main chips with each other, with gas concentration collection, data information processing, wireless transmission, processing result display, and the ability to work with other nodes. The block diagram of the system structure is shown in Figure 1. WSN nodes are used to realize wireless transmission during acquisition. One of the sensor arrays is connected and placed in the container where the apple is stored, and the other node is connected to the host computer to transmit data back and be processed by the host computer. Then multiple nodes can be added to form a wireless acquisition and prediction network. The actual self-made hardware is shown in Figure 2.

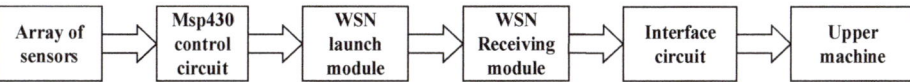

Figure 1. Block diagram of acquisition system.

Figure 2. Information processing center module (**left**), Information collection (**right**).

The sensor array is connected to the WSN transmitter module and placed in a gas environment containing the sample to be measured, which converts the gas concentration into an electrical signal packaged by the transmitter module. The receiving module is connected to the host computer, receives the data packets from the transmitting module, and uploads the data to the host computer through the serial port output for storage. The control chip of the whole acquisition system is Msp430F449, which has complete functions and can complete the preliminary data processing task. The ZigBee wireless transmission module adopts the CC2530 RF module circuit and supports the ZigBee2007Ztack protocol stack. The wireless transmission of data has the advantages of saving costs and making the system more convenient.

2.2. Choice of Sensor

According to the types of gases involved in the physiological action of apples after picking, the optimal sensor combination in the actual environment is selected to ensure the accuracy of gas concentration collection. According to the previous research and research results [14,15], the main response gases are selected as ethylene, ethanol, oxygen, and carbon dioxide gas sensors to form a sensor array, and the type of selection and performance indicators of the sensors are shown in Table 1. The sensor array composed of four types of sensors selected has a good response characteristic curve to the volatile gas of apples, and effectively improves the cross-sensitivity characteristics between the sensors during the detection process and the identification accuracy of apple odor in the experiment.

Table 1. Composition of sensor array.

Type of Sensor	Mainly Measured Gas	Measuring Range	Working Voltage/V
MQ3	ethanol	$(25\sim500) \times 10^{-6}$ ppm	5.0
MG811	Carbon dioxide	$(0\sim10,000) \times 10^{-6}$ ppm	6.0
ME2-O2	oxygen	0~25% Vol	3.3
ME3-C2H4	ethylene	$(0\sim100) \times 10^{-6}$ ppm	5.0

The advantage of using the sensor module is that after the power sensor module is turned on and the sensor module starts to work, the communication and data transmission between the main control chip can be realized through the general IO port, In terms of the collected data, the concentration value, voltage, and current value can be collected through the program according to its own needs. In addition, the use of ready-made sensor modules can also reduce the volume of the main circuit board, and you only need to reserve different voltage power supply interfaces and IO ports required for communication according to different sensor power supply needs in the design work. It also provides convenience for replacing the sensor, as the operation of the sensor ages and different environmental needs change, so it is necessary to detect and replace the sensor in time to ensure the stability and correctness of the collected data.

3. Algorithm Framework and Principle

The collected apple odor data is divided into a training set and a test set and imported into the optimized neural network for modeling and training, and the subsequent concentration change is predicted according to the current gas concentration of apple samples. The apple freshness is identified and classified by the predicted concentration, so as to realize the prediction of apple freshness.

3.1. Sparrow Search Algorithm

The Sparrow Search Algorithm (SSA) [16] is a new type of meta-heuristic algorithm. In the algorithm, individuals are divided into discoverers, watchers, and followers, with each individual corresponding to a solution. In the process of algorithmic foraging, the positions of the three are continuously updated to complete the resource acquisition. Compared with other optimization algorithms, this one is easy to implement and has relatively few control parameters and strong local search ability. In order to avoid falling into the local optimum, the tent chaotic sequence is introduced for optimization, which increases the population diversity, thereby improving the search and exploitation performance of the algorithm and increasing its stability [17].

The specific implementation steps are shown in Table 2:

Table 2. Tent-SSA implementation steps.

Step Number	Step Content
1	Set the population size N, the number of discoverers p_{num}, the number of reconnaissance warnings s_{num}, the target function dimension D, the initial values of the upper and lower bounds ub and lb, respectively, the maximum number of iterations T, and the solution accuracy ε.
2	Tent is applied to generate N D-dimensional vectors Z_i, and within the range of values of the variable by the carrier $X_{new}^d = d_{min} + (d_{max} - d_{min})Z_i$, d_{min} and d_{max} are, respectively, the minimum and maximum values of the d-dimension vector X_{new}^d.
3	Calculate the fitness f_i and select the optimal fitness f_g, the worst fitness f_w, and the corresponding positions xb and xw, respectively.
4	Select the first p_{num} with good fitness to be the discoverer, and the rest as joiners; update the locations of the discoverer and joiner.
5	Randomly select s_{num} as the vigilant and update their position.
6	One iteration was performed to calculate the fitness f_i and average fitness f_{avg} for each animal. Perform an iteration to calculate the fitness and average fitness of each animal
7	If $f_i \geq f_{avg}$, the individual is chaotically disturbed, if the individual's performance is better after disturbance, it will replace the previous individual; otherwise, it will remain unchanged.
8	If $f_i < f_{avg}$, $mutation(x) = x(1 + N(0,1))$ was used to conduct Gaussian mutations on individuals. x represents the original parameter value, $N(0,1)$ represents the normally distributed random number, and $mutation(x)$ is the value after Gaussian mutation.
9	Update the optimal position xb and its fitness and the worst position xw and its fitness in the entire population.
10	Determine whether the maximum number of iterations or solution accuracy is reached, and if so, the loop ends and outputs the optimal parameter result.
11	If not, go back to Step 6 and iterate again.

3.2. Optimized BP Neural Network

According to the prediction of concentration change, the shelf life of apples is predicted, and the sparrow search algorithm is improved by the Tent chaotic sequence. SSA is used to optimize the BP neural network in order to complete the prediction of apple odor characteristics. Tent-SSA optimizes BP in two aspects: one is to optimize the weight threshold of BP by using the optimization function of the optimization algorithm; Second, the input layer is optimized, and the structure of the input matrix is changed by setting the expected value and the number of cycles to find the best input matrix suitable for prediction and the number of input nodes of BP, where the setting of the expected threshold is further calculated by the fitting function of the relationship between the coefficient of determination and the input vector. The optimized flowchart is shown in Figure 3.

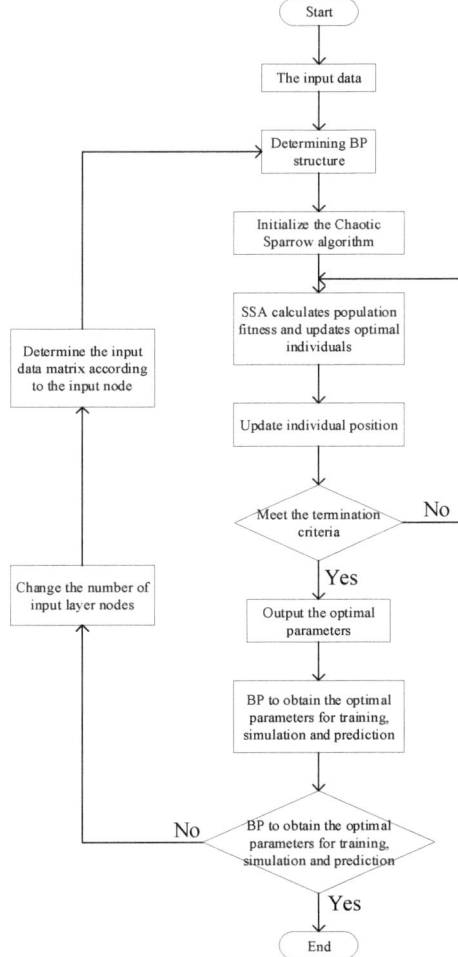

Figure 3. Flowchart of the algorithm.

The specific implementation steps are shown in Table 3.

Table 3. Optimized BP implementation prediction steps.

Step Number	Step Content
1	Initialize. According to the input matrix, determine the BP topology, initialize the maximum number of iterations T0, and determine the coefficient of determination ε.
2	Use an optimized sparrow search algorithm for iterative optimization through Tent.
3	The SSA algorithm is completed, the optimal parameters are output, and the values are assigned to the BP network for prediction and calculate ε.
4	According to $\varepsilon_0 = \begin{array}{l} 1.396 * 10^{-6}x^8 - 2.548 * 10^{-5}x^7 - 0.0003126x^6 + 0.01221x^5 \\ -0.1327x^4 + 0.701x^3 - 1.905x^2 + 2.461x - 0.2368 \end{array}$, calculate the quality threshold value ε_0 in this prediction result.
5	Make a judgement. If $\varepsilon < \varepsilon_0$, adjust the input matrix and return to Step 1 until the number of iterations T0 is reached or Step 6 is satisfied.
6	Make a judgement. If $\varepsilon > \varepsilon_0$, the loop ends, and the predicted result is output.

The pros and cons of the prediction model have a great relationship with the coefficient of determination ε_0, and according to multiple experiments and statistical analysis of data, there is a correlation between the prediction results of the model and the arrangement of the input matrix. After a fixed number of iterations, the fitting curve between ε_0 and the input vector obtained by multiple fitting using MATLAB is shown in Figure 4.

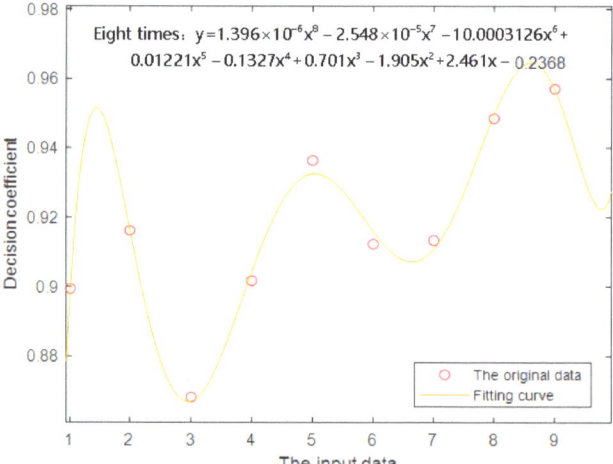

Figure 4. Relation diagram of fitting curve.

It can be seen from the figure that ε_0 shows a wave upward trend with the increase in the number of input vectors. Freshness prediction is the analysis of concentration change in days, and the number of input vectors should not be too large. The relationship between derived ε_0 and input vectors is shown in the following Equation (1), where x represents the number of input vectors.

$$\varepsilon_0 = \begin{array}{l} 1.396 \times 10^{-6}x^8 - 2.548 \times 10^{-5}x^7 - 0.0003126x^6 + 0.01221x^5 \\ -0.1327x^4 + 0.701x^3 - 1.905x^2 + 2.461x - 0.2368 \end{array} \quad (1)$$

The goodness-of-fit degree verified by the results is above 0.96, and ε_0 is used as the key parameter for model prediction. Through experimental testing, under indoor conditions of room temperature of 20 degrees Celsius and humidity of 50% RH, apples go from fresh to rotten in about 40 days. Each apple sample in the 800 mL container has a gas emission stability value as the characteristic value of the sample. By detecting the change in the characteristic value of apples within 40 days, a prediction model is established.

4. Freshness Classification Basis

In the process of storage, the physical and chemical changes of apples will lead to changes in the water content, freshness, and content and arrangement of some organic substances. The physiological tissue components inside apples can be regarded as an unconventional dielectric, and these physical and chemical changes will be further manifested in the internal biomolecules of apples in the change of charge arrangement response characteristics. Their macroscopic manifestation is the change of dielectric characteristics and its parameters, so the dielectric constant can be used to express the state of apple freshness.

Apple freshness feature detection technology based on dielectric properties has been relatively mature. In this experiment, a TH2822A handheld LCR instrument combined with a computer and a shielded box with plates was used to complete the detection of the dielectric constant of Fuji apples and classify the freshness of apples, and the relationship between the dielectric characteristics of Fuji apples and freshness is shown in the following table [10,11]. Among them, apples are divided into three categories: Fresh (apples without any wrinkles or shrinkage phenomena, not rotten); Not freshness (apples shrink after a period of time and do not decay); Decay (apples appear rotten).

5. Results and Analysis

5.1. Materials and Methods

Since the apples are picked, the cells in the fruit continue to respire, consuming oxygen while producing ethanol and carbon dioxide. In addition, ethylene is closely related to the ripeness of apples. Four gases: ethanol, ethylene, oxygen, and carbon dioxide, are selected to establish the apple freshness model. A number of Fuji apples picked from the same batch and purchased at the same time in the same market are selected and stored at 20 degrees Celsius indoors, and the volatile gas of each sample is sampled every 24 h to record the gas concentration change in the process from freshness to rot.

Based on the odor characteristics of apples, a BP neural network classification model was established. The eigenvalues of each apple sample were taken from the stable response values of the four sensors; the eigen input signal of the apple was four dimensions, and the result to be classified was three types, according to the Kolmogorov theorem combined with multiple experiments to verify that the number of optimal hidden layer nodes was 9, so the structure of the neural network was 4-9-3.

A total of 180 groups of pre-processed sample feature signals were selected, and 144 groups were randomly selected for network training. And it defined the expected output of each type of apple. For example, the expected output of a fresh apple sample is [1, 0, 0]; Not fresh is [0, 1, 0]; Decay is [0, 0, 1].

5.2. The Data Collection

Apples of varying degrees of freshness, from fresh to rotten, are tested separately. During the detection process, apple samples and fully preheated gas sensor arrays are put into the container, one sample at a time. The odor information concentration change of each sample in the container for 15 min is collected, and the relevant gas concentration change curve detected is shown in Figure 5.

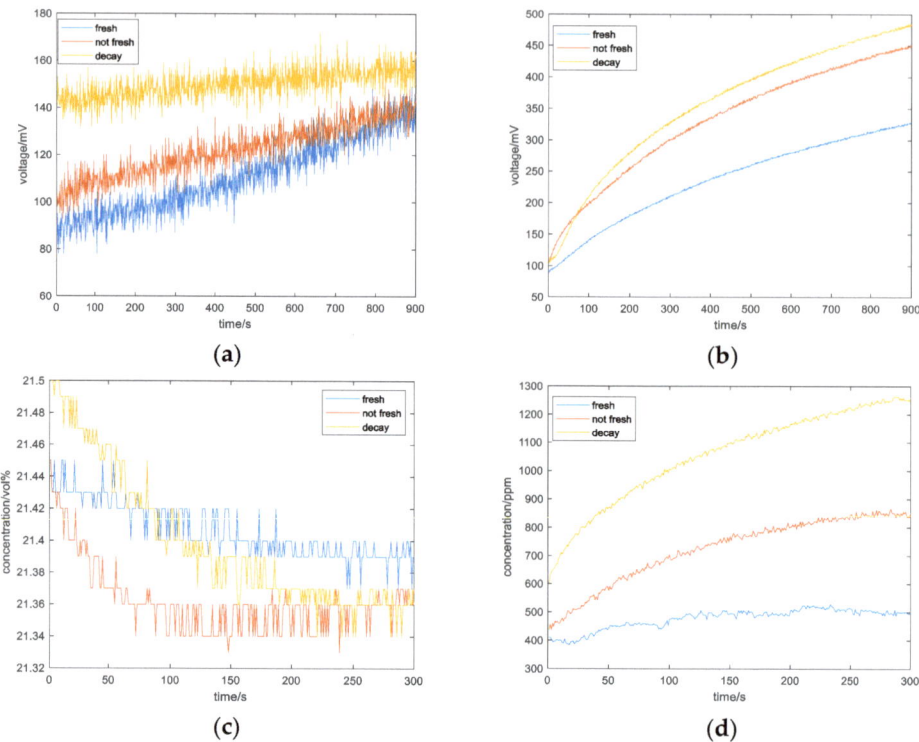

Figure 5. Comparison of gas concentrations of apples with different freshness: (**a**) ethylene; (**b**) ethanol; (**c**) oxygen; (**d**) carbon dioxide.

5.3. Data Pre-Processing

As can be seen from the data in the above figure, the concentration packets of ethylene and oxygen in the collected data contain a lot of noise and need to be filtered. Linear least squares filtering is selected, and the filtering effect is shown in Figure 6.

The characteristic values of the filtered noise data are extracted, and when the gas emitted by the apple tends to be stable, the stable, average, and maximum values are taken as the characteristic signal values for statistical analysis.

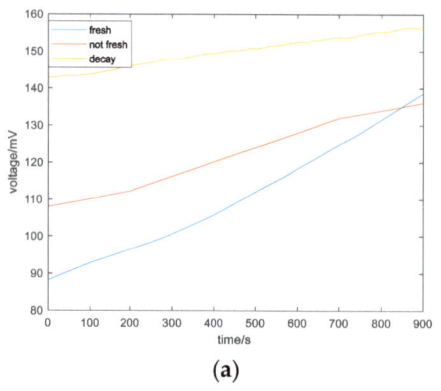

(**a**)

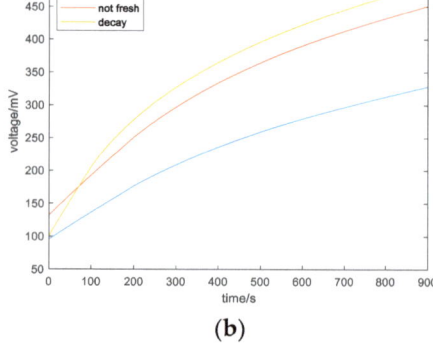

(**b**)

Figure 6. *Cont.*

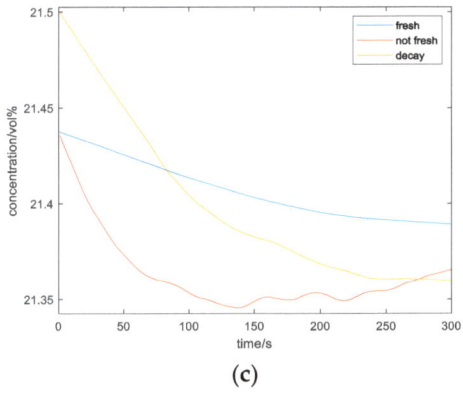
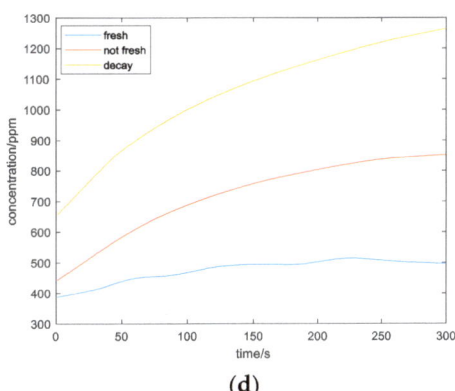

(c)   (d)

Figure 6. Gas concentration of apples with different freshness after filtering: (**a**) ethylene; (**b**) ethanol; (**c**) oxygen; (**d**) carbon dioxide.

5.4. Predicted Results

Figures 7 and 8 show the comparison of training errors before and after BP neural network optimization.

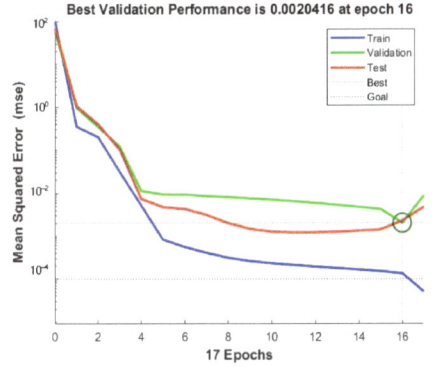
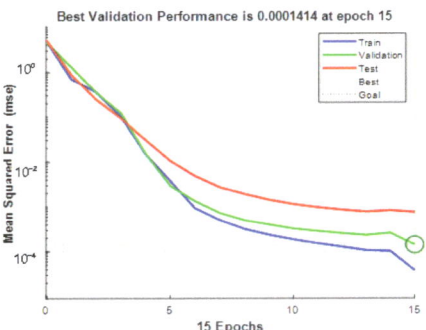

Figure 7. Training error before optimization (**left**), Training error after optimization (**right**).

As shown in Figure 7, the optimized neural network has a smaller prediction error. The 40-day concentration change curve of an apple sample during the detection process and the stable value concentration prediction curve of the apple sample before and after the optimization of the BP neural network are shown in Figure 8.

Based on the comparison of the above figures, it can be seen that the prediction errors of the prediction model before and after optimization are 0.002 and 0.0002, respectively, and the error after optimization is significantly reduced. The coefficient of determination is a key parameter used to reflect the reliability of model variables, and in order to evaluate the stability and reliability of the model more intuitively, the coefficient of determination is used as the evaluation index of the predictive model. Firstly, the sum of squares and total squares of the residuals are calculated according to Formulas (2) and (3), SS_{res} is the sum of residual squares, SS_{tot} is the total sum of squares, y_i represents the real data, $\overline{y}$ represents the average, and f_i is the predicted data. Then the determination coefficient is calculated according to Formula (4).

$$SS_{res} = \Sigma(y_i - f_i)^2 \tag{2}$$

$$SS_{tot} = \Sigma(y_i - \overline{y})^2 \tag{3}$$

$$R^2 = 1 - SS_{res}/SS_{tot} \qquad (4)$$

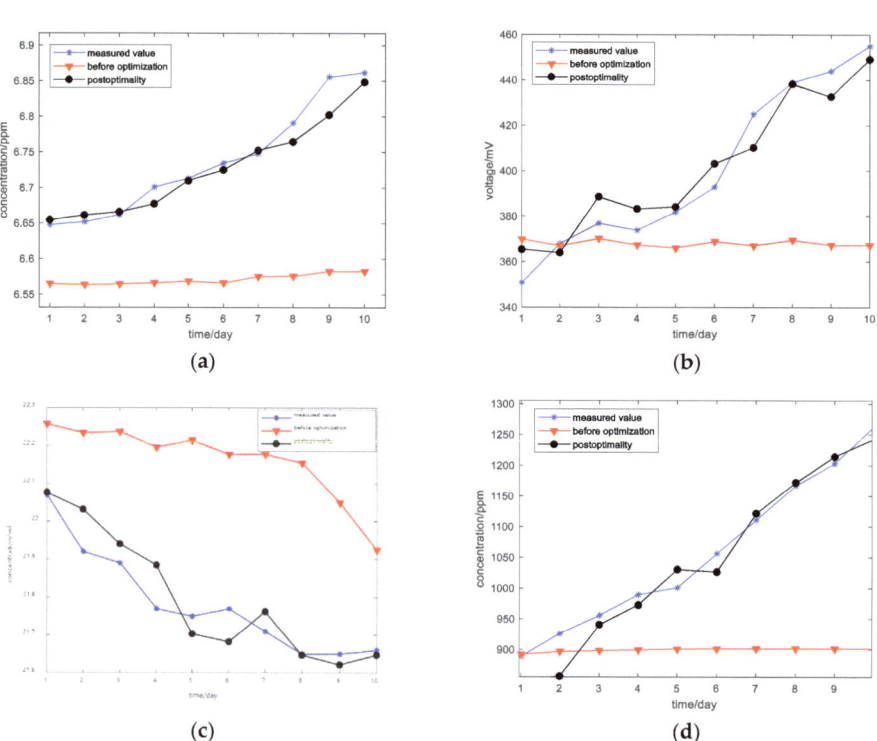

Figure 8. Comparison between the predicted value and the measured value of a sample before and after optimization: (**a**) ethylene; (**b**) ethanol; (**c**) oxygen; (**d**) carbon dioxide.

Finally, the coefficient of determination before optimization is 0.80057 and the coefficient of determination R^2 after optimization is 0.95851, which shows that the prediction stability of the optimized model is better than that before optimization, the prediction error is smaller, and the performance is stable.

5.5. The Classification Results

Through the pre-processed 180 sets of data, the ratio of training set to test set is 8:2 to divide, of which 144 groups are used to train the classification model and the other 36 groups are shuffled and sorted for recognition verification. We obtained some specific detection results as shown in both Tables 4 and 5, and the result change of the neural network for the freshness classification output of an apple sample within 30 days is shown in Figure 9, where the maximum value of the three types of output values on the same day is the predicted freshness result of the apple.

Table 4. The relationship between the dielectric properties and the freshness of Fuji apples [10,11].

Improvement of Characteristics	The Equivalent Capacitance $C_S/e^{-10}F$	Loss Factor D/e^{-2}	Relative Dielectric Constant ε/e^{-1}
fresh	2.0–2.5	6.1–6.8	5.0–5.5
not fresh	1.2–2.0	4.5–5.8	3.5–5.0
decay	>0.8	>2.8	>2.8

Table 5. Partial prediction results of an apple.

Time/d	The Actual Freshness	The Classification Results of Neural Network Output			Predicted Results
		Fresh	Not Fresh	Decay	
6	Fresh	0.8272	0.1970	−0.0242	Fresh
10	Fresh	0.6521	0.4064	−0.0585	Fresh
15	Not fresh	0.2219	0.8077	−0.0297	Not fresh
20	Not fresh	−0.0814	0.7105	0.3708	Not fresh
30	Decay	0.1381	0.1651	0.6967	Decay

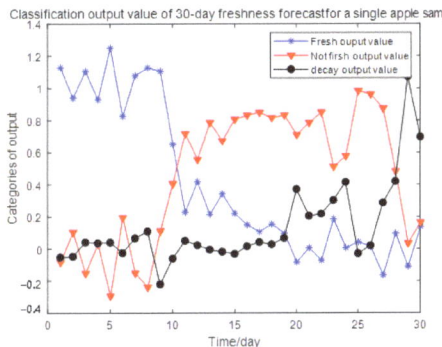

Figure 9. Freshness prediction output of a single apple.

According to the predicted concentration, the freshness of the apples was classified and identified. By comparing the recognition results, it was found that the accuracy rate of the sixth day was the highest, and the accuracy rate was 100%. The lowest accuracy rate was 80 percent on day 30. In practical application, the prediction days can be determined according to the accuracy requirements of the freshness prediction, and the accuracy of the freshness prediction will decrease with the increase in the number of days.

According to the predicted concentration, the freshness of apples was classified and identified. Then the identification results are compared, and the accuracy rate is the highest on the sixth day, which was 100%, and the lowest on the 30th day, which was 80%. In practical applications, the number of predicted days can be determined according to the requirements for freshness prediction accuracy, and the freshness prediction accuracy will decrease as the time goes on.

6. Conclusions

In this paper, a gas sensor array based on a wireless transmission module is used to collect the odor information of apples with different degrees of freshness, and a system model for apple odor information prediction is established by combining deep learning algorithms and intelligent senses by using the Tent-SSA-BP neural network prediction model. Compared with existing prediction models, the experimental results show that the model has strong optimization ability, high prediction accuracy, good stability, and a coefficient of determination of more than 0.95. Combined with the apple freshness classification system based on the gas sensor array, a complete apple freshness prediction system is formed that can accurately predict the freshness of apples in the next 30 days or so with the advantages of high accuracy, low cost, a small amount of data, convenient detection, and non-destructive testing. In the next phase of research, further research will be carried out on apples under practical application conditions, such as the shelf life of apples under refrigerated conditions and the change in shelf life of apples during transportation.

Author Contributions: Conceptualization, W.W. and W.Y.; methodology, W.Y.; software, W.Y.; validation, W.W., W.Y. and M.L.; formal analysis, W.Y.; investigation, W.D. and Z.Z.; resources, W.W. and W.Y.; data curation, W.W. and W.Y.; writing—original draft preparation, W.Y.; writing—review and editing, M.L.; visualization, W.Y.; supervision, W.W.; project administration, W.W. All authors have read and agreed to the published version of the manuscript.

Funding: This research was funded by Shanxi Provincial Natural Science Foundation General Project, China, Project, grant number 202203021221117. Grantee: Wei Wang.

Institutional Review Board Statement: Not applicable.

Informed Consent Statement: Not applicable.

Data Availability Statement: Publicly available datasets were analyzed in this study. This data can be found here: s2005131@st.nuc.edu.cn (W.Y.).

Conflicts of Interest: The authors declare no conflict of interest.

References

1. Guo, Z.; Wang, J.; Song, Y.; Zou, X.; Cai, J. Research progress on sensing detection and monitoring technology of quality deterioration of fruits and vegetables. *Smart Agric.* **2021**, *3*, 14–28.
2. Gómez, A.H.; Wang, J.; Hu, G.; Pereira, A.G. Monitoring storage shelf life of tomato using electronic nose technique. *J. Food Eng.* **2008**, *85*, 625–631. [CrossRef]
3. Guohua, H.; Yuling, W.; Dandan, Y.; Wenwen, D. Fuji Apple Storage Time Predictive Method Using Electronic nose. *Food Anal. Methods* **2013**, *6*, 82–88. [CrossRef]
4. Sanaeifar, A.; Mohtasebi, S.S.; Ghasemi-Varnamkhasti, M.; Ahmadi, H. Application of MOS based electronic nose for the prediction of banana quality properties. *Measurement* **2016**, *82*, 105–114. [CrossRef]
5. Wojnowski, W.; Kalinowska, K.; Majchrzak, T.; Płotka-Wasylka, J.; Namieśnik, J. Prediction of the Biogenic Amines Index of Poultry Meat Using an Electronic Nose. *Sensors* **2019**, *19*, 1580. [CrossRef] [PubMed]
6. Srinivasan, P.; Robinson, J.; Geevaretnam, J.; Rayappan, J.B.B. Development of electronic nose (Shrimp-Nose) for the determination of perishable quality and shelf-life of cultured Pacific white shrimp (Litopenaeus Vannamei). *Sens. Actuators B Chem.* **2020**, *317*, 128192. [CrossRef]
7. Lei, F. Intelligent Detection of Cucumber and Cherry Tomato Freshness Based on Electronic Nose and Low-Field NMR. Ph.D. Thesis, Jiangnan University, Wuxi, China, 2019.
8. Saoxia, C. Nondestructive Testing of Freshness of Baby Vegetable Based on Electronic Nose and Near-Infrared Spectroscopy. Master's Thesis, Nanjing Agricultural University, Nanjing, China, 2020.
9. Man, Z.; Zhigang, L.; Xinwu, L.; Xiaoshuan, Z. Prediction method of freshness of Chilled Mutton based on gas sensing information. *J. Shihezi Univ. (Nat. Sci. Ed.)* **2022**, *7*, 1–7.
10. Zhihui, G. *Research on Apple Freshness Detection Based on Odor Recognition*; North University of China: Taiyuan, China, 2019. (In Chinese with English abstract)
11. Yungang, L.; Wei, W. Apple freshness odor recognition system based on BP neural network optimized by SLA. *Sens. Microsyst.* **2020**, *39*, 96–99.
12. Zhuanhong, Y.; Wei, W. Research on apple sweetness recognition technology based on gas sensor array. *Foreign Electron. Meas. Technol.* **2021**, *40*, 71–76.
13. Wang, W.; Yang, W.; Liu, Y.; Wang, Z.; Yan, Z. A Research of Neural Network Optimization Technology for Apple Freshness Recognition Based on Gas Sensor Array. *Sci. Program.* **2022**, *2022*, 5861326. [CrossRef]
14. Jie, H.; Yan, L.; Fangyuan, Y.; Mengtian, Z.; Yan, J.; Feng, S.; Luye, W.; Guohua, H.; Yuquan, C. Prediction of Storage time of Tilapia cryopreservation using Electronic nose. *Chin. J. Sens. Sens.* **2013**, *26*, 1317–1322.
15. Dongjie, C.; Peihong, J.; Changfeng, Z.; Xiaobao, N. Prediction of freshness quality of sea bass based on electronic nose and statistical method. *Sci. Technol. Food Ind.* **2018**, *39*, 235–239.
16. Xu, L.; Zhang, Z.Y.; Chen, X.; Zhao, S.W.; Wang, L.Y.; Wang, T. Prediction of aero-optical imaging migration based on BP Neural Network Optimized by improved Sparrow Search Algorithm. *J. Optoelectron. Laser* **2021**, *32*, 653–658.
17. Huang, J.Y. Sparrow Search Algorithm Based on T Distribution and Tent Chaos Map. Master's Thesis, Lanzhou University, Lanzhou, China, 2021.

Disclaimer/Publisher's Note: The statements, opinions and data contained in all publications are solely those of the individual author(s) and contributor(s) and not of MDPI and/or the editor(s). MDPI and/or the editor(s) disclaim responsibility for any injury to people or property resulting from any ideas, methods, instructions or products referred to in the content.

Article

Swin Transformer-Based Edge Guidance Network for RGB-D Salient Object Detection

Shuaihui Wang, Fengyi Jiang and Boqian Xu *

Changchun Institute of Optics, Fine Mechanics and Physics, Chinese Academy of Sciences, Changchun 130033, China; wangshuaihui@ciomp.ac.cn (S.W.); jiangfengyi@ciomp.ac.cn (F.J.)
* Correspondence: xuboqian@ciomp.ac.cn

Abstract: Salient object detection (SOD), which is used to identify the most distinctive object in a given scene, plays an important role in computer vision tasks. Most existing RGB-D SOD methods employ a CNN-based network as the backbone to extract features from RGB and depth images; however, the inherent locality of a CNN-based network limits the performance of CNN-based methods. To tackle this issue, we propose a novel Swin Transformer-based edge guidance network (SwinEGNet) for RGB-D SOD in which the Swin Transformer is employed as a powerful feature extractor to capture the global context. An edge-guided cross-modal interaction module is proposed to effectively enhance and fuse features. In particular, we employed the Swin Transformer as the backbone to extract features from RGB images and depth maps. Then, we introduced the edge extraction module (EEM) to extract edge features and the depth enhancement module (DEM) to enhance depth features. Additionally, a cross-modal interaction module (CIM) was used to integrate cross-modal features from global and local contexts. Finally, we employed a cascaded decoder to refine the prediction map in a coarse-to-fine manner. Extensive experiments demonstrated that our SwinEGNet achieved the best performance on the LFSD, NLPR, DES, and NJU2K datasets and achieved comparable performance on the STEREO dataset compared to 14 state-of-the-art methods. Our model achieved better performance compared to SwinNet, with 88.4% parameters and 77.2% FLOPs. Our code will be publicly available.

Keywords: RGB-D salient object detection; edge guidance; transformer; cross-modal interaction

Citation: Wang, S.; Jiang, F.; Xu, B. Swin Transformer-Based Edge Guidance Network for RGB-D Salient Object Detection. *Sensors* **2023**, *23*, 8802. https://doi.org/10.3390/s23218802

Academic Editors: Man Qi and Matteo Dunnhofer

Received: 31 August 2023
Revised: 9 October 2023
Accepted: 24 October 2023
Published: 29 October 2023

Copyright: © 2023 by the authors. Licensee MDPI, Basel, Switzerland. This article is an open access article distributed under the terms and conditions of the Creative Commons Attribution (CC BY) license (https://creativecommons.org/licenses/by/4.0/).

1. Introduction

Salient object detection (SOD) is an important preprocessing method in computer vision tasks, with applications in video detection and segmentation [1], semantic segmentation [2], object tracking [3], etc.

CNN-based models for RGB SOD have yielded great performance in localizing salient objects [4–8]. However, it is still difficult to localize the salient object accurately in scenes such as those with low contrast or objects with a cluttered background. CNN-based RGB-D SOD models, which employ features from RGB images and depth maps, have attracted growing interest and presented promising performance [9–23]. However, some issues still limit the performance of existing CNN-based RGB-D SOD models.

The first issue is that CNN-based models cannot effectively capture long-range dependencies. Long-range semantic information plays an important role in identifying and locating salient objects [24]. Due to the intrinsic locality of convolution operations, CNN-based models cannot effectively extract global context information. In addition, the empirical receptive field of CNN is much smaller than the theoretical receptive field, especially on high-level layers [25].

The second issue is that depth maps are often noisy. The performance of RGB-D SOD models relies on reliable RGB images and depth maps. Misleading information in depth maps degrades the performance of RGB-D SOD models.

Global context information helps reduce errors created via poor depth maps. Transformers can extract features and model long-range dependencies, and Transformer-based methods have achieved outstanding performance in various computer vision tasks [26–29]. However, Transformers are less effective in capturing local features. The Swin Transformer [29], combining the advantages of Transformers and CNN, has been shown to have a powerful feature extraction ability. Considering the above challenges, the Swin Transformer is suitable as a feature extractor for RGB-D SOD tasks.

Swin Transformer-based models are relatively weak in their ability to model local context information. Therefore, Swin Transformer-based models should pay more attention to local feature information.

Based on the investigation above, we propose a novel Swin Transformer-based edge guidance network (SwinEGNet) that enhances feature locality to boost the performance of RGB-D SOD. We employed the Swin Transformer as the backbone to extract features from RGB images and depth maps for capturing long-range dependencies. We introduced a depth enhancement module (DEM) and a cross-modal interaction module to enhance local features. Unlike other methods, we employed edge clues to enhance depth features rather than edge clues as decoder guidance to directly refine the final prediction map. We designed the edge extraction module (EEM) to extract edge information and the depth enhancement module (DEM) to enhance depth features. Furthermore, we used a cross-modal interaction module to effectively integrate information from global and local contexts. To effectively explore the features of each layer, we employed a cascaded decoder to progressively refine our saliency maps.

Our main contributions are summarized as follows:

- A novel edge extraction module (EEM) is proposed, which generates edge features from the depth features.
- A newly designed edge-guided cross-modal interaction was employed to effectively integrate cross-modal features, where the depth enhancement module was employed to enhance the depth feature and the cross-modal interaction module was employed to encourage cross-modal interaction from global and local aspects.
- A novel Swin Transformer-based edge guidance network (SwinEGNet) for RGB-D SOD is proposed. The proposed SwinEGNet was evaluated with four evaluation metrics and compared to 14 state-of-the-art (SOTA) RGB-D SOD methods on six public datasets. Our model achieved better performance with less parameters and FLOPs than SwinNet, as shown in Figure 1. In addition, a comprehensive ablation experiment was also conducted to verify the effectiveness of the proposed modules. The experiment results showed the outstanding performance of our proposed method.

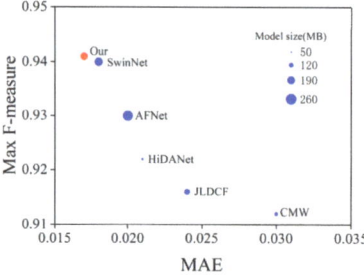

Figure 1. Max F-measure, MAE, and model size of different methods on the NLPR dataset. Our model achieves better performance with a smaller model size.

The remainder of this paper is structured as follows: The current status of RGB-D salient object detection is presented in Section 2. The overall architecture, detailed structure, and loss function of the proposed network are outlined in Section 3. The results of our experiments are provided in Section 4. Finally, our conclusions are presented in Section 5.

2. Related Work

CNN-based RGB-D salient object detection: Benefitting from the development of deep learning and depth sensors, many CNN-based RGB-D SOD methods have recently been proposed. Compared to RGB SOD methods, RGB-D SOD models employ depth clues as complementary information and have shown outstanding performance in salient object detection. Most RGB-D SOD models adopt CNN-based networks to extract features and focus on cross-modal fusion strategies to improve salient object detection performance. Various frameworks and fusion strategies have been proposed to effectively merge cross-modal cross-scale features [14,17,21–23,30,31]. Zhang et al. [30] designed an asymmetric two-stream network, where a flow ladder module is introduced to the RGB stream to capture global context information and DepthNet for the depth stream. Zhang et al. [17] proposed a multistage cascaded learning framework for RGB-D saliency detection, which minimizes the mutual information between RGB images and depth maps to model complementary information. Chen et al. [22] designed a triplet encoder network that processes RGB, depth, and fused features separately to suppress the background noise in the depth map and sharpen the boundaries of high-level features. Li et al. [14] designed a hierarchical alternate interaction module that progressively and hierarchically integrates local and global contexts. Wu et al. [21] proposed layer-wise, trident spatial, and attention mechanisms to fuse robust RGB and depth features against low-quality depths. Wu et al. [23] employed a granularity-based attention module to leverage the details of salient objects and introduced a dual-attention module to fuse the cross-modal cross-scale features in a coarse-to-fine manner.

To address the insufficiency of obtaining global semantic information of CNN-based networks, Liu et al. [7] proposed using a receptive field block to enhance feature discriminability and robustness by enlarging the receptive field. Dilated convolutions can enlarge the receptive field of CNN without loss of resolution. As a result, Yu et al. [32] presented modules based on dilated convolutions to aggregate multiscale information. Liu et al. [8] designed a global guidance module for RGB SOD that utilizes the revised pyramid pooling module to capture global semantic information.

Transformer-based RGB-D salient object detection: The Transformer was first employed for machine translation and gradually introduced in computer vision tasks. Dosovitskiy et al. [26] proposed the first Vision Transformer (ViT), Wang et al. [28] proposed a progressive shrinking pyramid Transformer (PVT), and Liu et al. [29] designed the Swin Transformer. Subsequently, researchers employed the Transformer as the backbone network to improve the detection performance of RGB-D SOD. Liu et al. [33] developed a unified model based on ViT for both RGB and RGB-D SOD. Zeng et al. [34] employed the Swin Transformer as the encoding backbone to extract features from RGB images and depth maps. Liu et al. [35] employed PVT as a powerful feature extractor to extract global context information and designed a lightweight CNN-based backbone to extract spatial structure information in depth maps. Pang et al. [36] proposed using a novel top-down information propagation path based on the Transformer to capture important global clues to promote cross-modal feature fusion. Liu et al. [37] proposed using a cross-modal fusion network based on SwinNet for RGB-D and RGB-T SOD. Roy et al. [38] employed the Swin Transformer as the encoder block to detect multiscale objects.

3. Methodologies

In this section, we present the proposed Swin Transformer-based edge guidance network (SwinEGNet). We provide an overview of our method and describe its main components in detail, including the feature encoder, edge extraction module, edge-guided cross-modal interaction module, cascaded decoder, and loss function.

3.1. The Overall Architecture

As illustrated in Figure 2, we present a Swin Transformer-based edge guidance network (SwinEGNet). Inspired by [37], we employed edge clues to guide salient object detection. However, unlike [37], edge clues were incorporated into cross-modal interaction blocks to enhance depth features rather than being employed as decoder guidance to refine the final prediction map. The proposed SwinEGNet adopts the encoder–decoder structure. As shown in Figure 2, SwinEGNet consists of a feature encoder, edge extraction module (EEM), edge-guided cross-modal interaction module (EGCIM), and cascaded decoder. Firstly, RGB images and depth maps are fed into two independent Swin Transformers for feature extraction, and an EEM is proposed to extract edge features. Then, these features are fed into EGCIM for depth feature enhancement and feature fusion, where the depth enhancement module (DEM) is responsible for depth feature enhancement and the cross-modal interaction module (CIM) is responsible for feature fusion. Finally, the fused features are fed into the decoder block for saliency maps. The cascaded decoder was employed to effectively explore the features of the four layers and progressively refine the saliency maps.

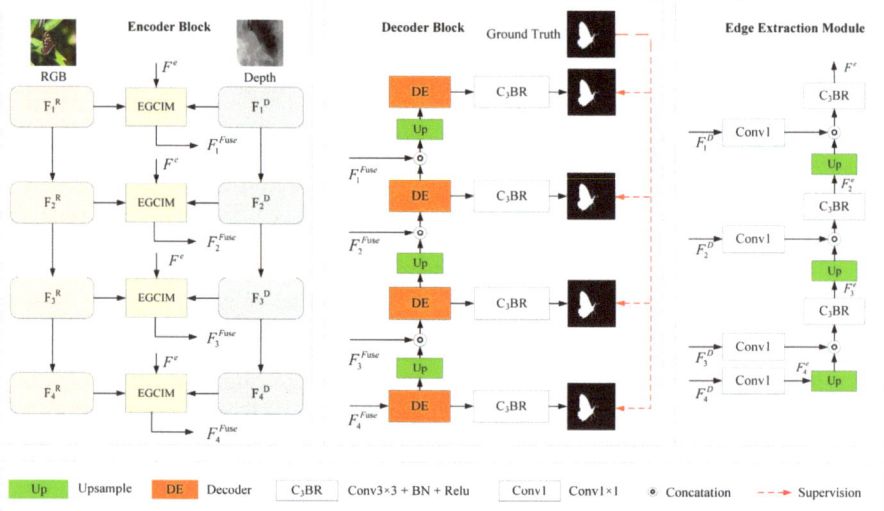

Figure 2. An overview of the proposed SwinEGNet. It consists of a feature encoder, an edge extraction module (EEM), an edge-guided cross-modal interaction module (EGCIM), and a cascaded decoder.

3.2. Feature Encoder

In contrast to other Transformers, the Swin Transformer computes multihead self-attention within a local window instead of the whole input to model locality relationships. Furthermore, it employs a shifted window operation to model long-range dependence across windows. Therefore, the Swin Transformer is suitable for feature extraction because it incorporates the merits of the Transformer and CNN. Considering the performance and computational complexity, we adopted the Swin-B Transformer as the backbone to extract features from RGB images and depth maps, which accept an input size of 384 × 384.

RGB images and depth maps are fed into two independent Swin Transformers for feature extraction. Considering the first layer contains redundant noisy information, the extracted features of the last four layers are employed for feature fusion. The features can be expressed as follows:

$$F_i^R = trans(I_R), i = 1, 2, 3, 4 \qquad (1)$$

$$F_i^D = trans(I_D), i = 1, 2, 3, 4 \qquad (2)$$

where F_i^R denotes the RGB feature; F_i^D denotes the depth feature, $trans(\cdot)$ denotes the Transformer; and I_R and I_D denote the input RGB image and depth image, respectively.

3.3. Edge Extraction Module

To extract edge features, we propose an edge extraction module (EEM). The extracted edge features are fed into EGCIM to enhance the depth feature. The details of the proposed EEM are illustrated in Figure 3.

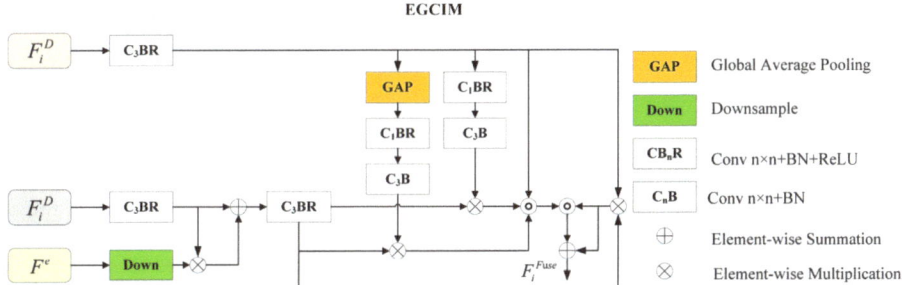

Figure 3. Details of the proposed edge-guided cross-modal interaction module (EGCIM).

Shallow layers contain low-level information such as structure clues, while deep layers contain global semantic information. They are all helpful in extracting edge information. In contrast to other methods that employ parts of the depth features for edge prediction, we employed all depth features for edge extraction, and the edge features were progressively refined in a coarse-to-fine manner.

In particular, the depth features $F_i^D (i = 1, 2, 3, 4)$ are fed into a 1×1 convolutional layer for channel reduction. Then, features $F_i^D (i = 2, 3, 4)$ perform the upsample operation to generate the same size features as F_{i+1}^D. The edge feature F_4^e can be expressed as follows:

$$F_4^e = Up\left(Conv_1(F_4^D)\right) \quad (3)$$

where $Up(\cdot)$ denotes the upsample operation.

Next, the edge feature performs a concatenation operation and a 3×3 convolutional layer with a BatchNorm and a ReLU function to generate the edge feature F_{i-1}^e, which can be expressed as follows:

$$F_i^e = C_3BR\left(Cat\left(Conv_1(F_i^D), F_{i+1}^e\right)\right), i = 1, 2, 3 \quad (4)$$

where $C_3BR(\cdot)$ denotes a 3×3 convolutional layer with a BatchNorm and a ReLU function, and $Cat(\cdot)$ denotes concatenation operation. The edge feature F_1^e is the final edge feature F^e. The final edge feature F^e will be fed into EGCIM for depth enhancement.

3.4. Edge-Guided Cross-modal Interaction Module

To enhance depth features and encourage cross-modal feature interaction, we designed an edge-guided cross-modal interaction module (EGCIM) to integrate features from both modalities, including a depth enhancement module (DEM) and a cross-modal interaction module (CIM).

Depth enhancement module: Though Transformer-based methods sufficiently capture global context information, they are relatively weak at capturing local context information compared to CNN-based methods. Therefore, it is necessary to utilize local clues like edge information to enhance the depth features. We designed a depth enhancement module (DEM) to enhance the depth features, which introduces edge information extracted from the depth features to these features for depth enhancement. The detailed structure of DEM is shown in Figure 3.

The depth features F_i^D and edge features F^e at a certain hierarchy $i = 1, 2, 3, 4$, F_i^D performs the convolution operation with a BatchNorm and a ReLU function for channel reduction, and F^e performs the downsample operation to gain the same size as F_i^D. Then, the depth features F_i^D and edge features of the same size are fused using multiplication and addition operations. The enhanced depth features can be expressed as follows:

$$F_i^{DE} = C_3BR\left(C_3BR(F_i^D) + C_3BR(F_i^D) \times Down(F^e)\right) \tag{5}$$

where + denotes the addition operation, and $Down(\cdot)$ denotes the downsample operation. The enhanced depth features F_i^{DE} will be fed into CIM for feature fusion.

Cross-modal interaction module: We used a cross-modal interaction module (CIM) to effectively combine RGB and depth modalities. The CIM contains a global attention branch and a local attention branch to enhance globality and locality. In addition, a residual connection is adopted to combine the fused features with RGB features for the preservation of the RGB images' original information. The local information of the depth features enhances the RGB features to sharpen the details of salient objects, and the global context information of the depth features enhances the RGB features to locate the salient object.

As shown in Figure 3, the RGB features are fed into a 3×3 convolutional layer with a BatchNorm and a ReLU activation function for channel reduction. There are three branches for feature fusion: the first branch employs global average pooling (GAP) to capture global context information, the second branch employs 1×1 convolution to obtain local information, and the third branch aims to keep the original information of RGB features. Then, we carry out multiplication, concatenation, and addition operations for fusion. The fused features can be expressed as follows:

$$F_i^{Fuse} = C_3BR\left(C_3BR\left(Cat\left(C_3BR(F_i^R), F_i^g, F_i^l, F_i^o\right)\right) + F_i^o\right) \tag{6}$$

$$F_i^{Fuse} = C_3BR(F_i^R) \times F_i^{DE} \tag{7}$$

$$F_i^g = C_3BR(F_i^R) \times C_1B\left(C_1BR\left(GAP(F_i^{DE})\right)\right) \tag{8}$$

$$F_i^l = C_3BR(F_i^R) \times C_1B\left(C_1BR(F_i^{DE})\right) \tag{9}$$

where $GAP(\cdot)$ represents the global average pooling operation, C_1B represents a convolution operation with a BatchNorm function, C_1BR represents a convolution operation with a BatchNorm function and a ReLU function, and F_i^{Fuse} represents the fused features.

3.5. Cascaded Decoder

The cascaded encoder can effectively leverage the multilevel features and eliminate the noise in low-level features, which improves the accuracy of salient maps. Moreover, deep-layer supervision performs better than single supervision [13]. Therefore, we employed a cascaded decoder for the final prediction map, as shown in Figure 3. The decoder has four decoding levels corresponding to the four-level cross-modal feature interaction. Consequently, the prediction map is refined progressively. Each decoder contains two 3×3 convolution layers with a BatchNorm and a ReLU function, a dropout layer, and an upsample layer. The initial prediction map S_4 is fed into the decoder and concatenates with the previous prediction map S_{n-1} for refinement. The prediction features S_i can be donated as follows:

$$S_i = \begin{cases} C_3BR(Up(S_{i+1}), S_i), i = 1,2,3 \\ C_3BR(F_i^{Fuse}), i = 4 \end{cases} \tag{10}$$

where $D(\cdot)$ represents the decoder operation, S_n represents the prediction map, and $Up(\cdot)$ represents the upsample operation. Next, features S_i perform convolution operations to obtain the prediction map, and S_1 is the final prediction map.

3.6. Loss Function

Detection loss is composed of the weighted binary cross-entropy (BCE) loss L_{BCE}^{ω} and the weighted intersection-over-union (IOU) loss L_{IoU}^{ω} [39], which has been invalidated in salient object detection. The detection loss can be formulated as follows:

$$L_d = L_{BCE}^{\omega} + L_{IoU}^{\omega} \qquad (11)$$

L_{IoU}^{ω} and L_{BCE}^{ω} pay more attention to the structure of SOD and the hard pixels to highlight the importance of the hard pixel. As illustrated in Figure 2, four-level supervisions are applied to supervise the four side-output maps. Each map S_i is upsampled to the same size as the ground truth map. Thus, the total loss function L can be expressed as follows:

$$L = \sum_{i=1}^{4} (L_d^i(S_i, G) \qquad (12)$$

4. Experiments

4.1. Datasets and Evaluation Metrics

Datasets: We evaluated the proposed method on six widely used benchmark datasets: STEREO (1000 image pairs) [40], NJU2K (2003 image pairs) [41], NLPR (1000 image pairs) [42], LFSD (100 image pairs) [43], SIP (929 image pairs) [44], and DES (135 image pairs) [45]. For a fair comparison, our training settings were the same as the existing works [12], which consisted of 1485 samples from the NJU2K dataset and 700 samples from the NLPR dataset. The remaining images from NLPR, DES, and NJU2K, and the whole SIP, STEREO, and LFSD were used for testing.

Evaluation metrics: We adopted four widely used evaluation metrics for quantitative evaluation, including S-measure (S_α, $\alpha = 0.5$) [46], maximum F-measure (F_m) [47], maximum E-measure (E_m) [48], and mean absolute error (MAE, M) [49]. S-measure evaluates the structural similarity between the saliency map and ground truth, which is defined as follows:

$$S = \alpha S_o + (1 - \alpha) S_r \qquad (13)$$

where α is a trade-off parameter set to 0.5, S_o represents the object perception, and S_r represents the regional perception. F-measure focuses on region-based similarity that considers precision and recall, which is defined as follows:

$$F_\beta = (1 + \beta^2) \frac{P \times R}{\beta^2 \times P + R} \qquad (14)$$

where P denotes precision, R denotes recall, and β^2 is a trade-off parameter set to 0.3. We used the maximum F-measure as the evaluation metric. MAE assesses the average absolute error at the pixel level, which is defined as follows:

$$MAE = \frac{1}{W \times H} \sum_{i=1}^{W} \sum_{j=1}^{H} |S(i,j) - G(i,j)| \qquad (15)$$

where W and H represent the width and height of the image, respectively. S represents the saliency maps, and G represents the ground truth. E-measure is employed to capture image-level statistics and local pixel matching, which is defined as follows:

$$E_m = \frac{1}{W \times H} \sum_{i=1}^{W} \sum_{j=1}^{H} \phi_{FM}(i,j) \qquad (16)$$

where ϕ_{FM} represents the enhanced alignment matrix. For a fair comparison, we used the evaluation tools provided by [15].

4.2. Implementation Details

We implemented our model on PyTorch with one NVIDIA A4000 GPU. The Swin Transformer that has been pretrained on ImageNet was employed as our backbone network. The parameters of the Swin-B model were initialized with the pretrained parameters, and the remaining parameters were initialized with PyTorch default settings. The Adam optimizer was employed to train the proposed model with a batch size of 5, a momentum of 0.9, and a weight decay of 0.1. The initial learning rate was 1×10^{-4}, which was then divided by 10 for every 60 epochs. All images were resized to 384×384 for training and testing. The single-channel depth image was replicated to a three-channel image, which was the same as the RGB image. Data augment strategies, including random flipping, rotating, and border clipping, were employed to augment the training data. The model was trained for 120 epochs.

4.3. Comparison with SOTAs

Quantitative comparison: We compared the proposed network with 14 SOTA CNN-based methods and Transformer-based methods, which were CMW [13], JLDCF [50], HINet [51], DSA2F [20], CFIDNet [52], C²DFNet [53], SPSNet [19], AFNet [22], HiDANet [23], MTFormer [54], VST [43], TANet [35], and SwinNet [37]. The compared saliency maps were directly provided by the authors or generated via their released codes. The quantitative comparison under four evaluation metrics on six datasets is shown in Table 1. As shown in Table 1, our SwinEGNet performed the best on LFSD, NLPR, and DES datasets and competitively performed on NJU2K, STEREO, and SIP datasets. In particular, SwinEGNet performed outstandingly on the LFSD dataset, which is considered a challenging dataset. Compared to the second model DSA2F, the improvements of S-measure, F-measure, E-measure, and MAE were about 0.011, 0.006, 0.005, and 0.002, respectively. On the NJU2K dataset, the performance of our method was comparable with SwinNet. On the STEREO dataset, our method performed the best in E_m.

Table 1. Quantitative comparison of SOTA methods under four evaluation metrics: S-measure (S_a), max F-measure (F_m), max E-measure (E_m), and MAE (M). ↑ denotes that higher is better, and ↓ denotes that lower is better. The best two results are shown in red and green fonts, respectively.

	Metric	CMW	JLDCF	HINet	HAINet	DSA2F	CFIDNet	C²DFNet	SPSNet	AFNet	HiDANet	MTFormer	VST	TANet	SwinNet	Our
LFSD	$S_m\uparrow$	0.876	0.854	0.852	0.854	0.882	0.869	0.863	-	0.89	-	0.872	0.89	0.875	0.886	0.893
	$F_m\uparrow$	0.899	0.862	0.872	0.877	0.903	0.883	0.89	-	0.9	-	0.879	0.903	0.892	0.903	0.909
	$E_m\uparrow$	0.901	0.893	0.88	0.882	0.920	0.897	0.899	-	0.917	-	0.911	0.918	-	0.914	0.925
	$M\downarrow$	0.067	0.078	0.076	0.08	0.054	0.07	0.065	-	0.056	-	0.062	0.054	0.059	0.059	0.052
NLPR	$S_m\uparrow$	0.917	0.925	0.922	0.924	0.918	0.922	0.928	0.923	0.936	0.93	0.932	0.931	0.935	0.941	0.941
	$F_m\uparrow$	0.912	0.916	0.915	0.922	0.917	0.914	0.926	0.918	0.93	0.929	0.925	0.927	0.943	0.94	0.941
	$E_m\uparrow$	0.94	0.962	0.949	0.956	0.95	0.95	0.957	0.956	0.961	0.961	0.965	0.954	-	0.968	0.969
	$M\downarrow$	0.03	0.022	0.026	0.024	0.024	0.026	0.021	0.024	0.02	0.021	0.021	A0.024	0.018	0.018	0.017
NJU2K	$S_m\uparrow$	0.903	0.903	0.915	0.912	0.904	0.914	0.908	0.918	0.926	0.926	0.923	0.922	0.927	0.935	0.931
	$F_m\uparrow$	0.913	0.903	0.925	0.925	0.916	0.923	0.918	0.927	0.933	0.939	0.923	0.926	0.941	0.943	0.938
	$E_m\uparrow$	0.925	0.944	0.936	0.94	0.935	0.938	0.937	0.949	0.95	0.954	0.954	0.942	-	0.957	0.958
	$M\downarrow$	0.046	0.043	0.038	0.038	0.039	0.038	0.039	0.033	0.032	0.029	0.032	0.036	0.027	0.027	0.026
STEREO	$S_m\uparrow$	0.913	0.903	0.892	0.915	0.898	0.91	0.911	0.914	0.918	0.911	0.908	0.913	0.923	0.919	0.919
	$F_m\uparrow$	0.909	0.903	0.897	0.914	0.91	0.906	0.91	0.908	0.923	0.921	0.908	0.915	0.934	0.926	0.926
	$E_m\uparrow$	0.93	0.944	0.92	0.938	0.939	0.935	0.938	0.941	0.949	0.946	0.947	0.939	-	0.947	0.951
	$M\downarrow$	0.042	0.043	0.048	0.039	0.039	0.042	0.037	0.035	0.034	0.035	0.038	0.038	0.027	0.033	0.031
DES	$S_m\uparrow$	0.937	0.929	0.927	0.939	0.917	0.92	0.924	0.94	0.925	0.946	-	0.946	-	0.945	0.947
	$F_m\uparrow$	0.943	0.919	0.937	0.949	0.929	0.937	0.937	0.944	0.938	0.952	-	0.949	-	0.952	0.956
	$E_m\uparrow$	0.961	0.968	0.953	0.971	0.955	0.938	0.953	0.974	0.946	0.98	-	0.971	-	0.973	0.98
	$M\downarrow$	0.021	0.022	0.22	0.017	0.023	0.022	0.018	0.015	0.022	0.013	-	0.017	-	0.016	0.014
SIP	$S_m\uparrow$	0.867	0.879	0.856	0.879	0.861	0.881	0.871	0.892	0.896	0.892	0.894	0.903	0.893	0.911	0.9
	$F_m\uparrow$	0.889	0.885	0.88	0.906	0.891	0.9	0.895	0.91	0.919	0.919	0.902	0.924	0.922	0.936	0.93
	$E_m\uparrow$	0.9	0.923	0.888	0.916	0.909	0.918	0.913	0.931	0.931	0.927	0.932	0.935	-	0.944	0.935
	$M\downarrow$	0.063	0.051	0.066	0.053	0.057	0.051	0.052	0.044	0.043	0.043	0.043	0.041	0.041	0.035	0.04

Qualitative comparison: We qualitatively compared seven representative methods on challenging scenes. The first scene had a similar foreground and background (first row), the second scene had poor depth map (second row and third row), the third scene had a complex background (fourth row and fifth row), the fourth scene had a small object (sixth row), the fifth scene had multiple objects (seventh row and eighth row), and the sixth scene had a fine structure (ninth row). As shown in Figure 4, our method obtained the best detection results. For the first scene, the foreground and background of the RGB image were similar, but the depth map provided correct information. Our method located salient objects better than other methods thanks to the power of EEM and EGCIM. For the second scene, though the depth map provided incorrect information, our method successfully located salient objects by eliminating misleading information of the poor depth map. For the fourth scene, our method fused the RGB feature and depth feature the best. For the fifth scene, our method not only located the salient objects but also maintained the sharp boundaries. These all indicate the effectiveness of our model.

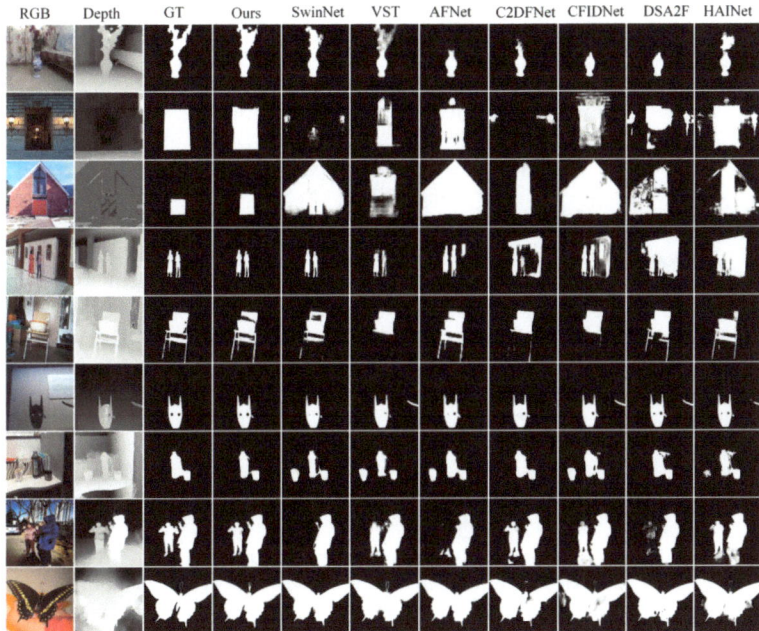

Figure 4. Visual comparison of our method and seven SOTAS, including CMW, DSA2F, CFIDNet, C²DFNet, AFNet, VST, and SwinNet.

4.4. Ablation Study

We conducted comprehensive ablation studies on LFSD and STEREO datasets to evaluate the effectiveness of the proposed modules in our proposed model.

Effectiveness of Swin Transformer backbone: We replaced the feature encoder with ResNet50 to verify the effectiveness of the Swin Transformer backbone. As shown in Table 2, the Transformer-based model showed better performance in all the evaluation benchmarks and metrics, especially on the LFSD dataset. We show the visual comparison of ResNet50 and Swin Transformer in Figure 5. The ResNet50 was inferior to the Swin Transformer. This validates the effectiveness of the Swin Transformer backbone for the RGB-D SOD.

Table 2. Effective analysis of the proposed modules on two datasets. The best results are shown in bold.

Models	LFSD				STEREO			
	$M\downarrow$	$S_m\uparrow$	$F_m\uparrow$	$E_m\uparrow$	$M\downarrow$	$S_m\uparrow$	$F_m\uparrow$	$E_m\uparrow$
Ours	**0.052**	**0.893**	**0.909**	**0.925**	**0.031**	**0.919**	**0.926**	**0.951**
ResNet50	0.084	0.835	0.864	0.868	0.044	0.893	0.0.9	0.927
w/o EGCIM	0.067	0.87	0.887	0.902	0.035	0.913	0.922	0.946
w/o DEM	0.064	0.875	0.893	0.906	0.032	0.917	0.925	0.949
w/o CIM	0.066	0.869	0.887	0.901	0.033	0.914	0.923	0.947

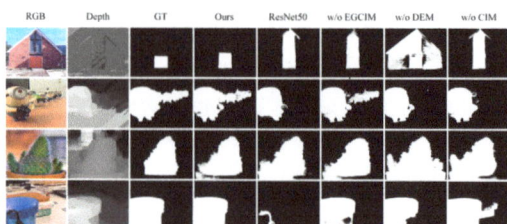

Figure 5. Visual comparison of the ablation study.

Effectiveness of EGCIM: To explore the effectiveness of EGCIM, we replaced EGCIM with a multiplication operation. In Table 2, we quantitatively demonstrate the contribution of the EGCIM. The performance of our model degraded without the help of EGCIM. This validates the effectiveness of the edge-guided cross-modal interaction module.

Effectiveness of DEM in EGCIM: To verify the effectiveness of DEM in EGCIM, we removed DEM from our full model. In Table 2, we quantitatively demonstrate the contribution of DEM. As shown in Table 2, the depth enhancement module improved the performance of the proposed model, especially on the LFSD dataset. The MAE, S-measure, F-measure, and E-measure are improved by about 0.012, 0.018, 0.013, and 0.009 in the LFSD dataset, respectively.

Effectiveness of CIM in EGCIM: We replaced CIM with a multiplication operation to verify the effectiveness of CIM in EGCIM. In Table 2, we quantitatively demonstrate the contribution of CIM. As shown in Table 2, the performance degradation caused by removing CIM supports our claim that the cross-modal interaction module can effectively fuse the RGB and depth features.

4.5. Complexity Analysis

We conducted a complexity comparison with the other five models on the number of parameters and FLOPs, as shown in Table 3. The performance of the CNN-based models was relatively poor compared to the Transformer-based models. Our model performed better with fewer parameters and lower computational costs compared to SwinNet. The parameters and FLOPs of our model were 175.6 M and 96 G, respectively. Our model achieved comparable performance to SwinNet, yielding 88.4% parameters and 77.2% FLOPs.

Table 3. Complexity comparison and performance on LFSD and NLPR datasets. The best two results are shown in red and green fonts, respectively.

Backbone	Model	Num_Parameters ↓	FLOPs ↓	LFSD F_m ↑	NLPR F_m ↑
CNN	CMW	85.7 M	208 G	0.899	0.912
	HiDANet	59.8 M	73.6 G	0.877	0.922
	JLDCF	143.5 M	211.1 G	0.862	0.916
Transformer	AFNet	242 M	128 G	0.902	0.93
	SwinNet	198.7 M	124.3 G	0.903	0.94
	Ours	175.6 M	96 G	0.909	0.941

4.6. Failure Cases

We show some failure cases on the challenging scenes in Figure 6: the first scene with multiple objects (first row and second column), and the second scene with poor depth map (third row and fourth row). As shown in the first scene, our model could not accurately locate multiple objects with complex backgrounds. Global feature relations are important for locating multiple salient objects. Multihead self-attention within a local window enhanced the locality, but it also limited the long-range model ability of the Swin Transformer. The second scene indicates that our model could not locate salient objects well in some scenes with poor depth maps. In addition to the low quality of depth maps, there were misalignments between RGB images and depth maps at the pixel level. It is difficult to effectively fuse features for direct pixel-wise fusion. We will conduct further research in the future.

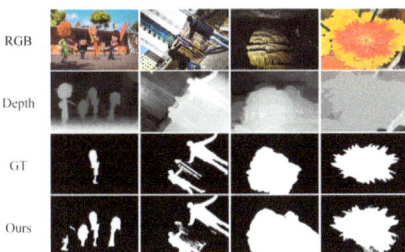

Figure 6. Visualization of failure cases in challenging scenes.

5. Conclusions

In this paper, we propose a novel Swin Transformer-based edge guidance network for RGB-D SOD. We employed the Swin Transformer as the backbone to extract features from RGB images and depth maps for capturing the long-range dependencies. Additionally, we proposed using the edge extraction module (EEM), the depth enhancement module, and the cross-modal interaction module (CIM) to enhance the local features. The EEM extracts edge features from the depth features, and the DEM employs edge information to enhance the depth features. The CIM effectively fuses RGB features and depth features from global and local contexts. With all these modules working together, our SwinEGNet model can accurately localize salient objects in various complex scenarios with sharp boundaries. Countless comparison studies and ablation experiments demonstrated that the proposed SwinEGNet showed outstanding performance on six widely used RGB-D SOD benchmark datasets. As an independent module, EEM can be applied to related tasks. In the future, we will extend our model to RGB-T salient object detection.

Author Contributions: Conceptualization, B.X.; methodology, S.W.; software, S.W.; validation, F.J.; supervision, B.X.; data curation, S.W.; formal analysis, S.W.; investigation, S.W.; resources, B.X.; writing—original draft, S.W.; writing—review and editing, F.J. and B.X; visualization, F.J. All authors have read and agreed to the published version of the manuscript.

Funding: This work is funded by the National Natural Science Foundation of China under grant number 62205334.

Institutional Review Board Statement: Not applicable.

Informed Consent Statement: Not applicable.

Data Availability Statement: Not applicable.

Conflicts of Interest: The authors declare no conflict of interest.

References

1. Fan, D.-P.; Wang, W.; Cheng, M.; Shen, J. Shifting more attention to video salient object detection. In Proceedings of the IEEE/CVF Conference on Computer Vision and Pattern Recognition (CVPR), Long Beach, CA, USA, 15–20 June 2019.
2. Shimoda, W.; Yanai, K. Distinct class-specific saliency maps for weakly supervised semantic segmentation. In Proceedings of the European Conference on Computer Vision (ECCV), Amsterdam, The Netherlands, 11–14 October 2016.
3. Mahadevan, V.; Vasconcelos, N. Saliency-based discriminant tracking. In Proceedings of the IEEE/CVF Conference on Computer Vision and Pattern Recognition(CVPR), Miami, FL, USA, 20–25 June 2009.
4. Ma, C.; Huang, J.B.; Yang, X.K.; Yang, M.H. Hierarchical convolutional features for visual tracking. In Proceedings of the 2015 IEEE International Conference on Computer Vision, Santiago, Chile, 7–13 December 2015; pp. 3074–3082.
5. Lin, T.Y.; Dollár, P.; Girshick, R.; He, K.; Hariharan, B.; Belongie, S. Feature pyramid networks for object detection. In Proceedings of the IEEE/CVF Conference on Computer Vision and Pattern Recognition (CVPR), Honolulu, HI, USA, 21–26 July 2017.
6. Wang, X.; Ma, H.; Chen, X.; You, S. Edge preserving and multiscale contextual neural network for salient object detection. *IEEE Trans. Image Process.* **2018**, *27*, 121–134. [CrossRef] [PubMed]

7. Liu, S.; Huang, D.; Wang, Y. Receptive field block net for accurate and fast object detection. In Proceedings of the European Conference on Computer Vision (ECCV), Munich, Germany, 8–14 September 2018; pp. 385–400.
8. Liu, J.; Hou, Q.; Cheng, M.; Feng, J.; Jiang, J. A simple pooling-based design for real-time salient object detection. In Proceedings of the IEEE/CVF Conference on Computer Vision and Pattern Recognition (CVPR), Long Beach, CA, USA, 15–20 June 2019.
9. Zhao, J.; Cao, Y.; Fan, D.-P.; Cheng, M.; Li, X.; Zhang, L. Contrast prior and fluid pyramid integration for RGBD salient object detection. In Proceedings of the IEEE/CVF Conference on Computer Vision and Pattern Recognition (CVPR), Long Beach, CA, USA, 15–20 June 2019.
10. Piao, Y.; Rong, Z.; Zhang, M.; Ren, W.; Lu, H. A2dele: Adaptive and attentive depth distiller for efficient RGB-D salient object detection. In Proceedings of the IEEE/CVF Conference on Computer Vision and Pattern Recognition (CVPR), Seattle, WA, USA, 13–19 June 2020.
11. Chen, S.; Fu, Y. Progressively guided alternate refinement network for RGB-D salient object detection. In Proceedings of the European Conference on Computer Vision (ECCV), Virtual, 23–28 August 2020; pp. 520–538.
12. Fan, D.-P.; Zhai, Y.; Borji, A.; Yang, J.; Shao, L. BBS-Net: RGB-D salient object detection with a bifurcated backbone strategy network. In Proceedings of the European Conference on Computer Vision (ECCV), Virtual, 23–28 August 2020; pp. 275–292.
13. Li, G.; Liu, Z.; Ye, L.; Wang, Y.; Ling, H. Cross-modal weighting network for RGB-D salient object detection. In Proceedings of the European Conference on Computer Vision (ECCV), Virtual, 23–28 August 2020; pp. 665–681.
14. Li, G.; Liu, Z.; Chen, M.; Bai, Z.; Lin, W.; Ling, H. Hierarchical Alternate Interaction Network for RGB-D Salient Object Detection. *IEEE Trans. Image Process.* **2021**, *30*, 3528–3542. [CrossRef] [PubMed]
15. Zhou, T.; Fu, H.; Chen, G.; Zhou, Y.; Fan, D.-P.; Shao, L. Specificity-preserving RGB-D Saliency Detection. In Proceedings of the IEEE International Conference on Computer Vision (ICCV), Montreal, QC, Canada, 10–17 October 2021; pp. 4681–4691.
16. Chen, H.; Li, Y.; Su, D. Multi-modal fusion network with multi-scale multi-path and cross-modal interactions for RGB-D salient object detection. *Pattern Recognit.* **2019**, *86*, 376–385. [CrossRef]
17. Zhang, J.; Fan, D.-P.; Dai, Y.; Yu, X.; Zhong, Y.; Barnes, N.; Shao, L. RGB-D saliency detection via cascaded mutual information minimization. In Proceedings of the IEEE International Conference on Computer Vision (ICCV), Montreal, QC, Canada, 10–17 October 2021; pp. 4318–4327.
18. Ji, W.; Li, J.; Yu, S.; Zhang, M.; Piao, Y.; Yao, S.; Bi, Q.; Ma, K.; Zheng, Y.; Lu, H.; et al. Calibrated rgb-d salient object detection. In Proceedings of the IEEE/CVF Conference on Computer Vision and Pattern Recognition. In Proceedings of the IEEE/CVF Conference on Computer Vision and Pattern Recognition (CVPR), Virtual, 19–25 June 2021; pp. 9471–9481.
19. Lee, M.; Park, C.; Cho, S.; Lee, S. SPSN: Superpixel prototype sampling network for RGB-D salient object detection. In Proceedings of the European Conference on Computer Vision (ECCV), Tel Aviv, Israel, 23–27 October 2022.
20. Sun, P.; Zhang, W.; Wang, H.; Li, S.; Li, X. Deep RGB-D Saliency Detection with Depth-Sensitive Attention and Automatic Multi-Modal Fusion. In Proceedings of the IEEE/CVF Conference on Computer Vision and Pattern Recognition (CVPR), Virtual, 19–25 June 2021.
21. Wu, Z.; Gobichettipalayam, S.; Tamadazte, B.; Allibert, G.; Paudel, D.P.; Demonceaux, C. Robust RGB-D fusion for saliency detection. In Proceedings of the 2022 International Conference on 3D Vision (3DV), Prague, Czechia, 12–15 September 2022; pp. 403–413.
22. Chen, T.; Xiao, J.; Hu, X.; Zhang, G.; Wang, S. Adaptive fusion network for RGB-D salient object detection. *Neurocomputing* **2023**, *522*, 152–164. [CrossRef]
23. Wu, Z.; Allibert, G.; Meriaudeau, F.; Ma, C.; Demonceaux, C. HiDAnet: RGB-D Salient Object Detection via Hierarchical Depth Awareness. *IEEE Trans. Image Process.* **2023**, *32*, 2160–2173. [CrossRef] [PubMed]
24. Pang, Y.; Zhao, X.; Zhang, L.; Lu, H. Caver: Cross-modal view mixed transformer for bi-modal salient object detection. *IEEE Trans. Image Process.* **2023**, *32*, 892–904. [CrossRef] [PubMed]
25. Zhao, H.; Shi, J.; Qi, X.; Wang, X.; Jia, J. Pyramid scene parsing network. In Proceedings of the IEEE/CVF Conference on Computer Vision and Pattern Recognition (CVPR), Honolulu, HI, USA, 21–26 July 2017; pp. 2881–2890.
26. Vaswani, A.; Shazeer, N.; Parmar, N.; Uszkoreit, J.; Jones, L.; Gomez, A.N.; Kaiser, Ł.; Polosukhin, I. Attention is all you need. *Adv. Neural Inform. Process. Syst.* **2023**, *30*, 5998–6008.
27. Liu, Y.; Zhang, Y.; Wang, Y.; Hou, F.; Yuan, J.; Tian, J.; Zhang, Y.; Shi, Z.; Fan, J.; He, Z. A Survey of visual transformers. *IEEE Trans. Neural Netw. Learn. Syst.* **2023**, *early access*. [CrossRef] [PubMed]
28. Wang, W.; Xie, E.; Li, X.; Fan, D.-P.; Song, K.; Liang, D.; Lu, T.; Luo, P.; Shao, L. Pyramid vision transformer: A versatile backbone for dense prediction without convolutions. In Proceedings of the IEEE/CVF International Conference on Computer Vision (ICCV), Montreal, QC, Canada, 10–17 October 2021; pp. 568–578.
29. Liu, Z.; Lin, Y.; Cao, Y.; Hu, H.; Wei, Y.; Zhang, Z.; Lin, S.; Guo, B. Swin transformer: Hierarchical vision transformer using shifted windows. In Proceedings of the IEEE/CVF Conference on Computer Vision and Pattern Recognition, Virtual, 19–25 June 2021; pp. 10012–10022.
30. Zhang, M.; Fei, S.; Liu, J.; Xu, S.; Piao, Y.; Lu, H. Asymmetric two-stream architecture for accurate rgb-d saliency detection. In Proceedings of the European Conference on Computer Vision (ECCV), Virtual, 23–28 August 2020; pp. 374–390.
31. Jiang, B.; Chen, S.; Wang, B.; Luo, B. MGLNN: Semi-supervised learning via multiple graph cooperative learning neural networks. *Neural Netw.* **2022**, *153*, 204–214. [CrossRef] [PubMed]
32. Yu, F.; Koltun, V. Multi-scale context aggregation by dilated convolutions. *arXiv* **2015**, arXiv:1511.07122. [CrossRef]

33. Liu, N.; Zhang, N.; Wan, K.; Han, J.; Shao, L. Visual Saliency Transformer. In Proceedings of the IEEE/CVF Conference on Computer Vision and Pattern Recognition. In Proceedings of the IEEE/CVF Conference on Computer Vision and Pattern Recognition (CVPR), Virtual, 19–25 June 2021; pp. 4722–4732.
34. Zeng, C.; Kwong, S. Dual Swin-Transformer based Mutual Interactive Network for RGB-D Salient Object Detection. *arXiv* **2022**, arXiv:2206.03105. [CrossRef]
35. Liu, C.; Yang, G.; Wang, S.; Wang, H.; Zhang, Y.; Wang, Y. TANet: Transformer-based Asymmetric Network for RGB-D Salient Object Detection. *IET Comput. Vis.* **2023**, *17*, 415–430. [CrossRef]
36. Pang, Y.; Zhao, X.; Zhang, L.; Lu, H. Transcmd: Cross-modal decoder equipped with transformer for rgb-d salient object detection. *arXiv* **2021**, arXiv:2112.02363. [CrossRef]
37. Liu, Z.; Tan, Y.; He, Q.; Xiao, Y. Swinnet: Swin transformer drives edge-aware RGB-D and RGB-T salient object detection. *IEEE Trans. Circ. Syst. Video Technol.* **2021**, *32*, 4486–4497. [CrossRef]
38. Roy, A.M.; Bhaduri, J. DenseSPH-YOLOv5: An automated damage detection model based on DenseNet and Swin-Transformer prediction head-enabled YOLOv5 with attention mechanism. *Adv. Eng. Inform.* **2023**, *56*, 102007. [CrossRef]
39. Wei, J.; Wang, S.; Huang, Q. F^{3}net: Fusion, feedback and focus for salient object detection. In Proceedings of the AAAI Conference on Artificial Intelligence, New York, NY, USA, 7–12 February 2020; pp. 12321–12328.
40. Niu, Y.; Geng, Y.; Li, X.; Liu, F. Leveraging stereopsis for saliency analysis. In Proceedings of the IEEE/CVF Conference on Computer Vision and Pattern Recognition (CVPR), Province, RI, USA, 16–21 June 2012; pp. 454–461.
41. Ju, R.; Ge, L.; Geng, W.; Ren, T.; Wu, G. Depth saliency based on anisotropic center-surround difference. In Proceedings of the IEEE International Conference on Image Processing (ICIP), Paris, France, 27–30 October 2014; pp. 1115–1119.
42. Peng, H.; Li, B.; Xiong, W.; Hu, W.; Ji, R. RGBD salient object detection: A benchmark and algorithms. In Proceedings of the European Conference on Computer Vision (ECCV), Zurich, Switzerland, 5–12 September 2014; pp. 92–109.
43. Li, N.; Ye, J.; Ji, Y.; Ling, H.; Yu, J. Saliency detection on light field. In Proceedings of the IEEE/CVF Conference on Computer Vision and Pattern Recognition (CVPR), Zurich, Switzerland, 5–12 September 2014; pp. 2806–2813.
44. Fan, D.-P.; Lin, Z.; Zhang, Z.; Zhu, M.; Cheng, M. Rethinking RGB-D salient object detection: Models, data sets, and large-scale benchmarks. *IEEE Trans. Neural Netw. Learn. Syst.* **2021**, *32*, 2075–2089. [CrossRef] [PubMed]
45. Cheng, Y.; Fu, H.; Wei, X.; Xiao, J.; Cao, X. Depth enhanced saliency detection method. In Proceedings of the International Conference on Internet Multimedia Computing and Service, Xiamen, China, 10–12 July 2014; pp. 23–27.
46. Fan, D.-P.; Cheng, M.M.; Liu, Y.; Li, T.; Borji, A. Structure-measure: A new way to evaluate foreground maps. In Proceedings of the IEEE international conference on computer vision (CVPR), Honolulu, HI, USA, 21–26 July 2017; pp. 4548–4557.
47. Achanta, R.; Hemami, S.; Estrada, F.; Susstrunk, S. Frequency-tuned salient region detection. In Proceedings of the 2009 IEEE Conference on Computer Vision and Pattern Recognition (CVPR), Miami, FL, USA, 22–25 June 2009; pp. 1597–1604.
48. Fan, D.-P.; Gong, C.; Cao, Y.; Ren, B.; Cheng, M.; Borji, A. Enhanced-alignment measure for binary foreground map evaluation. In Proceedings of the International Joint Conference on Artificial Intelligence (IJCAI), Stockholm, Sweden, 13–19 July 2018; pp. 698–704.
49. Perazzi, F.; Krähenbühl, P.; Pritch, Y.; Hornung, A. Saliency filters: Contrast based filtering for salient region detection. In Proceedings of the IEEE/CVF Conference on Computer Vision and Pattern Recognition (CVPR), Providence, RI, USA, 16–21 June 2012; pp. 733–740.
50. Fu, K.; Fan, D.-P.; Ji, G.; Zhao, Q. JL-DCF: Joint learning and densely-cooperative fusion framework for rgb-d salient object detection. In Proceedings of the IEEE/CVF conference on computer vision and pattern recognition (CVPR), Seattle, WA, USA, 13–19 June 2020; pp. 3052–3062.
51. Bi, H.; Wu, R.; Liu, Z.; Zhu, H. Cross-modal Hierarchical Interaction Network for RGB-D Salient Object Detection. *Pattern Recognit.* **2023**, *136*, 109194. [CrossRef]
52. Chen, T.; Hu, X.; Xiao, J.; Zhang, G.; Wang, S. CFIDNet: Cascaded Feature Interaction Decoder for RGB-D Salient Object Detection. *Neural Comput. Applic.* **2022**, *34*, 7547–7563. [CrossRef]
53. Zhang, M.; Yao, S.; Hu, B.; Piao, Y.; Ji, W. C^{2}DFNet: Criss-Cross Dynamic Filter Network for RGB-D Salient Object Detection. *IEEE Trans. Multimed.* **2022**. early access. [CrossRef]
54. Wang, X.; Jiang, B.; Wang, X.; Luo, B. Mutualformer: Multimodality representation learning via mutual transformer. *arXiv* **2021**, arXiv:2112.01177. [CrossRef]

Disclaimer/Publisher's Note: The statements, opinions and data contained in all publications are solely those of the individual author(s) and contributor(s) and not of MDPI and/or the editor(s). MDPI and/or the editor(s) disclaim responsibility for any injury to people or property resulting from any ideas, methods, instructions or products referred to in the content.

Article

Lightweight Detection Methods for Insulator Self-Explosion Defects

Yanping Chen [1], Chong Deng [1], Qiang Sun [1], Zhize Wu [1], Le Zou [1], Guanhong Zhang [1] and Wenbo Li [2,*]

[1] School of Artificial Intelligence and Big Data, Hefei University, Hefei 230601, China
[2] Institute of Intelligent Machines, Chinese Academy of Sciences, Hefei 230001, China
* Correspondence: wbli@iim.ac.cn

Citation: Chen, Y.; Deng, C.; Sun, Q.; Wu, Z.; Zou, L.; Zhang, G.; Li, W. Lightweight Detection Methods for Insulator Self-Explosion Defects. *Sensors* **2024**, *24*, 290. https://doi.org/10.3390/s24010290

Academic Editors: Hossam A. Gabbar, Man Qi and Matteo Dunnhofer

Received: 29 September 2023
Revised: 7 November 2023
Accepted: 14 December 2023
Published: 3 January 2024

Copyright: © 2024 by the authors. Licensee MDPI, Basel, Switzerland. This article is an open access article distributed under the terms and conditions of the Creative Commons Attribution (CC BY) license (https://creativecommons.org/licenses/by/4.0/).

Abstract: The accurate and efficient detection of defective insulators is an essential prerequisite for ensuring the safety of the power grid in the new generation of intelligent electrical system inspections. Currently, traditional object detection algorithms for detecting defective insulators in images face issues such as excessive parameter size, low accuracy, and slow detection speed. To address the aforementioned issues, this article proposes an insulator defect detection model based on the lightweight Faster R-CNN (Faster Region-based Convolutional Network) model (Faster R-CNN-tiny). First, the Faster R-CNN model's backbone network is turned into a lightweight version of it by substituting EfficientNet for ResNet (Residual Network), greatly decreasing the model parameters while increasing its detection accuracy. The second step is to employ a feature pyramid to build feature maps with various resolutions for feature fusion, which enables the detection of objects at various scales. In addition, replacing ordinary convolutions in the network model with more efficient depth-wise separable convolutions increases detection speed while slightly reducing network detection accuracy. Transfer learning is introduced, and a training method involving freezing and unfreezing the model is employed to enhance the network's ability to detect small target defects. The proposed model is validated using the insulator self-exploding defect dataset. The experimental results show that Faster R-CNN-tiny significantly outperforms the Faster R-CNN (ResNet) model in terms of mean average precision (mAP), frames per second (FPS), and number of parameters.

Keywords: target detection; lightweight; self-explosion defects in insulators; EfficientNet; small target defects

1. Introduction

Insulators, as important components of high-voltage transmission lines, serve the functions of electrical separation and support for conductors [1]. Due to their long-term outdoor exposure to sunlight, rain, climate changes, and chemical corrosion, insulators often suffer from self-exploding defects, causing the disconnection of insulator strings and interfering with their performance, thus affecting the safety and stability of power systems [2]. Insulator detection methods are generally divided into two types. The first is manual inspection, where workers directly observe insulators to identify defective parts. However, this method is time-consuming and not safe. The second is intelligent inspection, which can effectively locate defective parts by carrying edge detection equipment on drones for regular inspection of insulators. This is also the current mainstream inspection method.

Currently, the implementation of insulator defect detection mainly relies on traditional methods and deep learning methods. Traditional detection methods primarily differentiate insulators from the background based on features such as size, texture, and color of the images [3]. For example, Tan et al. [4] takes a fusion algorithm based on insulator contour features and grayscale similarity matching. It can extract the contours of insulator pieces, accurately separate them, and construct a defect detection model based on the spacing between insulator pieces and grayscale similarity matching. Liu et al. [5] proposed an edge-based segmentation method for insulator strings. It uses a multi-scale morphological gradient algorithm to extract the edges of insulator strings, determine the largest connected region, and provide guidance for addressing the problem of mis-segmentation of iron caps and umbrella discs caused by edge loss in infrared images of insulator strings. However, these traditional detection methods have low efficiency in feature extraction, poor generalization capabilities, and difficulty in recognizing small-scale and high-likelihood objects in images [6].

To enhance the feature extraction capability and anti-interference ability of insulator detection, traditional detection methods are no longer able to meet modern needs. Many scholars have turned their attention to deep learning methods. For example, Guo et al. [7] used a lightweight target detection network called CenterNet-GhostNet to address the issue of the large number of parameters in the insulator defect detection model, which makes it difficult for unmanned aerial vehicles to deploy on the edge. This network significantly reduces the number of network parameters while achieving a slight increase in detection accuracy, thereby improving the detection speed of the network. Jia et al. [8] considered a lightweight detection method called MDD-YOLOv3. The improved YOLOv3 can quickly and accurately recognize and locate insulator defects in complex backgrounds. Li et al. [9] proposed a method that utilizes multiple-scale feature encoding and dual attention fusion to improve the accuracy and speed of detecting insulator defects in transmission lines. It has a certain reference value for accurate insulator defect detection by unmanned aerial vehicles. In summary, compared with the traditional manual feature extraction of insulators, deep learning-based detection methods can automatically and accurately extract target features and have stronger generalization capabilities.

In recent years, due to the development of "Intelligentization" in the power system, the combination of using drones to collect insulator defect data and computer vision technology has become a popular method for intelligent inspection [10,11]. However, deep learning-based object detection networks usually require a large number of computational resources and parameters for training and inference, which limits their deployment and usage in practical applications. Therefore, the construction of lightweight detection models has become crucial [12–14].

The existing deep learning detection methods can be mainly divided into two categories. One is the two-stage detection model represented by R-CNN, Faster R-CNN, and Mask R-CNN. These algorithms require two-stage processing: (1) candidate region acquisition and (2) classification and regression of candidate regions. The other is the single-stage detection model represented by the YOLO series, which simultaneously obtains candidate regions and categories through joint decoding. Among them, the Faster R-CNN model, as a representative of two-stage networks, exhibits a more pronounced advantage when it comes to handling high-precision, multi-scale, and small object detection tasks. However, the original Faster R-CNN (ResNet) model suffers from significant drawbacks in terms of detection speed performance. Firstly, its feature extraction capability is relatively poor. This is because the original ResNet cannot effectively extract high-level semantic information and low-level fine-grained features from images, making it difficult for deeper feature maps to learn information about small objects. Secondly, the network's inference speed is slow. The original model contains a lot of redundant information, resulting in a slow detection speed. Finally, the network parameters are not well optimized. For instance, the original model's learning rate can easily get stuck in local optima, leading to a decline in the overall model performance.

In this paper, we have lightweighted the original Faster R-CNN (ResNet) and constructed a new detection model (Faster R-CNN-tiny), as shown below:

(1) We use the lightweight EfficientNet [15] as the backbone network to capture multi-scale detailed features of faulty object. These features serve as inputs to the Feature Pyramid Network (FPN) [16], enhancing the network's capability to extract characteristics from defects of various scales.

(2) A feature fusion module is added to effectively combine high-level semantic information with low-level detail information, enhancing the accuracy of defect detection. Ordinary convolutions in the network are replaced with depth-wise separable convolutions (DSConv), which improve the detection speed to some extent.
(3) Transfer learning methods [17] are employed in the network training process, combining freezing and unfreezing training strategies to enhance the network's detection performance for defects in complex environments.
(4) The proposed lightweight Faster R-CNN-tiny object detection model can effectively locate the defects of insulators by learning a large number of features from self-made defect images of insulators. This is a crucial step towards the edge detection of defects in the next step. We have also introduced a new dataset for insulator defects called Tiny-Insulator.

The rest of this paper is structured as follows: In Section 2, we introduce the structure and principles of the original Faster R-CNN model. In Section 3, we provide a detailed description of the target model, Faster R-CNN-tiny, analyzing the functionality and flow of each component. In Section 4, we validate the impact of different network structures on experimental performance using the insulator defect dataset. In Section 5, we summarize this research and discuss future work.

2. Related Works

2.1. Faster R-CNN

Defect detection involves the following two tasks: defect classification and localization. This paper chooses two-stage Faster R-CNN [18] as the lightweight base network structure, which exhibits a high accuracy in object detection tasks. Its working principle is to first identify and locate defective insulators in an image, then select them with rectangles, and, finally, mark their belonging categories near the rectangles.

Faster R-CNN is a two-stage object detection network proposed by Ross B. Girshick, building upon the foundations of R-CNN and Fast R-CNN. As shown in Figure 1, the Faster R-CNN network structure consists of four parts: the backbone network, the Region Proposal Network (RPN), the Region of Interest (RoI) pooling, and the detection network. The backbone network is a ResNet network stacked with multiple 7×7 convolutions of stride 2 and 3×3 convolutions of stride 2. The RPN is a feature-processing part composed of two parallel 1×1 convolutions by 3×3 deep separable convolutions (DWConv). The detection network consists of two parallel fully connected layers (FC).

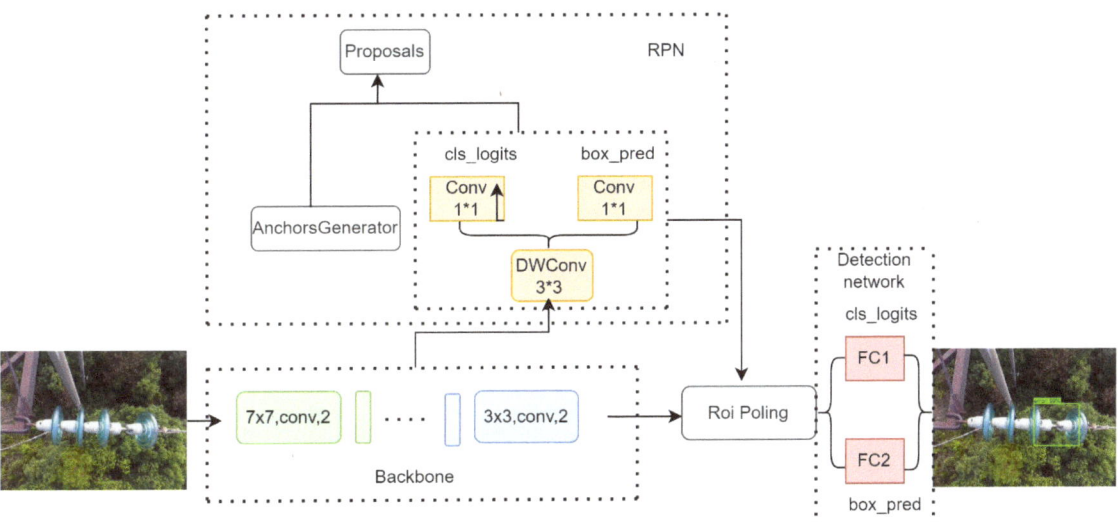

Figure 1. The original Faster R-CNN network structure.

The entire algorithm process is divided into several parts. First, the backbone network extracts features from preprocessed images by capturing multi-scale information with inter-channel interactions. Then, these features are used as input for the RPN, which generates candidate boxes. The candidate boxes are mapped to the feature map output by the backbone network. The obtained feature matrix is passed through the RoI Pooling layer, resulting in a 7×7 feature map. Finally, the detection network utilizes the feature map to obtain class information and bounding box regression parameters. The candidate boxes are adjusted using the bounding box regression parameters to obtain the final target position.

To address the low accuracy and slow speed issues of the original model in insulator defect detection, we propose a lightweight defect detection model based on Faster R-CNN-tiny. The aim is to make the original detection model more suitable for future edge deployment requirements.

2.2. ResNet

ResNet, which stands for Deep Residual Network, is a landmark convolutional neural network (CNN) that uniquely solves the problems of gradient disappearance and explosion in deep neural networks.

In 2015, ResNet won the ILSVRC (ImageNet Large Scale Visual Recognition Challenge) championship and significantly improved error accuracy in the ImageNet classification task. This is mainly due to ResNet's "shortcut connections", also known as "skip connections". Through this connection method, the output of the deep network can be directly added to some layers of the shallow network, which helps the gradient to be directly transmitted to the shallow network. This design allows the network to train deep networks with dozens or even hundreds of layers.

2.3. EfficientNet

EfficientNet, proposed by Google in 2019, constructs models through compound scaling to improve model efficiency. It is composed of one ordinary convolutional layer and sixteen mobile inverted bottleneck convolution modules (MBConv). Among them, the MBConv module is its core component, which mainly draws inspiration from the residual structure of MobileNetv3 [19]. As shown in Figure 2, it has the following functional features: firstly, a Swish activation function [20] is used instead of a ReLU activation function, and Swish performs better on deep models. Secondly, a squeeze-and-excitation networks

(SENet) [21] attention mechanism is added to each MBConv module to strengthen the extraction of small-scale target features and suppress useless feature information. Thirdly, dropout layers are introduced. When there are shortcut branches (shortcuts), the main branch of the whole module will be randomly dropped, leaving only the shortcut branch, making the network lighter and improving the detection speed of the model.

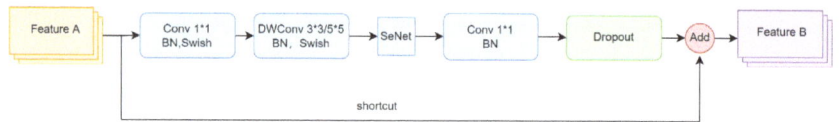

Figure 2. The MBConv module in the EfficientNet network.

The main difference between the two lies in their network structure and optimization strategies. EfficientNet adopts a deeper and wider network structure, while using compound scaling to adjust the depth, width, and resolution of the network. This makes EfficientNet reduce the number of parameters and computations, thereby improving the efficiency of the model. On the other hand, ResNet mainly solves the vanishing and exploding gradient problems using residual blocks, with a relatively simple network structure.

3. Methodology

In order to make the model more suitable for the detection of small targets and reduce the number of model parameters, this paper proposes a new object detection model called Faster R-CNN-tiny. The Faster R-CNN-tiny model only improves the backbone part of the original Faster R-CNN model, as shown in Figure 3. The input image first goes through a feature extraction layer (EfficientNetB0) to obtain feature maps at different resolutions (C2, C3, C4, C5), then enters a feature fusion layer (Feature Pyramid Network), and, finally, the resulting different feature maps (P2, P3, P4, P5) are further processed in the RPN.

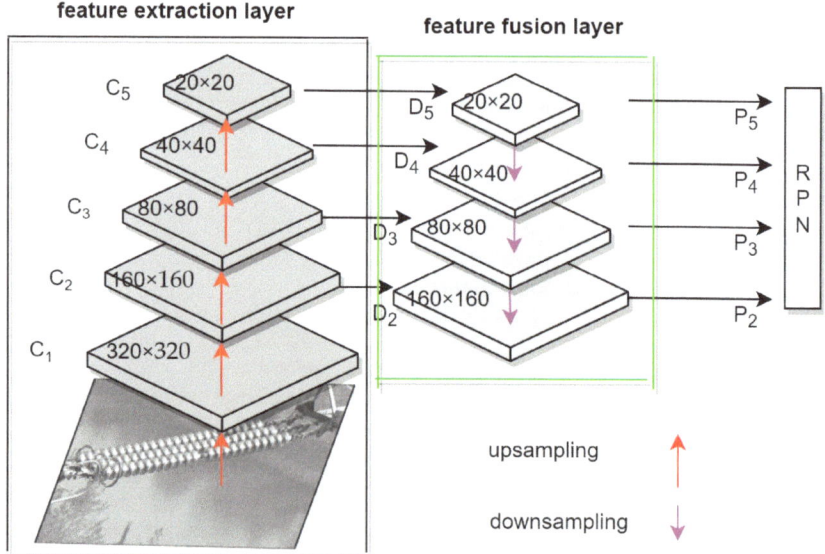

Figure 3. The backbone of the Faster R-CNN-tiny model.

To enable Faster R-CNN-tiny to detect more small object features, we have added a D2 object detection layer to detect shallower features. This can be specified as follows: First, the use of EfficientNet with attention mechanisms as the backbone network for feature extraction from input images, addressing the issue of partial feature information loss in the generation of multi-resolution feature maps by the backbone network [22]. After that, a lightweight feature fusion module is proposed and added to the backbone network. This module effectively integrates low-level positional information with high-level semantic information, ensuring that the fused feature maps retain sufficient detailed information. Finally, DSConv are employed to replace regular convolutions in the FPN and RPN. This not only reduces the network's parameter count but also enhances its detection speed.

To ensure that the target detection algorithm has a good scale invariance, the original Faster-RCNN algorithm generates anchor boxes with ratios of (1:1, 1:2, 2:1) and sizes of (64, 128, 256) when traversing the feature map. However, the original Faster-RCNN algorithm's anchor boxes are not suitable for detecting small targets or actual-scale defective insulator targets. To obtain better anchor box ratios, this paper statistically analyzes the length-to-width ratios of the defective insulators in the dataset. The length-to-width ratio of the insulators is approximately 60% for 1:1, 26% for 2:1, 11% for 3:1, and 3% for 4:1. Therefore, in this paper, the anchor box ratios are set as (1:1, 2:1, 3:1) with sizes of (16, 32, 64, 128, 256).

After a series of consecutive convolutional and pooling operations on the input image, the information on the feature map gradually diminishes. In Figure 3, the C2 feature map layer contains more object information than the C3 feature map layer. Therefore, this paper introduces detection in the C2 feature layer, which contains more feature information. In the original Faster R-CNN algorithm, the feature extraction part only utilized ResNet, whereas the new model incorporates EfficientNet and FPN. When the input image size is 640×640, the detection layer corresponding to C3 has a size of 80×80, suitable for detecting objects larger than 8×8; the detection layer corresponding to C4 has a size of 40×40, suitable for detecting objects larger than 16×16, and the detection layer corresponding to C5 has a size of 20×20, suitable for detecting objects larger than 32×32.

3.1. Feature Extraction Layer

Traditional object detection algorithms have many similarities between the feature layers in their feature extraction networks. While these similar feature layers improve accuracy, they also introduce a lot of redundant information, making the network models large and difficult to deploy on small mobile devices [23,24]. Therefore, research on lightweight networks has become a hot topic. Currently, the mainstream lightweight networks include SqueezeNet, ShuffleNet, EfficientNet, RegNet, and MobileNet. Considering that the detection targets are small objects and that the purpose is to build a lightweight and efficient feature extraction network, EfficientNetB0 is chosen as the backbone network. EfficientNet is a network model obtained through Neural Architecture Search (NAS), and it achieves EfficientNet B0~B7 network models by rationalizing the configuration of the following three parameters: image input resolution (r), network depth (d), and network width (w), as shown in Figure 4.

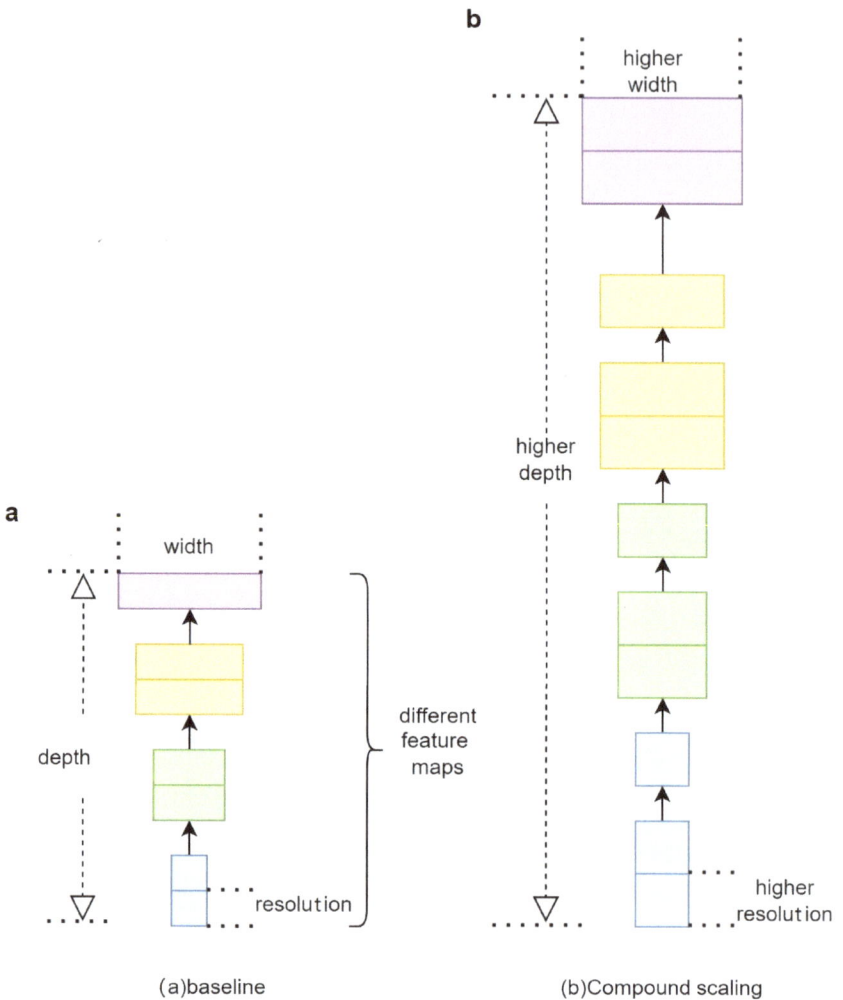

Figure 4. Illustrates the NAS search process within EfficientNet. (**a**) baseline and (**b**) compound scaling.

Then, the NAS search is used to obtain the best r, d, and w factors for the EfficientNet network.

$$s.t: N(d,w,r) = \bigodot_{i=1...s} \hat{F}_i^{d \cdot \hat{L}_i}\left(X_{\langle r \cdot \hat{H}_i, r \cdot \hat{W}_i, w \cdot \hat{C}_i \rangle}\right) \quad (1)$$

$$\underset{d,w,r}{MAX}[Accuracy(N(d,w,r))] \quad (2)$$

$$\text{Memory}(N) \leq \text{Target\_memory} \quad (3)$$

$$\text{FLOPs}(N) \leq \text{Target\_flops} \quad (4)$$

In Equations (1)–(4), $\bigodot_{i=1...s}$ represents the multiplication operation. $\hat{F}_i^{d \cdot \hat{L}_i}$ represents that the $\hat{F}_i$ operation is repeated $d \cdot \hat{L}_i$ times in the i-th stage. $\hat{F}_i$ represents an operation. Here, $X_{\langle r \cdot \hat{H}_i, r \cdot \hat{W}_i, w \cdot \hat{C}_i \rangle}$ represents the feature matrix of the *i*-th stage's input; $<\hat{H}_i, \hat{W}_i, \hat{C}_i>$ represents the height, width, and number of channels of X. Moreover, d, r, and w are used for scaling, respectively, $\hat{L}_i$, $\hat{H}_i$, and $\hat{C}_i$. Memory and FLOPs represent the limitations of the hardware's memory and maximum computational load.

By utilizing the MBConv module mentioned above (Figure 2), a lightweight network structure called EfficientNet can be constructed. Depending on the network's different stages, the MBConv structure can be modified in various ways. Firstly, the number of channels in the input features is expanded through 1×1 convolutions. Depending on the needs of each stage, DSConv can be selected in either a 3×3 or 5×5 size to integrate the extracted features and reduce noise. Then, SENet modules are used to enhance the ability to extract features at small target scales, followed by dimensionality reduction using 1×1 convolutions. Finally, shortcut connections only exist when the shape of the input feature matrix and the final output feature matrix are the same, allowing for the superposition of two types of features to enhance the network's feature extraction capabilities.

3.2. Feature Fusion Layer

The original Faster-RCNN object detection network only uses a single feature layer with high-order semantic information to predict target information. For small object detection, the high-order semantic feature maps lack the underlying information about details and have an impact on the accuracy of the detection results [25,26]. The new model leverages the multi-resolution feature maps from the feature extraction layer; these are merged through the feature fusion layer to retain the panoramic information on the targets.

To provide a more intuitive understanding of the principle behind the feature fusion layer, we have separately illustrated the feature fusion layer in Figure 5, as shown. Different resolution feature maps of 20×20, 40×40, 80×80, and 160×160 obtained from the feature extraction layer serve as inputs to the feature fusion layer. These four feature maps of varying scales contain rich semantic and positional information, and their effective fusion results in the final full-sized feature map, greatly enhancing the network's detection performance. Firstly, a step-1 1×1 convolution operation is applied to the four feature layers—L2, L3, L4, and L5—obtained from the feature extraction layer, resulting in dimensions of 112 and providing the necessary conditions for the subsequent feature fusion. Then, the higher-level feature maps are upsampled using bicubic interpolation and are pixel-wise added to the corresponding lower-level feature maps from the next layer to produce the initial feature fusion map. Finally, a DSConv operation is performed on the initial feature fusion map to integrate the extracted features and reduce noise interference on the detection results.

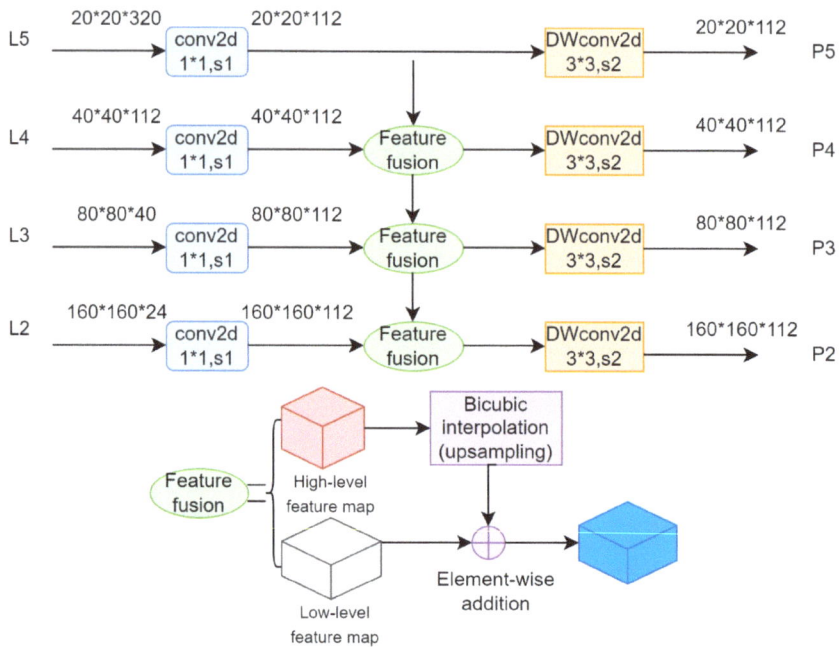

Figure 5. Feature fusion layer construction.

3.3. Depth-Wise Separable Convolution

Addressing the issues of large network size and slow detection speed, a lightweight network structure is proposed. The common convolutions in the FPN and RPN are replaced with DSConv [27], which reduces network parameters and computation while increasing detection speed.

As shown in Figure 6, DSConv consists of two main steps: channel-wise convolution (DWConv) and point-wise convolution (PoConv). DWConv produces feature maps with the same number of channels as the input by applying a convolutional kernel to each channel in the feature map after the convolution procedure. PoConv establishes dimensional links between the feature maps created by DWConv by further convolving them using a 1×1 convolution.

The parameter volume of a normal convolution is $D_w \times D_h \times M \times N$, while that of a channel–wise convolution is $D_w \times D_h \times M \times 1$ and that of point–wise convolution is $D_w \times D_h \times 1 \times 1$. The ratio of the parameter volume of DSConv to that of a normal convolution is the following:

$$\frac{D_w \times D_h \times M \times 1 + 1 \times 1 \times M \times N}{D_w \times D_h \times M \times N} = \frac{1}{N} + \frac{1}{D_w \times D_h} \tag{5}$$

For feature maps in which the scale does not change after processing, the calculation volume of a normal convolution is $D_w \times D_h \times F_w \times F_h \times M \times N$. The calculation volume of DWConv is $D_w \times D_h \times F_w \times F_h \times M \times 1$ and that of PoConv is $1 \times 1 \times F_w \times F_h \times M \times N$. The ratio of the calculation volume of DSConv to that of a normal convolution is as follows:

$$\frac{D_w \times D_h \times F_w \times F_h \times M \times 1 + 1 \times 1 \times F_w \times F_h \times M \times N}{D_w \times D_h \times F_w \times F_h \times M \times N} = \frac{1}{N} + \frac{1}{D_w \times D_h} \tag{6}$$

Among them, D_w and D_h represent the width and height of the convolution kernel; M represents the dimension of the input feature map; N represents the number of convolutional kernels; F_w and F_h represent the width and height of the feature map.

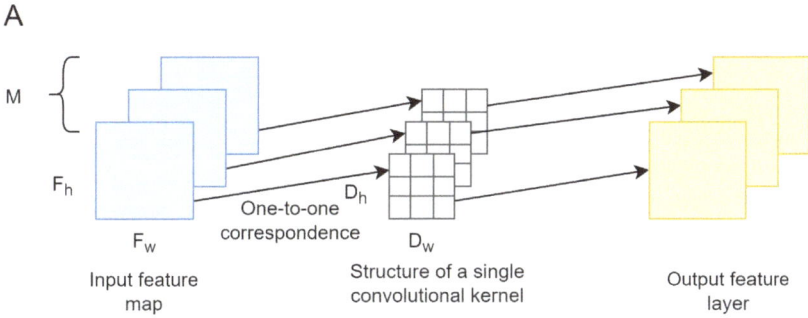

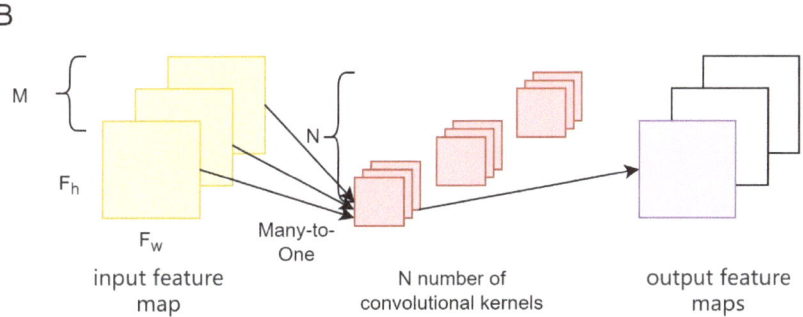

Figure 6. Structure of DSConv: (**A**) the process of DWConv and (**B**) the process of PoConv.

Due to the fact that, in normal circumstances, the number of convolutional kernels (N) is much greater than the size of the convolution kernel $D_w \times D_h$, when the depth-separable convolution kernel size is $3 \times 3 \times M$, Equations (1) and (2) are approximately equal to 1/9. Therefore, it can be seen that depth-separable convolution greatly reduces the parameter and computation amount of network models, thereby improving the detection speed of the network models.

3.4. Training Network Models with Transfer Learning

ImageNet [28] is an authoritative benchmark for evaluating network performance, with more than 1.2 million images which are finely classified into 1000 categories. In addition, ImageNet provides a large number of pre-trained weights that can be used in current object detection tasks, while revealing key features such as edges, corners, textures, and other characteristics in natural images, laying the foundation for visual tasks.

Due to safety concerns in power systems, it is difficult for drones to obtain data on defective insulators in transmission lines, and using small datasets to train models can easily lead to the slow or non-convergence of networks, resulting in poor detection performance [29,30]. To improve the detection performance of networks, transfer learning methods are introduced into the model training process, combining common feature knowledge of objects with the target object, thereby improving the detection performance of the target object.

This paper first uses the ImageNet dataset to pre-train the main network EfficientNet and obtain new model weights for the main network. Then, in order to save hardware costs, a combined model training strategy of freezing and unfreezing is used, using a homemade small dataset of insulator defects to re-fine-tune the network so that the network model can quickly adapt to small sample insulator defect datasets [31].

4. Experimental Results and Additional Requirements of the Analysis

4.1. Experimental Setup and Evaluation Metrics

The software environment used in all the experiments in this paper includes Python 3.6, Pycharm Community Edition 2022.2.1, and the Pytorch (1.10.2) deep learning framework. The CPU model is an Intel Core i7-12700H @ 2.30 GHz; the GPU model is an NVIDIA GeForce RTX 3070 with 8GB of memory and 16 GB of RAM, and the experiments were conducted on a PC. The detection performance of the model on self-exploding insulators was measured using the following three evaluation metrics: mean average precision (AP), frames per second (FPS), and parameters (Para).

4.2. Dataset

Since there is currently no publicly available insulator defect dataset, this paper proposes a new Tiny-Insulator dataset consisting of 400 insulator defect images and 400 annotation files. The initial dataset is divided into a 3:1 ratio for obtaining training and validation sets, with the training set being used for augmentation and network training and the validation set being used to evaluate network performance. These images are primarily sourced from a well-known power station in the region. To enhance the model's generalization and prevent overfitting, data augmentation techniques are applied to the training dataset, including the addition of noise and image transformations like flips. Brightness and contrast adjustments are also employed to simulate various insulator environments, thereby improving the robustness of model training. As depicted in Figure 7, the final training dataset is augmented to include 600 images and 600 annotation files.

Figure 7. Data augmentation.

To address the significant differences in resolution among the self-made insulator images, all images were resized to 640 × 640 pixels to create the final dataset. The defects on the insulators were labeled using the Labelimg (1.8.6) image annotation software, and all label files were organized in the PASCAL VOC dataset format.

4.3. Network Training

The batch size was set to four, and Adam was selected as the optimizer. The number of iterations for model training was 300, measured in epochs, and there were two categories (one background). In order to prevent the neural network model from getting stuck in local optima, a cosine annealing learning rate decay is used, with an initial learning rate set to 10×10^{-2} and a decay multiplier factor of 10×10^{-3}. To avoid a drop in model performance when training on a self-made small dataset, a transfer learning approach is employed. First, the backbone network is pretrained on a large-scale dataset like ImageNet, and its pretrained weights are used as the initial weights for the entire network. The model is trained using a combination of frozen and unfrozen layers to prevent issues arising from large discrepancies between the initial network weights and the actual data distribution. This approach also saves hardware resources and speeds up network convergence. During training, the weights of the backbone network are frozen for the first 30 epochs, and the remaining network structure is fine-tuned. The remaining 210 epochs involve unfreezing and training the entire network, resulting in increased GPU memory usage and changes to all model parameters. The loss and learning rate change process of the lightweight network model is shown in Figure 8.

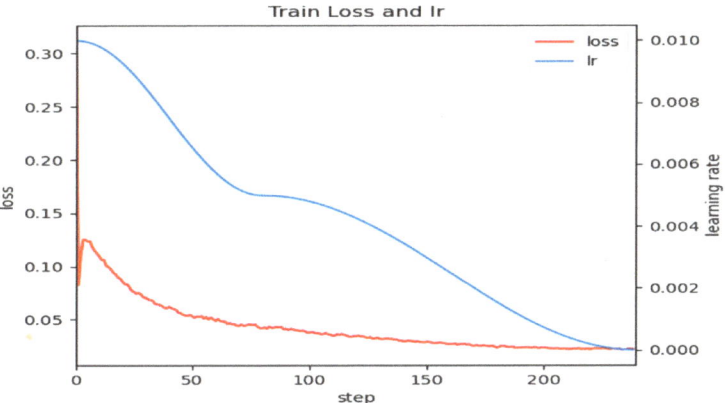

Figure 8. Change in training loss and learning rate curve.

As shown in Figure 8, transfer learning can significantly improve the convergence speed of the network.

There are three stages in the change in model loss. Firstly, the loss value decreases rapidly during the first 50 steps. Then, when training is between 50 and 150 steps, the loss decreases slowly. Finally, after the training iterations reach 150 steps, the loss gradually tends to be stable around 0.015. The learning rate decreases gradually along with the increase in iteration steps and, eventually, drops to a level close to the loss value.

4.4. Performance Analysis of Faster R-CNN-Tiny Models

4.4.1. Main Network Selection

According to the evaluation results generated using the Faster R-CNN target detection framework and by combining it with five different lightweight backbone networks in Table 1, the following points can be seen. The mean average precision (mAP) metric using EfficientNetB0 as the backbone network reaches 85.1%, which is 22.7% higher than ShuffleNetV2, 9.6% higher than SqueezeNet, 6.2% higher than MobileNetV3, and 5.6% higher than RegNetY800MF. In addition, the parameter count (Params) of EfficientNetB0 as the backbone network is lower or equal to that of the other four backbone networks, which is only 5.3M.

Table 1. Experimental results of different backbone networks.

Object Detection Framework	Backbone Network	mAP$_{0.5}$ (%)	Params (M)
Faster R-CNN	ShuffleNetV2	62.4	5.3
	SqueezeNet	75.5	6.9
	MobileNetV3	78.9	5.8
	RegNetY800MF	79.5	6.3
	EfficientNetB0	85.1	5.3

These results show that EfficientNetB0 not only has stronger feature extraction capabilities for defective insulators but also requires fewer parameters. Therefore, this research selects EfficientNetB0 as the feature extraction network.

4.4.2. Ablation Experiment

To verify the effectiveness of the proposed lightweight network structure, we conducted an ablation experiment in the same experimental environment to evaluate the impact of the improved modules on target detection. We chose the original Faster R-CNN with ResNet50 as the backbone network and trained it for 240 epochs. The results are shown in Table 2. The main evaluation metrics used were $mAP_{0.5}$ (which measures the target detection accuracy at an IoU threshold of 0.5), Params (which represents the size of the network), GFLOPs (which represents the model's computational complexity), and the symbol "✓" represents the use of this network structure.

Table 2. Evaluation results of ablation experiments.

Network Framework	Improvement				Evaluation Results		
	Feature Processing				$mAP_{0.5}$	Params	GFLOPs
	EfficientNetB0	Feature Fusion	DSConv	New Anchor Box	(%)	(M)	(G)
Faster R-CNN					83.0	70.5	88.8
	✓				85.1	7.6	6.3
	✓	✓			90.4	10.9	19.8
	✓	✓	✓		90.0	10.4	11.8
	✓	✓	✓	✓	90.3	10.4	11.8

As shown in Table 2, the first row shows the original Faster_rcnn algorithm without any detection results from any improved modules. Each subsequent row gradually adds different improved modules. The last row of the table shows the proposed Faster R-CNN-tiny model.

The following conclusions can be drawn. Firstly, when comparing the original target detection network Faster R-CNN with the replaced lightweight backbone network EfficientNetB0, the network parameters are reduced by 62.9 M and GFLOPs by 82.5 G, while the detection accuracy is improved, indicating that replacing the backbone network with EfficientNetB0 significantly improves the detection performance of the network. Secondly, the addition of the feature fusion module (FPN) leads to a significant improvement in model $mAP_{0.5}$ but also results in an increase in parameters, proving that fully integrating high-level semantic information and low-level position information into the model is indeed useful for improving its detection accuracy. Thirdly, replacing ordinary convolutional layers with DSConv makes the model lighter and more complex, resulting in a slight increase in detection speed under slightly lower detection accuracy, demonstrating the effectiveness of DSConv. And, finally, adding new anchor boxes on top of these experiments results in a slight increase in the model's detection accuracy, as adding small-scale detection anchor boxes better adapts to insulator defect detection in this paper. To sum up, this paper performs lightweight improvements on the Faster R-CNN algorithm, which not only greatly improve the detection accuracy of the model but also ensure real-time detection, demonstrating the effectiveness of the improved modules on model performance.

4.4.3. Detection Result Visualization

To further verify the detection effectiveness of the improved algorithm on self-exploding insulator defects in practical power inspection, the detection results are visualized and compared. As shown in Figure 9, Line a represents the real inspection chart; Line b represents the original Faster R-CNN detection result, and Line c represents the Faster R-CNN model detection result after the lightweight optimization outlined in this paper.

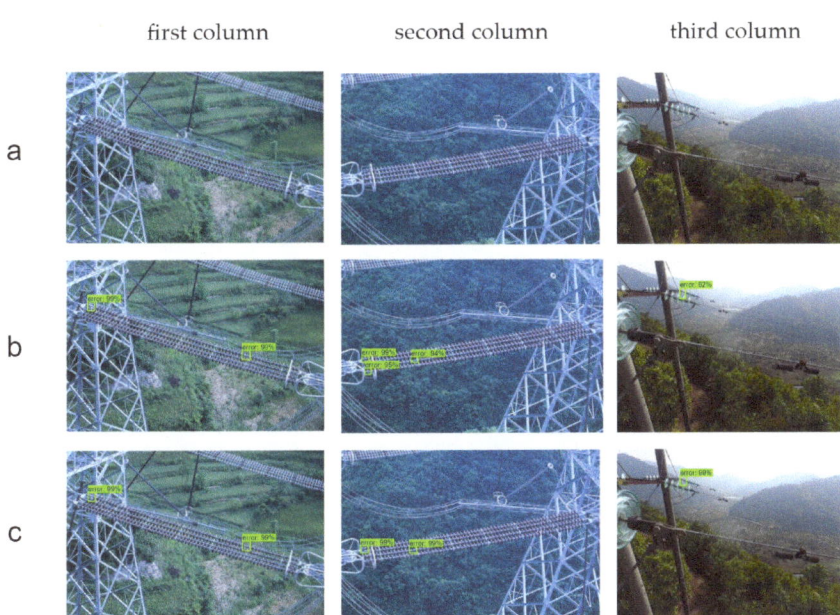

Figure 9. Visualization results of model improvement before and after.

As can be seen from the visualization results in Figure 9, for the detection of normal small targets, as shown in the first column, both the original Faster R-CNN algorithm and the improved algorithm can be recognized. For the detection of multiple targets in complex environments, as shown in the second column, the original algorithm has false positives, while the improved algorithm can accurately detect them. In the third column, although both algorithms can be recognized, the improved algorithm has a higher confidence. Based on the comprehensive evaluation metrics and image visualization results, the model proposed in this paper can effectively locate defective insulators in different environments.

4.4.4. Comparison Experiment on Different Object Detection Algorithms

To further validate the superiority of the algorithm proposed in this paper, a comparative experiment is conducted on the same hardware environment setup and power transmission line self-destructing insulator dataset with current mainstream object detection algorithms, including Faster R-CNN, RetinaNet, YOLOv5s, and YOLOv6s. The evaluation metrics used in the experiment are the number of parameters (Params), the average precision (AP), and the frames per second (FPS). The FPS metric represents the number of images the detection model can process per second and is commonly used to assess the detection speed of object detection networks. The experimental results are shown in Table 3.

Table 3. Comparison of different models.

Network Model	Backbone Network	Params (M)	$mAP_{0.5}$ (%)	FPS (F/S)
Faster R-CNN	ResNet50	70.5	83.0	21.8
RetinaNet	ResNet50 + FPN	32.2	86.5	22.3
YOLOv5s	CSPDarknet	7.0	87.3	50.0
YOLOV6s	EfficientRep	17.1	87.7	63.6
Ours	EfficientNetB0	10.4	90.3	35.2

As can be seen from Table 3, with a certain level of detection speed, the detection accuracy of Faster R-CNN-tiny algorithm surpasses the other algorithms. Compared with the original Faster R-CNN and RetinaNet, the average precision (mAP$_{0.5}$) is improved by 7.3% and 12.9%, respectively. The frames per second (FPS) are also increased by 13.4 and 12.9, indicating that this algorithm exhibits high detection efficiency for self-destructing insulator defects. Moreover, the parameter count of this paper's algorithm is significantly smaller than those of the other two algorithms, indicating a smaller model size. Compared to the YOLO algorithm, although the detection speed of this paper's algorithm is slower than YOLOv5s and YOLOv6s by 50.0 frames per second and 63.6 frames per second, respectively, it possesses certain advantages in terms of detection accuracy and model size. In summary, this paper's algorithm enhances network detection accuracy while considering network detection speed and model size, meeting the real-time and efficient detection requirements for defective insulators.

5. Conclusions

To address the issues of large volume and slow detection speed of the original Faster R-CNN model, an insulator defect detection model based on Faster R-CNN-tiny is proposed. The experimental results show that, compared with the original algorithm, the AP and FPS of this algorithm have improved by 7.3% and 13.4 frames/second, respectively, and the model parameter has decreased from 70.5 M to 10.4 M. This model can effectively identify insulator defects in transmission lines and has a certain reference significance for maintaining national line safety.

In the next stage of research, we will attempt to combine the new model with an embedded platform to prepare for future edge detection of the model. At the same time, we will also use other datasets in the power grid for testing. This will enable the model to have a better network structure and a higher detection accuracy.

Author Contributions: Conceptualization, Y.C.; methodology, C.D.; software, Z.W. and L.Z.; validation, Q.S.; formal analysis, C.D., Z.W. and G.Z.; project administration, Y.C., W.L. and L.Z.; funding acquisition, W.L. All authors have read and agreed to the published version of the manuscript.

Funding: This work was supported by the grant of the Anhui Provincial Natural Science Foundation under Grant (No.2108085J19) and in part by the National Nature Science Foundation of China under Grant (No.41871302) and the grant of Scientific Research and Talent Development Foundation of the Hefei University (No.21-22RC15), the Program for Scientific Research Innovation Team in Colleges and Universities of Anhui Province (No.2022AH010095), the Anhui Provincial Natural Science Foundation (No.2108085MF223, 2308085MF213).

Institutional Review Board Statement: Not applicable.

Informed Consent Statement: Not applicable.

Data Availability Statement: Data are contained within the article.

Conflicts of Interest: The authors declare no conflict of interest.

References

1. Feng, S.; Sheng, H.; Yong, L.; Lu, T. RSIn-Dataset: An UAV-Based Insulator Detection Aerial Images Dataset and Benchmark. *Drones* **2023**, *7*, 125.
2. Law, F.C.; Niu, L.L.; Wang, S.H.; Zhu, Y.P. Evaluation of the Severity of Discharge in Porcelain Suspension Insulator Based on Ultraviolet Imaging and Improved YOLOv3. *High Volt. Technol.* **2021**, *47*, 377–386. (In Chinese)
3. Miao, X.R.; Liu, X.Y.; Chen, J.; Zhuang, S.; Fan, J.; Jiang, H. Insulator Detection in Aerial Images for Transmission Line Inspection Using Single Shot Multibox Detector. *IEEE Access* **2019**, *7*, 79945–79956. [CrossRef]
4. Tan, P.; Li, X.F.; Xu, J.M.; Wang, F.J.; Ding, J.; Fang, Y.T.; Ning, Y. Catenary insulator defect detection based on contour features and gray similarity matching. *J. Zhejiang Univ. Sci. A* **2020**, *21*, 64–73. [CrossRef]
5. Liu, Y.; Lu, Y.P.; Gao, S.; Bi, X.T.; Yin, Q.G.; Zhu, X.Q.; Yao, J.G. Application of Edge Detection in Infrared Images of Coil-type Suspension Porcelain Insulator. *Electro-Ceram. Light. Prot.* **2020**, 198–203. (In Chinese) [CrossRef]
6. Tao, X.; Zhang, D.P.; Wang, Z.H.; Liu, X.; Zhang, H.; Xu, D. Detection of Power Line Insulator Defects Using Aerial Images Analyzed with Convolutional Neural Networks. *IEEE Trans. Syst. Man Cybern. Syst.* **2020**, *50*, 1486–1498. [CrossRef]
7. Guo, J.N.; Du, S.S.; Wang, S.D.; Zhang, X.Y. CenterNet Self-Exploding Detection of Insulators on Transmission Lines Based on Lightweight Feature Fusion. *J. Beijing Univ. Aeronaut. Astronaut.* **2022**, 1–13. (In Chinese) [CrossRef]

8. Jia, X.F.; Yu, Y.Q.; Guo, Y.C.; Huang, Y.R.; Zhao, B.T. Lightweight Detection Method for Insulator Self-Explosion Defects in Aerial Photography. *High Volt. Technol.* **2023**, *49*, 294–300. (In Chinese)
9. Li, L.R.; Chen, P.; Zhang, Y.L.; Zhang, K.; Xiong, W.; Gong, P.C. Insulator Defect Detection Based on Multi-Scale Feature Coding and Dual Attention Fusion. *Prog. Laser Photonics* **2022**, *59*, 81–90. (In Chinese)
10. Li, X.; Zhang, J.; Zhang, L.; Chen, X. Layout Optimization and Multi-scenes Intelligent Inspection Scheme Design Based on Substation Video Monitoring. *J. Phys. Conf. Ser.* **2023**, *2560*, 012022. [CrossRef]
11. Wu, J.; Liu, Z.B.; Ren, Q. Detection of Defects in Power Grid Inspection Images Based on Multi-scale Fusion. *J. Phys. Conf. Ser.* **2022**, *2363*, 012013. [CrossRef]
12. Zhao, H.; Wan, F.; Lei, G.B.; Xiong, Y.; Xu, L.; Xu, C.; Zhou, W. LSD-YOLOv5: A Steel Strip Surface Defect Detection Algorithm Based on Lightweight Network and Enhanced Feature Fusion Mode. *Sensors* **2023**, *23*, 6558. [CrossRef]
13. Li, Q.; Sun, B.Q.; Bir, B. Lite-FENet: Lightweight multi-scale feature enrichment network for few-shot segmentation. *Knowl.-Based Syst.* **2023**, *278*, 110887. [CrossRef]
14. Song, K.C.; Wang, H.; Zhao, Y.; Huang, L.; Dong, H.; Yan, Y. Lightweight multi-level feature difference fusion network for RGB-D-T salient object detection. *J. King Saud Univ.-Comput. Inf. Sci.* **2023**, *35*, 101702. [CrossRef]
15. Tan, M.; Le, Q.V. EfficientNet: Rethinking model scaling for convolutional neural networks. *Proc. Int. Conf. Mach. Learn.* **2019**, *97*, 6105–6114.
16. Lin, T.; Dollár, P.; Girshick, B.R.; He, K.; Hariharan, B.; Belongie, S. Feature Pyramid Networks for Object Detection. In Proceedings of the IEEE Conference on Computer Vision and Pattern Recognition, Honolulu, HI, USA, 21–26 July 2017; pp. 2117–2125.
17. Zhang, Y.W.; Zhang, Y.; Ding, Z.H.; Wang, Z. Classification and Recognition Method of Non-Cooperative Object Based on Transfer Learning. *Opt. Laser Technol.* **2024**, *169*, 110005. [CrossRef]
18. Ren, S.Q.; He, K.M.; Girshick, R.; Sun, J. Faster R-CNN: Towards Real-Time Object Detection with Region Proposal Networks. *IEEE Trans. Pattern Anal. Mach. Intell.* **2017**, *39*. [CrossRef] [PubMed]
19. Liu, Y.M.; Wang, Z.L.; Wang, R.J.; Chen, J.; Gao, H. Flooding-based MobileNet to identify cucumber diseases from leaf images in natural scenes. *Comput. Electron. Agric.* **2023**, *213*, 108166. [CrossRef]
20. Taro, S.; Kohei, T.; Shohei, O.; Ito, S. Introducing Swish and Parallelized Blind Removal Improves the Performance of a Convolutional Neural Network in Denoising MR Images. *Magn. Reson. Med. Sci.* **2021**, *20*, 410–424.
21. Hu, J.; Shen, L.; Albanie, S.; Sun, G.; Wu, E.H. Squeeze-and-Excitation Networks. *IEEE Trans. Pattern Anal. Mach. Intell.* **2019**, *42*. [CrossRef]
22. Zhao, X.; Zhao, J.; He, Z. A multiple feature-maps interaction pyramid network for defect detection of steel surface. *Meas. Sci. Technol.* **2023**, *34*, 055401. [CrossRef]
23. Ma, S.H.; Sun, H.C.; Gao, S.; Zhou, G. A real-time mechanical fault diagnosis approach based on lightweight architecture search considering industrial edge deployments. *Eng. Appl. Artif. Intell.* **2023**, *123*, 106433. [CrossRef]
24. Anu, J.; Jithin, J. Multi-task learning approach for modulation and wireless signal classification for 5G and beyond: Edge deployment via model compression. *Phys. Commun.* **2022**, *54*, 101793.
25. Liu, B.; Jia, Y.X.; Liu, L.Y.; Dang, Y.; Song, S. Skip DETR: End-to-end Skip connection model for small object detection in forestry pest dataset. *Front. Plant Sci.* **2023**, *14*, 1219474. [CrossRef] [PubMed]
26. Mirzaei, B.; Nezamabadi-pour, H.; Raoof, A.; Derakhshani, R. Small Object Detection and Tracking: A Comprehensive Review. *Sensors* **2023**, *23*, 6887. [CrossRef]
27. Chollet, F. Xception: Deep learning with depthwise separable convolutions. In Proceedings of the IEEE Conference on Computer Vision and Pattern Recognition, Honolulu, HI, USA, 21–26 July 2017; pp. 1251–1258.
28. Deng, J.; Dong, W.; Socher, R.; Li, L.J.; Li, K.; Li, F.F. ImageNet: A large-scale hierarchical image database. In Proceedings of the 26th IEEE Conference on Computer Vision and Pattern Recognition, Miami, FL, USA, 20–25 June 2009.
29. Zhang, H.P.; Zhang, X.Y.; Meng, G.; Guo, C.; Jiang, Z. Few-Shot Multi-Class Ship Detection in Remote Sensing Images Using Attention Feature Map and Multi-Relation Detector. *Remote Sens.* **2022**, *14*, 2790. [CrossRef]
30. Lu, Y.; Chen, X.Y.; Wu, Z.X.; Yu, J. Decoupled Metric Network for Single-Stage Few-Shot Object Detection. *IEEE Trans. Cybern.* **2022**, *53*, 514–525. [CrossRef]
31. Zou, L.; Wang, K.; Wang, X.; Zhang, J.; Li, R.; Wu, Z. Automatic Recognition Reading Method of Pointer Meter Based on YOLOv5-MR Model. *Sensors* **2023**, *23*, 6644. [CrossRef]

Disclaimer/Publisher's Note: The statements, opinions and data contained in all publications are solely those of the individual author(s) and contributor(s) and not of MDPI and/or the editor(s). MDPI and/or the editor(s) disclaim responsibility for any injury to people or property resulting from any ideas, methods, instructions or products referred to in the content.

Article

The Impact of Noise and Brightness on Object Detection Methods

José A. Rodríguez-Rodríguez [1], Ezequiel López-Rubio [1,2], Juan A. Ángel-Ruiz [1] and Miguel A. Molina-Cabello [1,2,*]

1. Department of Computer Languages and Computer Science, University of Málaga, 29071 Málaga, Spain; joseantoniorodriguez@uma.es (J.A.R.-R.); ezeqlr@lcc.uma.es (E.L.-R.); juanan7999@uma.es (J.A.Á.-R.)
2. Instituto de Investigación Biomédica de Málaga y Plataforma en Nanomedicina-IBIMA Plataforma BIONAND, 29009 Málaga, Spain
* Correspondence: miguelangel@lcc.uma.es

Abstract: The application of deep learning to image and video processing has become increasingly popular nowadays. Employing well-known pre-trained neural networks for detecting and classifying objects in images is beneficial in a wide range of application fields. However, diverse impediments may degrade the performance achieved by those neural networks. Particularly, Gaussian noise and brightness, among others, may be presented on images as sensor noise due to the limitations of image acquisition devices. In this work, we study the effect of the most representative noise types and brightness alterations on images in the performance of several state-of-the-art object detectors, such as YOLO or Faster-RCNN. Different experiments have been carried out and the results demonstrate how these adversities deteriorate their performance. Moreover, it is found that the size of objects to be detected is a factor that, together with noise and brightness factors, has a considerable impact on their performance.

Keywords: deep learning; object detection; noise; brightness

Citation: Rodríguez-Rodríguez, J.A.; López-Rubio, E.; Ángel-Ruiz, J.A.; Molina-Cabello, M.A. The Impact of Noise and Brightness on Object Detection Methods. *Sensors* **2024**, *24*, 821. https://doi.org/10.3390/s24030821

Academic Editors: Man Qi, Yun Zhang and Matteo Dunnhofer

Received: 12 December 2023
Revised: 12 January 2024
Accepted: 24 January 2024
Published: 26 January 2024

Copyright: © 2024 by the authors. Licensee MDPI, Basel, Switzerland. This article is an open access article distributed under the terms and conditions of the Creative Commons Attribution (CC BY) license (https://creativecommons.org/licenses/by/4.0/).

1. Introduction

The constant presence of Gaussian noise, among other noises, in data processed by electronic devices is a phenomenon inherent to our highly technological society. This noise, closely linked to electromagnetic radiation, inevitably infiltrates any device that requires electrical communication, posing a constant challenge to the integrity of the data we process. In particular, image processing is significantly affected by this type of noise, since the vast majority of sensors used in this field are exposed to its influence. These sensors play a crucial role in a wide range of applications, especially in the life sciences. From dental imaging to digital mammography and ophthalmological examinations to assess eye health and analyze pathologies, electronic sensors are essential. However, the real problem arises when the images captured by these sensors are affected by Gaussian Noise, which can lead to alterations in the conclusions drawn from their processing [1].

This challenge is compounded in the field of Deep Learning, a discipline that has gained increasing prominence in the field of Artificial Intelligence. In this context, neural networks emerge as the fundamental pillars of this technological revolution. These networks appear as powerful tools that allow complex tasks to be tackled more efficiently in terms of time and resources, offering highly accurate results. Their application is diverse, ranging from event prediction to the simulation of complex systems, pattern recognition and classification, and system monitoring [2].

Among the rich variety of neural network models available, Convolutional Neural Networks (CNNs) stand out for their effectiveness in detecting objects in images and videos, analyzing medical images, processing natural language and creating recommendation systems, among other applications. These networks, specifically designed to process data in image format, are an essential tool in fields as diverse as medical diagnosis, autonomous driving and traffic management [3]. In particular, the object detection problem is a field where many applications are being developed [4,5].

However, the relationship between images processed by electronic sensors and neural networks brings with it a major challenge: the introduction of noise in the data, which can affect the quality of the inferences and results obtained by the networks. This interference may have critical implications in a wide range of applications, from medical misdiagnosis to safety decisions in autonomous vehicles, for example. Understanding and addressing this issue is critical to ensure the reliability of these technologies in real-world situations.

The specific problem to be addressed in this study concerns understanding the impact of certain factors on the performance of neural networks based on convolutional models. As mentioned above, when information is acquired from an electronic sensor to digitize an image, this image is inevitably affected to a greater or lesser extent by noise, such as Gaussian noise. This interference, combined with the actual lighting conditions in which the image was captured, can have a more significant impact than might be imagined on the performance of various neural network models. This can lead to the networks generating incorrect or insufficiently accurate results [6].

In addition to exploring the implications of noise on the performance of CNNs, this study aims to closely analyze the impact that different lighting conditions can have on their performance. In fact, illumination variability is a critical factor in many real-world applications, where images may be captured in environments with varying or insufficient illumination [7].

Through the evaluation of the networks under various brightness conditions, solid conclusions can be drawn about their ability to adapt and produce accurate results in challenging lighting situations. This information will be essential to understanding how these network models can be used effectively in real-world applications where lighting conditions can vary considerably. By considering both noise and lighting conditions in this study, we seek to provide a comprehensive view of the robustness and versatility of these CNNs in various circumstances.

Despite the growing presence of CNNs in our daily lives, there is a surprising paucity of studies that thoroughly analyze the impact of factors such as noise and others on their performance and efficiency. This underscores the pressing need to investigate and better understand how these systems respond under adverse conditions. Improving the robustness of neural networks in the presence of adverse conditions would not only be beneficial from an academic point of view, but could also have a significant impact in critical sectors such as healthcare and the automotive industry, where accuracy and reliability are imperative [8].

The main contribution of this work is to analyze the impact of noise and changes in the brightness of images on the object detection performance of deep convolutional neural networks. This analysis is based on an accurate physical model of noise and brightness for imaging CMOS sensors. Qualitative and quantitative assessments of the relative object detection performance of the deep networks are carried out, along with a discussion of the results.

The remainder of this paper is organized as follows. Related works are considered in Section 2. After that, Section 3 describes the physical model of noise and brightness for imaging CMOS sensors. The experiments that have been carried out are presented in Section 4. Finally, the conclusions of this work are detailed in Section 5.

2. Related Works

The images can be hindered by several factors, particularly by the influence of varied types of sensor noise. One of the possible reasons can be found in the constraints that image obtainment devices manifest [9–11]. Consequently, the input pixel values may be altered, and this behaviour may have an impact on the performance achieved by those methods that accomplish different tasks, such as background segmentation or classification. In order to analyze the impact of these noises on the performance of methods of different kinds, several studies have been carried out.

For example, the performance of different foreground object detection methods was studied when the input images were affected under diverse quantities of Gaussian and uniform noise [12]. Moreover, a certain number of situations were stated where the addition of noise to the input image might be beneficial in order to alleviate the constraints of a method. The performance yielded by a total of nine methods from the state-of-the-art where the input images were corrupted with both noises were analyzed in that paper.

Several factors may hinder the result of the classification provided by a CNN. In particular, one of these drawbacks is the sensor noise. An analysis of the effects of noise in these kinds of neural networks was presented in [7]. The methodology detailed two noise models for current CMOS vision sensors that allow entering Poisson, Gaussian, salt-and-pepper, speckle and uniform noise as an origin of imperfections in the image acquisition devices. This way, synthetic noise can be added to an image by using the proposed methodological framework to imitate usual sources of image distortion. With that suggested framework, each kind of noise type was combined with a bright scale factor to simulate images with low lighting conditions, and their impact on the classification performance of several state-of-the-art CNNs pretrained models was studied. The results showed that Gaussian and uniform noise have a moderate effect; speckle and salt-and-pepper noise, together with the level of brightness, could significantly decrease the classification performance; while Poisson noise did not have a substantial impact on the performance.

Another of the possible noises is the linear motion blur. Its impact in the performance of CNNs was evaluated by proposing a realistic vision sensor model to generate a linear motion blur effect in raw input images. By using this methodology, the classification performance of different pretrained CNNs was studied and the obtained results demonstrated that the more the displacement the more the degradation of the performance as expected. However, although the angle of displacement does not have as much impact as the length, the performance is slightly deteriorated. It is interesting to observe how higher values of motion length produce a higher drop in CNNs performances and make it more sensitive to the motion angle. Moreover, angles close to odd multiples of 45° imply a more relevant drop of the performance. Regarding opposite angles, they achieve the same performance; however, conjugate angles do not provide the same performance [13].

3. Methodology

This section outlines the methodology developed for this study. It involves the creation of an accurate model for an imaging CMOS sensor (Section 3.1), the process for introducing synthetic noise into digital images based on this model (Section 3.2), and the description of how the performance of an object detection deep neural network degrades as the noise level increases (Section 3.3).

3.1. Sensor Noise Model

We have defined two realistic noise models for a CMOS vision sensor, drawing inspiration from the European Machine Vision Association (EMVA) Standard 1288, which characterizes image sensors and cameras [14]. These models serve as the foundation for subsequent experiments to assess the performance of Convolutional Neural Networks (CNNs) under different noise sources. Figures 1 and 2 represent these models, known as Model A and Model B, respectively:

- Model A simulates an imaging CMOS sensor with a single source of noise, namely Poisson type.
- Model B accounts for additional noise types, including Gaussian, speckle, salt-and-pepper, and uniform noises.

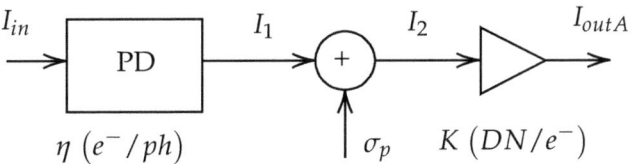

Figure 1. Model A. A conceptual model of a CMOS vision sensor, comprising a photodiode (PD) followed by a Poisson noise source (illustrated as a circular element). Subsequently, a conversion gain is represented as a triangular element. This model exclusively accounts for Poisson noise.

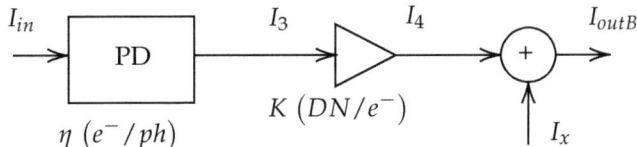

Figure 2. Model B. A conceptual model of a CMOS vision sensor, where a photodiode (PD) is succeeded by a conversion gain, represented as a triangular element, and followed by the introduction of a specific type of noise I_x (depicted as a circular element).

Both models are employed to maintain a realistic image creation process while isolating the sources of noise. The operation of the sensor, common to both models, is described based on Figures 1 and 2. Images enter the device associated to a given amount of photons (ph) and interact with the photodiode (PD), which transforms the photons to electrons with a quantum efficiency factor (η (e^-/ph)). The result of this transformation carried out by the photodiode can be quantified as accumulated electrons. The Full Well Capacity (FWC) is defined as the maximum accumulation capacity for any given pixel. The equation that governs the transformation of photons into pixels reads:

$$I_1 = I_3 = \eta I_{in} \tag{1}$$

where I_{in} is in photons (ph), and I_1 is in electrons (e^-).

Shot noise, associated with photon counting errors, is assumed to distribute according to the Poisson distribution, and it is only considered by Model B. The signal corrupted by Poisson type noise (measured in electrons) features a mean parameter of the Poisson distribution equal to the noiseless signal I_1:

$$I_2 \sim \text{Poisson}(\lambda = I_1) \tag{2}$$

The subsequent steps include conversion gain, presented in both Model A and Model B, where accumulated electrons are converted to voltage (in microvolts, μV). In this step, the gain factor is noted χ ($\mu V/e^-$). The conversion from analog information to digital data converts the voltage signal into digital numbers (DN). This time the gain factor is noted ξ ($DN/\mu V$). These two gains are merged into an overall gain factor ($K = \chi \xi$). Therefore, the computation of the theoretical result of the conversion from analog information to digital data reads as follows:

$$I_4 = KI_3 \tag{3}$$

$$I_{outA} = KI_2 \tag{4}$$

However, the two conversions from electrons to a voltage signal, and then from analog information to digital data, are subject to additive noise, which is only considered in Model B. A signal level (I_x) is added before the output. If I_x is measured in digital numbers

(DN), then Model B indicates that the observed image can be expressed (in digital numbers DN) as follows:

$$I_{outB} = I_4 + I_x \tag{5}$$

3.2. Synthetic Noise Emulation

This subsection outlines a procedure for obtaining a noisy digital image based on the accurate models of noise discussed in Section 3.1. The noiseless pixel value in a digital image will be noted φ, measured in digital numbers (DN).

The simulation of Poisson noise is carried out by application of Model A. This involves the modification of the original noiseless signal by Poisson noise, where the signal is expressed in electrons (6). To achieve this, a division of the pixel value expressed in digital numbers by the overall gain K must be carried out, in order to yield the pixel value expressed in electrons. Moreover, low illumination situations may be modeled by applying a brightness scale factor b. Afterwards, a subsequent conversion to digital numbers is performed. The noisy pixel value obtained by the imaging device ($\hat{\varphi}_{Poisson}$) can be calculated considering the original pixel value φ, as indicated next:

$$\hat{\varphi}_p = K \operatorname{Poisson}\left(\frac{b\varphi}{K}\right) \tag{6}$$

Model B is utilized to generate synthetic noise. Various types of synthetic noise representing common degradation mechanisms in digital images, including Gaussian (g), salt-and-pepper (sp), and uniform (u) noise, are considered. The resulting noisy pixel value $\tilde{\varphi}$ is determined based on the type of synthetic noise introduced.

Gaussian noise is expressed as:

$$\tilde{\varphi}_g = b\varphi + \operatorname{Gauss}\left(0, \sigma'_g\right) \tag{7}$$

where b stands for the brightness scale, while σ'_g denotes the standard deviation. Common sources of Gaussian noise include intrinsic circuit noise and an elevated operating temperature.

A probability mass function is employed to model salt-and-pepper noise:

$$P(\tilde{\varphi}_{sp}) = \begin{cases} p_0 & \text{if } \tilde{\varphi}_{sp} = 0 \\ p_1 & \text{if } \tilde{\varphi}_{sp} = 255 \\ 1 - p_0 - p_1 & \text{if } \tilde{\varphi}_{sp} = b\varphi \end{cases} \tag{8}$$

with probabilities p_0 for black pixels and p_1 for saturated pixels, respectively. The function can be simplified by assuming that p_0 and p_1 are equal, i.e., $p_0 = p_1$.

Uniform noise is defined as:

$$\tilde{\varphi}_u = b\varphi + \operatorname{Uniform}(-\Delta, \Delta) \tag{9}$$

where Δ specifies the extremes of the valid values for the corrupted image, drawn uniformly at random from the interval $[-\Delta, +\Delta]$.

These Equations (6)–(9) must be employed for each channel (red, green, and blue) of the original image in order to obtain its corrupted version.

3.3. Object Detection Performance Degradation

Here, we describe how the performance of an object detection deep neural network degrades as the amount of noise present in the input image increases. Let us note ζ the noise level. This corresponds to a different parameter depending on the noise type, $\zeta \in \{K, \sigma'_g, p_0, \Delta\}$ for Poisson, Gaussian, salt-and-pepper, and uniform noise, respectively. Also, let us remember that the brightness scale factor is noted b.

Then the performance of an object detection neural network can be expressed as a function of the noise level ζ and the brightness scale factor b, $\mathcal{A}(\zeta, b)$. Maximum performance should be attained for zero noise:

$$\mathcal{A}(0, b) \geq \mathcal{A}(\zeta, b), \ \forall \zeta \geq 0 \tag{10}$$

Moreover, the performance should decrease as the noise level increases:

$$\zeta_0 \leq \zeta_1 \Rightarrow \mathcal{A}(\zeta_0, b) \geq \mathcal{A}(\zeta_1, b) \tag{11}$$

The specific characteristics of the performance function $\mathcal{A}$ must be determined by experimentation for various object detection networks.

4. Experimental Results

Figure 3 offers a graphical abstract of how the study proposed in this work was conducted. First, the raw image is contaminated with a source of noise. Additionally, a brightness scale factor is applied to emulate low-illumination conditions. Therefore, after both noise and brightness processes, a noisy image is obtained. Next, that image is supplied to a detector method in order to locate and classify the objects presented in that image. Once the detections of all tuned configurations have been performed, a fair comparison has been carried out by using different well-known metrics. Finally, the obtained results from that comparison can be discussed.

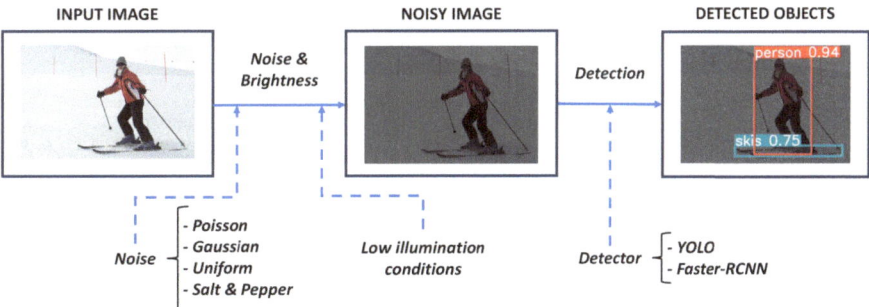

Figure 3. Schema of the proposed methodology. A raw image is contaminated with a source of noise and low illumination conditions are applied. The obtained noisy image from that process is supplied to a detector method in order to locate and classify the objects presented in that image.

This study aims to analyze the effect of the most relevant sources of sensor noises. With this intent, a set of experiments was carried out, and the obtained results are shown in this section. First of all, the considered methods are described in Section 4.1. Next, the selected dataset is detailed in Section 4.2. Then, the parameter configuration is presented in Section 4.3. At last, results are depicted in Sections 4.4 and 4.5.

4.1. Methods

YOLO (You Only Look Once) and Faster R-CNN (Faster Region-based Convolutional Neural Network) are the neural network models that we will use for the battery of experiments. These architectures excel in their ability to identify and locate objects in a variety of scenarios, even in situations with brightness variations and Gaussian noise.

However, it is critical to understand that YOLO and Faster R-CNN, while sharing the purpose of object detection, differ in their approach and operation [15]. While YOLO excels in speed and efficiency in addressing real-time detection, Faster R-CNN offers a higher level of accuracy at the cost of greater computational complexity [16]. This distinction in performance and efficiency will be a key element in our evaluation, as it will determine which of the two architectures is more suitable for object detection in images with variations in brightness and Gaussian noise.

For the different experiments, different versions of these models have been used. As for YOLO, the pre-trained versions of YOLOv5nu, YOLOv5mu, YOLOv5xu [17], as well as other more current versions such as YOLOv8n, YOLOv8m, and YOLOv8x [18] have been used. Among these versioned versions of the YOLO model, YOLOv8 outperforms YOLOv5 in accuracy, achieving 54.2% average accuracy on the COCO dataset compared to 50.5% for YOLOv5. Both are suitable for real-time applications, with YOLOv5 offering higher FPS on the CPU, but YOLOv8 being preferable on some GPUs. The 'n' version of YOLOv8 is optimal for embedded devices such as Jetson Nano. In summary, YOLOv8 is more accurate, while YOLOv5 is faster on the CPU and YOLOv8 is preferable on some GPUs and embedded devices. Both mentioned YOLO versions v5 (https://github.com/ultralytics/yolov5, accessed on 11 December 2023) and v8 (https://github.com/ultralytics/ultralytics, accessed on 11 December 2023) are extracted from the Ultralytics library (https://github.com/ultralytics, accessed on 11 December 2023).

For the Faster R-CNN [19] models we have used the Pytorch-Torchvision library (https://pytorch.org/vision/main/models/faster_rcnn.html, accessed on 11 December 2023), making use of the Faster R-CNN ResNet50 FPN V2 (https://pytorch.org/vision/main/models/generated/torchvision.models.detection.fasterrcnn_resnet50_fpn_v2.html, accessed on 11 December 2023) and Faster R-CNN Mobilenet V3 Large FPN (https://pytorch.org/vision/main/models/generated/torchvision.models.detection.fasterrcnn_mobilenet_v3_large_fpn.html, accessed on 11 December 2023) versions, both with pre-trained weights. As for the differences between these versions, Faster R-CNN ResNet50 FPN V2 stands out for its high accuracy in object detection, thanks to the complex ResNet50 FPN architecture. However, it requires more computational resources, which may affect its speed. On the other hand, Faster R-CNN MobileNet V3 Large FPN focuses on speed and is ideal for real-time applications on resource-constrained devices, although its accuracy may be slightly lower due to its lighter architecture. In summary, ResNet50 offers accuracy, while MobileNet V3 is fast and efficient on resource-constrained devices.

4.2. Dataset

The COCO dataset, with its diversity of natural scenarios, detailed labels, and divisions for detection tasks, represents an essential resource in our computer vision research. COCO contains an extensive collection of about 300,000 images, meticulously selected to represent diverse and realistic natural settings. Of the more than 200,000 images available, more than 80 different object categories have been thoroughly labeled. This accurate labeling allows for detailed analysis of a wide range of objects, improving the robustness of our experiment. This provides a rich and varied dataset for our research. COCO dataset offers different versions of the dataset [20] and it has been divided into essential subsets for detection tasks. These subsets include training, validation, and test sets, each accompanied by corresponding annotations.

In this work, we have used the 2017 Val images dataset (http://images.cocodataset.org/zips/val2017.zip, accessed on 11 December 2023), which is composed of 5000 images with a size of approximately 1 GB, and their annotations (http://images.cocodataset.org/annotations/annotations_trainval2017.zip, accessed on 11 December 2023).

4.3. Parameter Selection

In order to establish a fair comparison between YOLO and Faster R-CNN models, their parameters were fixed to the same values. This way, the parameter object confidence threshold for detection $conf$ for YOLO models and box_score_thresh for Faster R-CNN models were fixed to 0.5.

Regarding the images, each color channel of each pixel will have a value within the interval [0, 255], so 8-bit encoded images are assumed.

Respecting the source of noise and brightness, the number of possible tuned configurations is enormous, as can be deduced from Section 3. This way, a parameter analysis has been established to obtain realistic results and a set of enough experiments that allow us to deduce solid conclusions. The tuned parameters and their description are as follow:

- b: The brightness scale factor emulates illumination conditions by controlling the minimum and maximum values of the image. The tuned values for this parameter b have been selected from the interval 0.1 to 1.0 with a step of 0.1. With this configuration, the lower the value of b, the darker the noisy image.
- K: The image sensor gain that converts electrons into digit values is represented by this parameter, which is exclusively dependent on the sensor performance. With the aim of analysing realistic scenarios, different commercial vision sensor data-sheets have been collected [21–24], where K goes from 0.01 to 0.1 DN/e^-. Furthermore, in order to study those more complex situations, higher values for K have been considered.
- σ'_g: The standard deviation modulates the quantity of Gaussian noise. The higher the value of σ'_g, the noisier the image. While the read-out noise of most commercial vision sensors is less than 1 DN when 8-bit encoding is used, values from 0 to 22.5 DN with a step of 2.5 DN have been considered.
- Δ: The limit range value establishes the minimum and maximum values that define the uniform noise that can be reached. The parameter Δ manages both values by considering the range $[-\Delta, \Delta]$ to introduce additive noise. The chosen values for this parameter are in the interval from 0 to 22.5 DN with a step of 2.5 DN.
- p: This parameter represents the probability of having a pixel affected by salt-and-pepper noise. It has been considered that the likelihood for salt is precisely the same for pepper, so that, $p_0 = p_1 = p$ (see Section 3.2). The selected values for this parameter go from 0.00 to 0.27 with a step of 0.03.

Table 1 summarizes the parameter values which form the set of tuned configurations.

Table 1. Considered parameters and their possible values to study the performance of the different selected methods for diverse noises and illumination conditions.

Parameter		Value
Bright scale factor, b	=	{0.1, 0.2, 0.3, 0.4, 0.5, 0.6, 0.7, 0.8, 0.9, 1.0}
Image sensor gain, Poisson noise, K	=	{0.01, 0.03, 0.05, 0.07, 0.09, 0.2, 0.4, 0.6, 0.8, 1.0}
Standard deviation, Gaussian noise, σ'_g	=	{0, 2.5, 5.0, 7.5, 10.0, 12.5, 15.0, 17.5, 20.0, 22.5}
Probability salt and pepper noise, p	=	{0.00, 0.03, 0.06, 0.09, 0.12, 0.15, 0.18, 0.21, 0.24, 0.27}
Limit range, uniform noise, Δ	=	{0, 2.5, 5.0, 7.5, 10.0, 12.5, 15.0, 17.5, 20.0, 22.5}

4.4. Qualitative Results

The different types of noise considered in this work have their nature. In order to provide a better comprehension of their incidence and low illumination conditions on the images, Figure 4 details an example of the effect they produce on an input raw image.

The performances yielded by the considered methods are compared from a qualitative point of view in this subsection. It aims to better comprehend the performance deterioration of the selected approaches. Without loss of generality, YOLOv5nu detector and Gaussian noise have been chosen for this purpose. The image selected is shown in Figure 5. It exhibits a room with plenty of objects that the methods may detect. The objects presented in the image are varied, such as chairs, tables, televisions, vases, potted plants, or people. As can be demonstrated, a noisier image does not have to provide a worse detection. In fact, a low quantity of noise can even be beneficial to enhance the detection. This remark can be observed with the clock: it is not detected in the raw image, but it is well detected when low-illumination conditions are applied.

However, in general, a more noisy image produces fewer detections and lower confidence in the model. As shown, the television is well detected in all images; nevertheless, the higher the quantity of noise, the lower the confidence of the method in that detection.

This effect is more visible in the case of the vase, which is recognized in the absence of noise or with a low quantity of noise, but it is not detected when the noise is much higher.

This same behavior of the detections occurs when low illumination conditions are presented in the image. As can be observed, the person is better detected when brightness has not been modified.

These observations are presented similarly for the rest of the different considered models. Depending on the intrinsic characteristics of the selected model, the noise and the brightness of the input image, the model can detect a specific object in that image well or not.

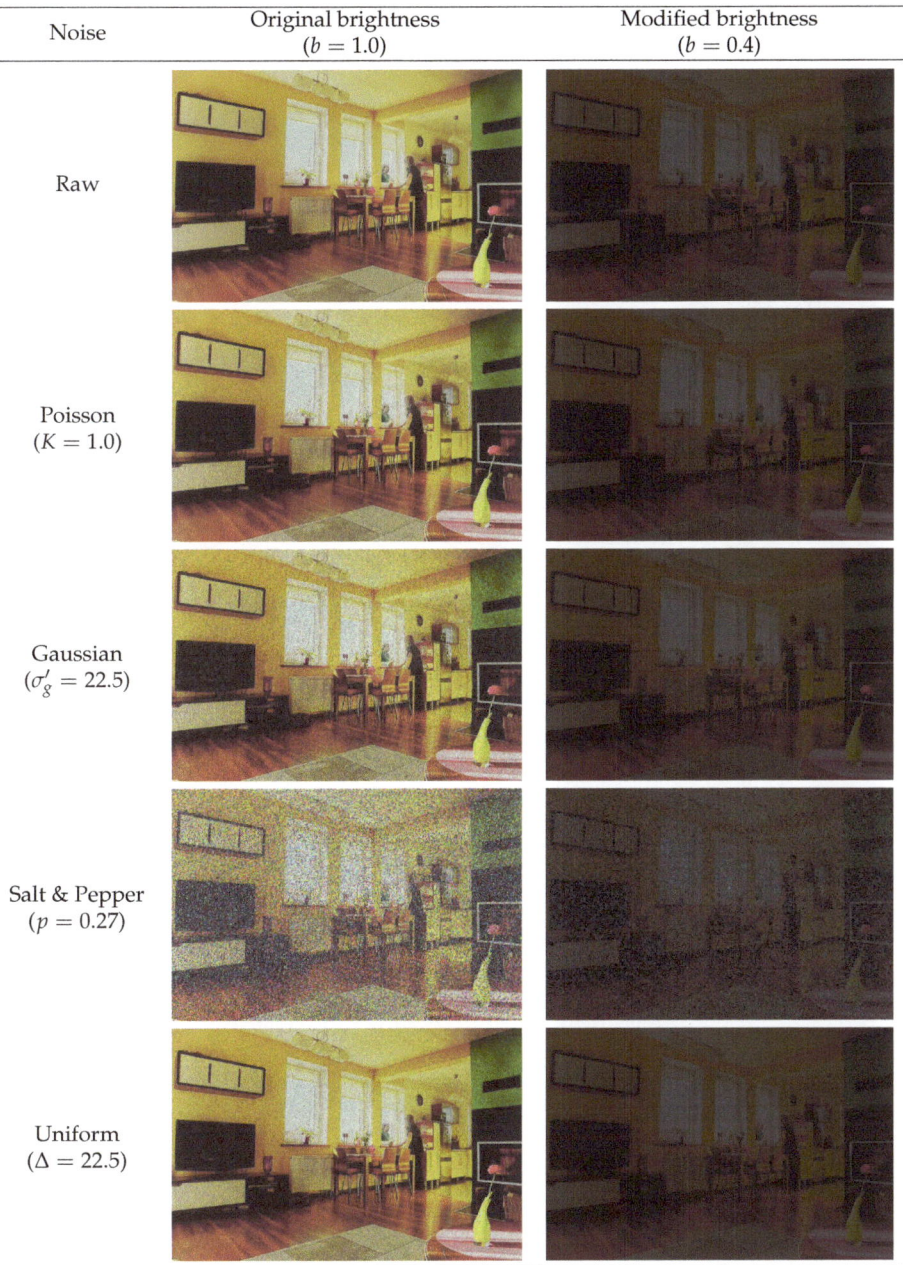

Figure 4. Image 139 without and with several quantities of different noises and bright scale configurations.

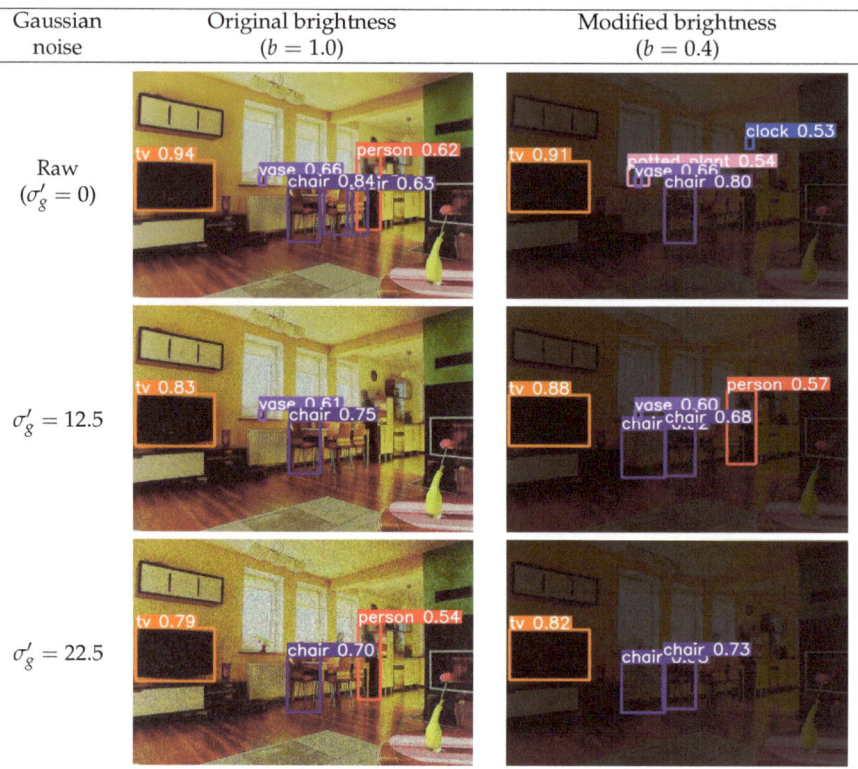

Figure 5. Image 139 without and with several quantities of **Gaussian noise** and bright scale configurations. Bounding boxes represent the object detections that the YOLOv5nu method has predicted. Each bounding box shows the class of the object and the confidence that the method has performed in that detection.

4.5. Quantitative Results

In order to measure the performance of each method and establish a fair comparison between them, several well-known metrics have been considered for that purpose. In the context of the detection of objects presented in images, the utilization of Average Precision (AP) metric is useful to evaluate the effectiveness of the predictions.

Before going into the details of the evaluation, it must be highlighted what AP entails. In the field of object detection. AP is a performance metric used to measure the ability of a model to detect and locate objects in an image [25] and it considers two crucial aspects:

- Detection accuracy: This component evaluates how many of the detected objects are actually relevant. It is essential to determine the model's ability to identify objects of interest under varying conditions, such as noise and brightness.
- Location accuracy: Accuracy in the location and size of detected objects is another essential element of the AP calculation. This is crucial to evaluate the ability of the model to not only detect objects but also to accurately localize them.

The AP metric provides a complete and detailed view of the model's performance in terms of object detection and localization, which is essential to understanding how it performs against modifications introduced in the experiments.

Within the field of object detection, in addition to the AP metric, the mean Average Precision (mAP) metric is frequently used, which is a well-known metric that provides a more comprehensive evaluation by averaging the AP values obtained on different classes or categories of objects. The behaviour of mAP can be defined as:

- Calculation of AP by class: First, the AP value is calculated for each class of object being detected. This involves measuring the detection and localization accuracy specifically for that category.
- Average AP: Once the AP has been calculated for each class, these values are averaged to obtain the mAP. This average takes into account the detection and localization efficiency across all object categories, providing an overall assessment of the model.

The results obtained in the different experiments of this work are based on these metrics AP and mAP mentioned.

The AP measures the accuracy of the model by calculating the average accuracy value for the recovery value from 0 to 1, based on the Intersection over Union (IoU), which is a measure that evaluates the overlap between the predicted area and the true annotation area, i.e., how much the boundary predicted by our model overlaps with the boundary of the real object in the image.

Another concept that we should comment on, and to which we have made reference, is the accuracy itself. The accuracy simply measures how accurate the predictions made by our model are based on the IoU obtained from our detection, i.e., the percentage of predictions on object detections that are correct. In this work, the percentage of detections whose IoUs are 50% (0.5) or higher have been considered. This IoU threshold can be varied according to how strict we want the evaluation of the detections of our model to be, since all the detections that have an IoU lower than the established threshold will be discarded as possible true predictions.

Our evaluation code disaggregates between true positive (TP), true negative (TN), false positive (FP), and false negative (FN) cases. Then, for each object category, we calculate the precision at different IoU thresholds as mentioned above and average it based on the number of thresholds, obtaining the AP.

$$Precision = \frac{TP}{TP + FP} \qquad (12)$$

The advantage of using mAP lies in its ability to evaluate an object detection model more comprehensively, taking into account performance on multiple classes of objects rather than considering only one. This is especially relevant in practical applications where it is common to detect and locate various objects in an image. The use of mAP provides a global assessment of how the model performs in both conditions (detection accuracy and localization accuracy) across multiple object categories. This further enriches the understanding of the effectiveness of the model. Additionally, ground-truth objects are categorized into *small, medium,* and *large* according to their area measured as the number of pixels in the segmentation mask. In this way, a better understanding of the performance detection may be reported. More details about detection evaluation can be found in the COCO dataset website (https://cocodataset.org/#detection-eval, accessed on 11 December 2023).

Next, an analysis and comparison have been performed for each type of considered noises. For each type of noise, the performance of each detector (methods are detailed in Section 4.1) is shown according to the quantity of noise and the brightness of the image (considered values are described in Section 4.3). The performance is reported in terms of mAP, which is considered the primary challenge metric by the COCO dataset. This measure provides values between 0 and 1, where higher is better. To better understand the effect of noise and brightness, the performance yielded by each method for each kind of noise is reported using a heatmap, where the performance for each configuration of noise and brightness is detailed. Two figures each from Figures 6–13 show the results for Poisson, Gaussian, salt-and-pepper, and uniform noises, respectively. For each type of noise, the first figure describes the performance of the methods by considering the overall size of the ground-truth objects, while the second one details the performance of a selected method across small, medium, and large sizes. With the aim of not overcharging this study with the performance across scales small, medium, and large sizes for all selected methods, only the most significant method for each noise is exhibited. Note that the configuration with $b = 1.0$ corresponds with no synthetic low-illumination conditions.

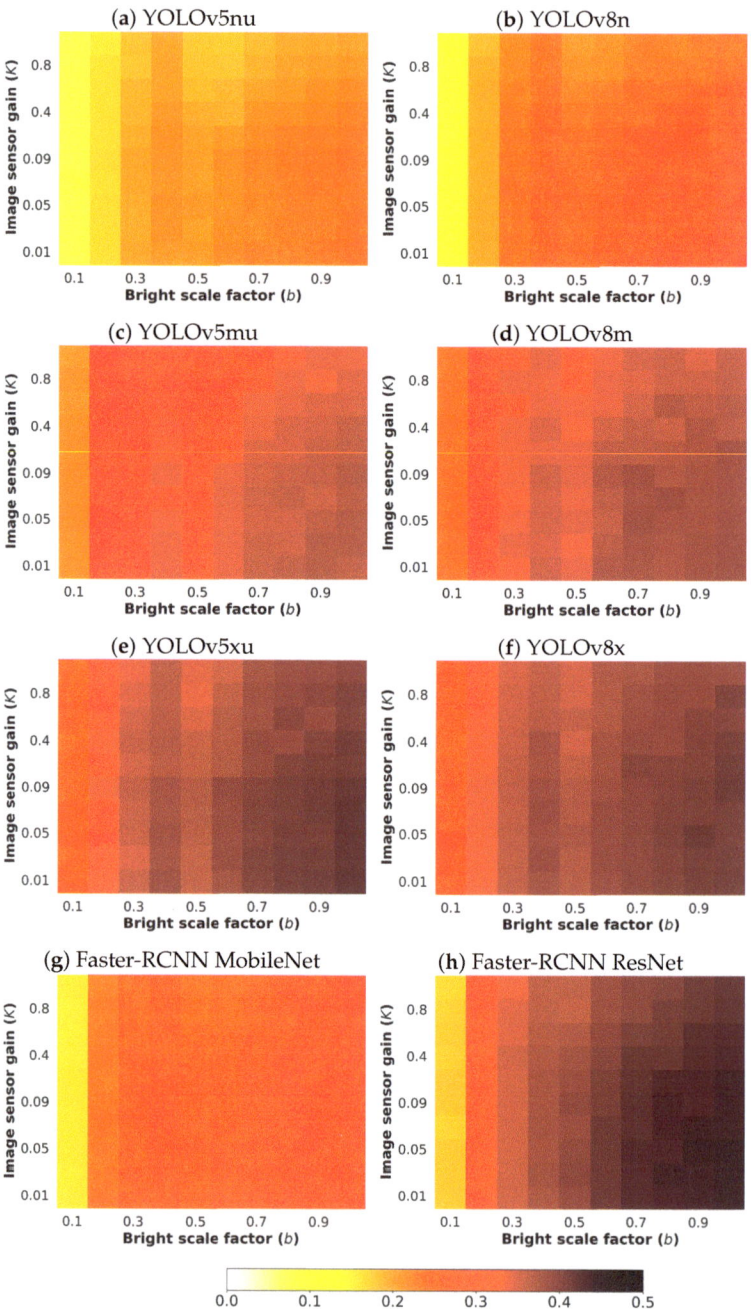

Figure 6. Heatmap of mAP for considered detectors where images have been degraded introducing different levels of bright scale and **Poisson noise**.

It must be highlighted that the first row, from bottom to top of each heatmap, shows the performance of that method when no noise is introduced, except for Poisson noise.

This way, this row will be the same for each analyzed noise. Moreover, note that the configuration with $b = 1.0$ (no synthetic low illumination conditions) and no noise matches the configuration where the raw original images are supplied to the detectors. In the case of Poisson noise, there is no raw without noise because this noise is multiplicative, not additive like the remaining noises.

In general terms, the results from Figures 6–13 demonstrate, as expected, the lower the illumination conditions (bright scale factor), the lower the performance. Also predictable is that the performance deterioration is proportional to the addition of noise for every bright scale value. Regarding the performance according to the size of the ground-truth objects, the size influences the efficiency of the methods: the larger the size of the objects, the better they are detected and well-classified. The difficulty detecting small objects must be highlighted, where the detector methods do not yield properly even in the absence of noise and adequate illumination conditions.

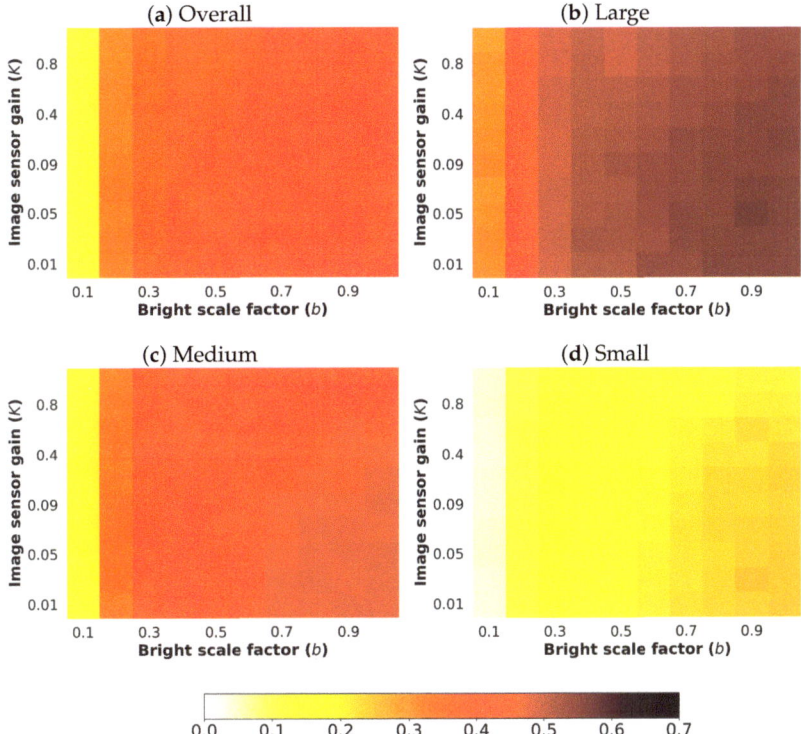

Figure 7. Heatmap of mAP for **Faster-RCNN ResNet** according to the size of the ground-truth objects where images have been degraded introducing different levels of bright scale and **Poisson noise**.

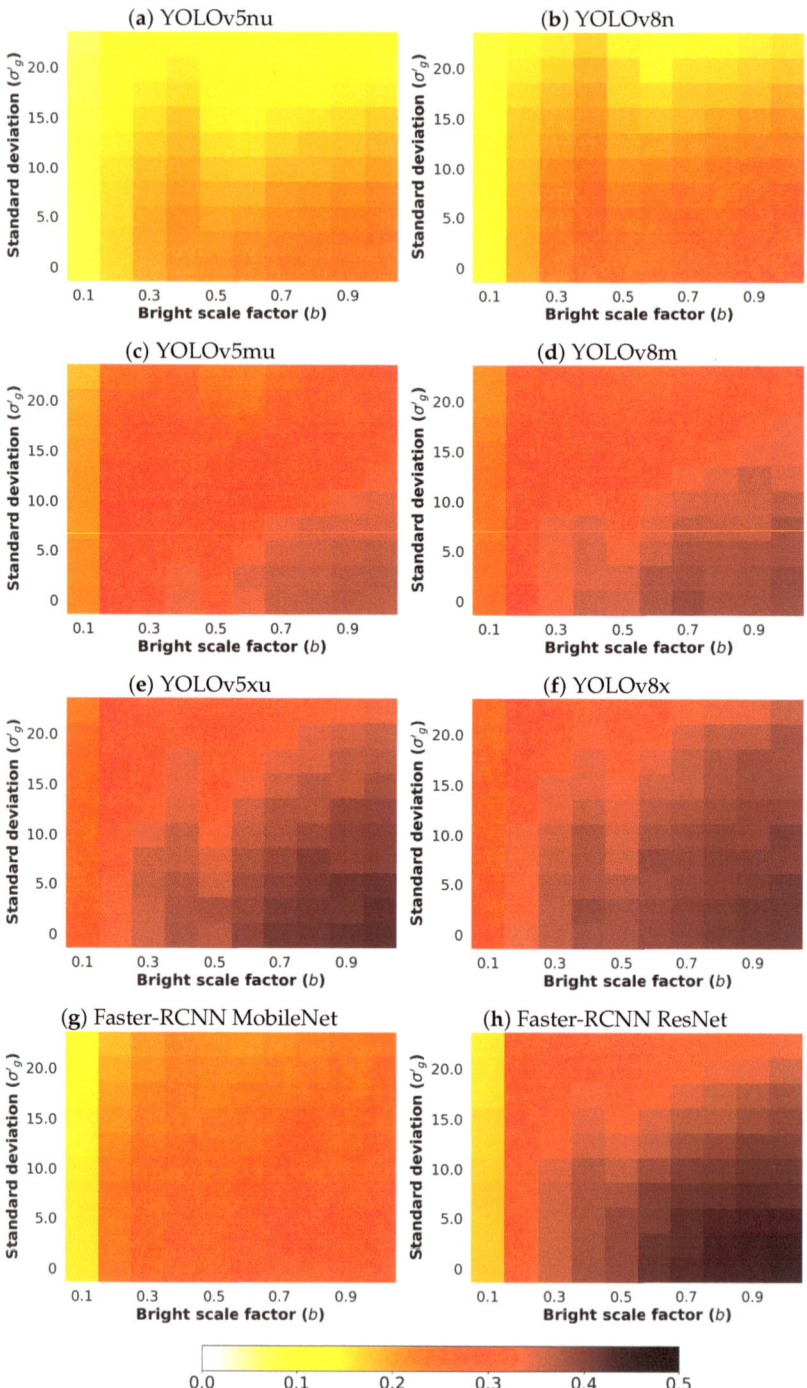

Figure 8. Heatmap of mAP for considered detectors where images have been degraded introducing different levels of bright scale and **Gaussian noise**.

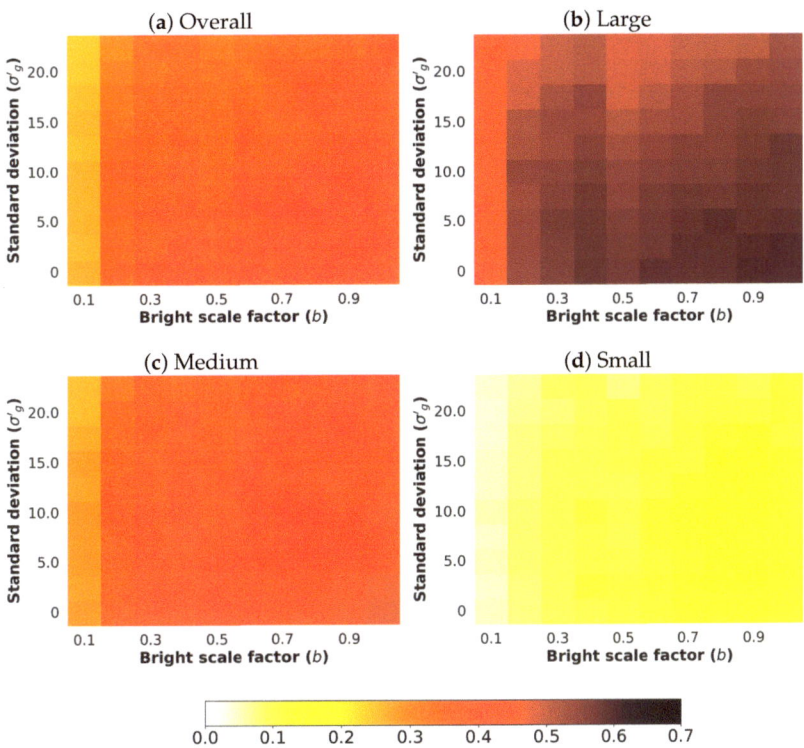

Figure 9. Heatmap of mAP for **YOLOv8x** according to the size of the ground-truth objects where images have been degraded introducing different levels of bright scale and **Gaussian noise**.

4.5.1. Poisson Noise

The overall results for Poisson noise are shown in Figure 6. From this figure, it can be deduced that the Faster-RCNN ResNet model achieves the best performance. Furthermore, this degradation is more perceptible for the lowest values of the bright scale. In these cases, YOLOv5xu performs better.

The performance of the Faster-RCNN ResNet model, taking into account the categorization of the objects by their size, can be observed in Figure 7.

4.5.2. Gaussian Noise

Figure 8 shows the results for Gaussian noise. As can be observed, YOLOv8x has the best performance. The performance degradation is proportional to the standard deviation σ'_g for every bright scale value. Moreover, this deterioration is more noticeable in the lowest-illumination conditions. This is because the quantity of noise introduced for each pixel is comparable with the maximum pixel value of the input image.

Regarding the performance according to the size of the ground-truth objects, the size influences the efficiency of the methods, as can be observed in Figure 9, where the performance of YOLOv8x is shown.

4.5.3. Salt & Pepper Noise

Figure 10 exhibits the performances yielded by the methods for salt-and-pepper noise. It is interesting to observe the great impact that this noise has on the detections: from a certain value of p, depending on the method, the performance of the detectors drops drastically. The impact of this noise on the performance surpasses the effect of the brightness, which is practically non-existent.

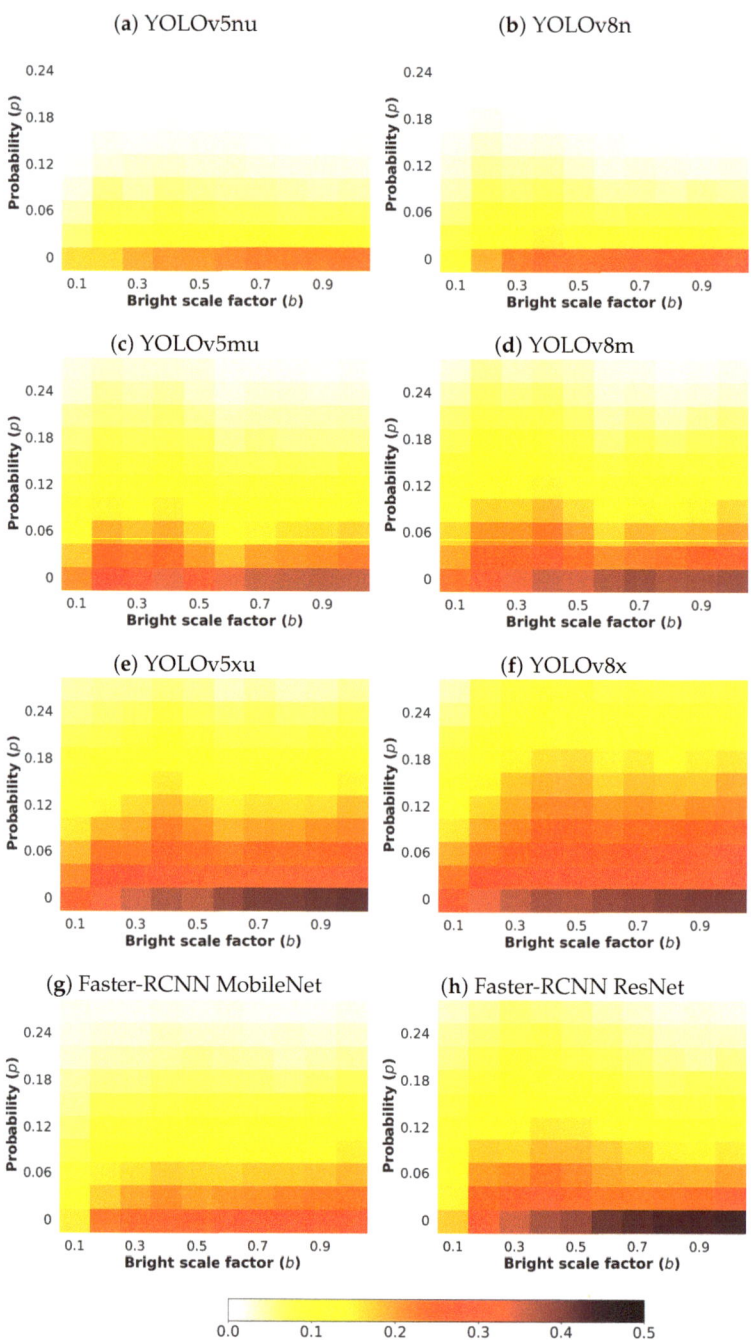

Figure 10. Heatmap of mAP for considered detectors where images have been degraded introducing different levels of bright scale and **salt-and-pepper noise**.

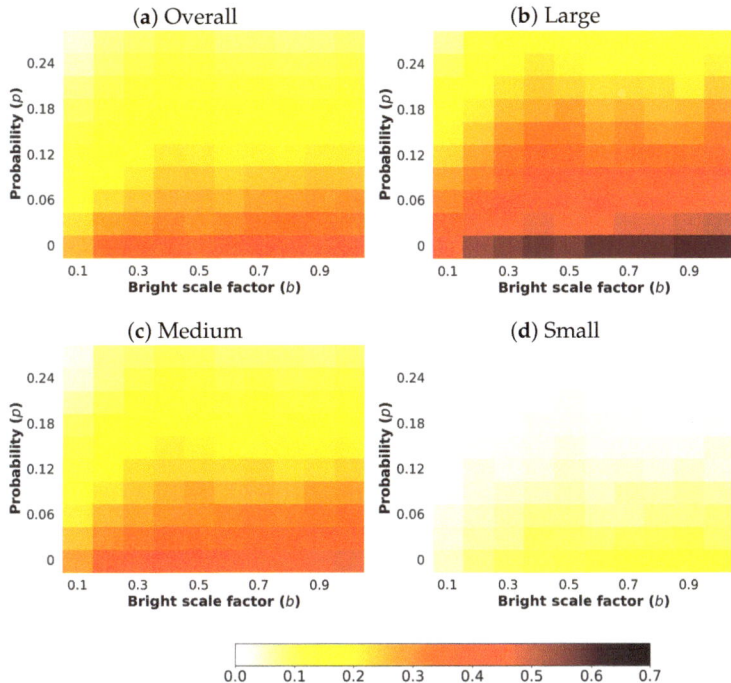

Figure 11. Heatmap of mAP for **YOLOv8x** according to the size of the ground-truth objects where images have been degraded introducing different levels of bright scale and **salt-and-pepper noise**.

YOLOv8x achieves the highest scores, and Figure 11 reports the results by size. As shown, the largest objects are well-detected for a considerable amount of salt-and-pepper noise, even under low illumination conditions; however, detecting the smallest objects involves serious difficulties.

4.5.4. Uniform Noise

The results for uniform noise are shown in Figure 12. The results are similar to those obtained for Gaussian noise, although the impact of the uniform noise is lesser than the Gaussian noise. Again, YOLOv8x has the best performance.

Figure 13 exhibits the performance of YOLOv5xu according to the size of the ground-truth objects.

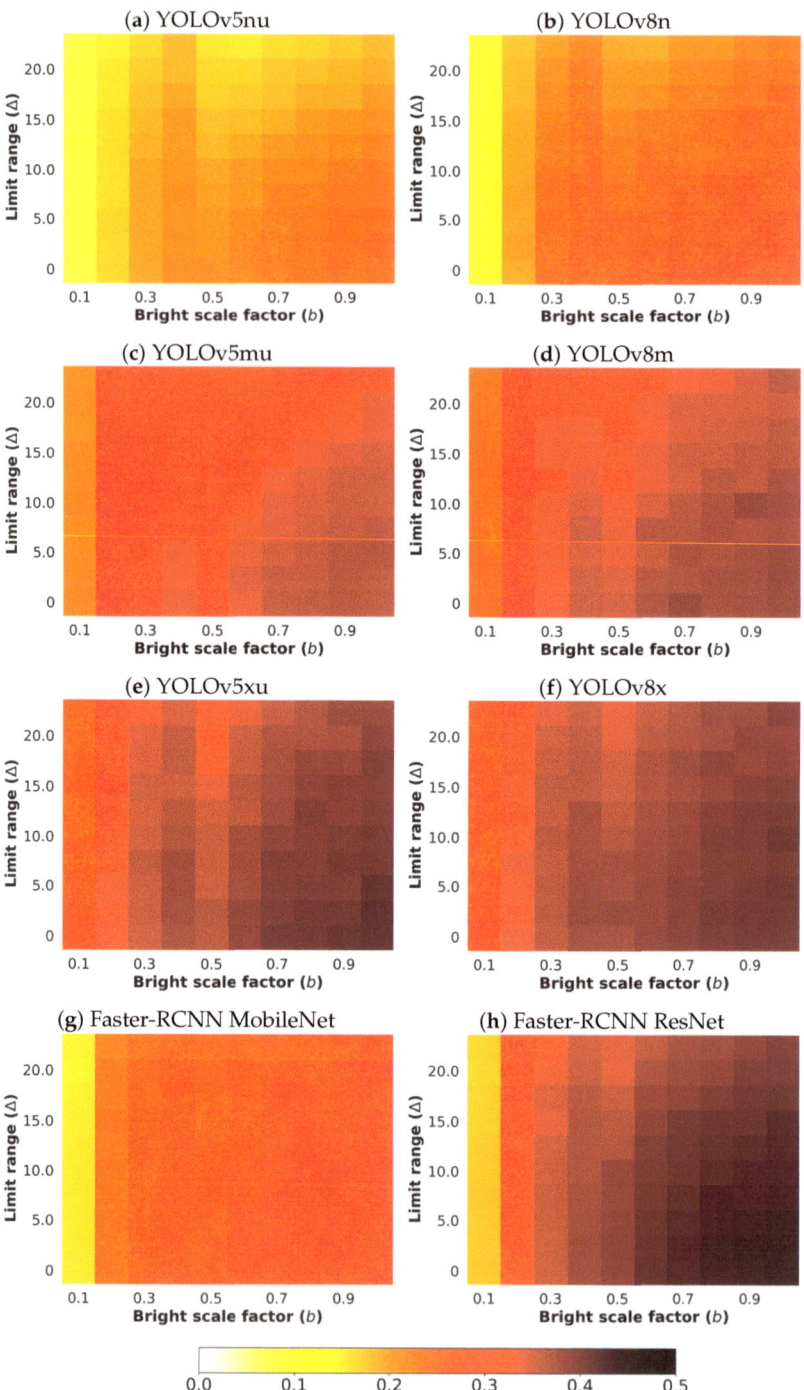

Figure 12. Heatmap of mAP for considered detectors where images have been degraded introducing different levels of bright scale and **uniform noise**.

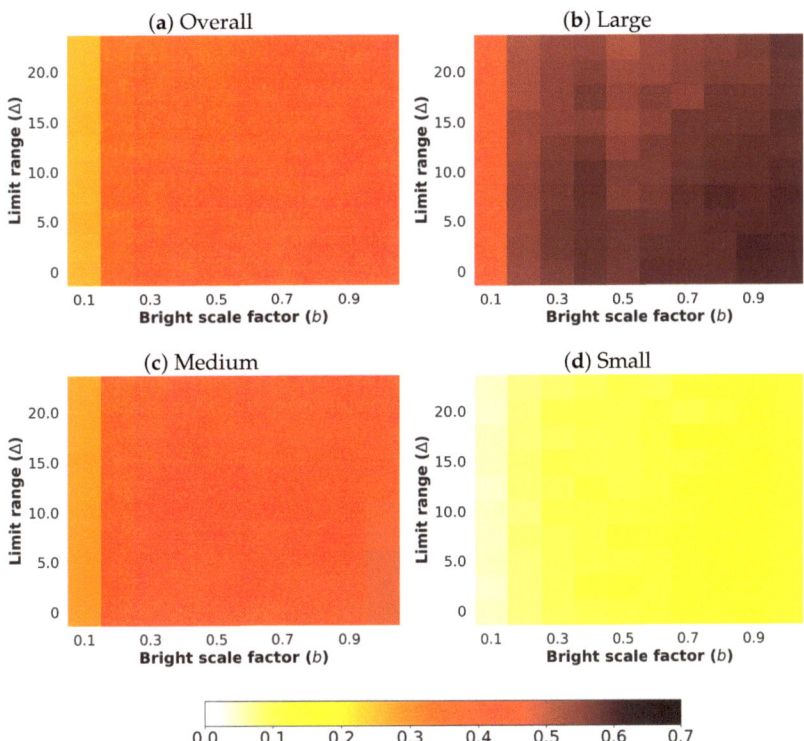

Figure 13. Heatmap of mAP for **YOLOv8x** according to the size of the ground-truth objects where images have been degraded introducing different levels of bright scale and **uniform noise**.

5. Conclusions

This paper presents a study of the impact of the most prevalent noise types and brightness alterations in images on the performance of object detectors. The exposed methodology proposes an accurate model of vision sensor noise in order to carry out the analysis of the noise types considered in this work: Poisson, Gaussian, salt-and-pepper, and uniform noise. The influence of the low illumination conditions accompanying each type of noise has also been studied. Several object detection methods have been selected such as different versions of YOLO v5 and v8, and Faster-RCNN.

Qualitative results demonstrate the need for a more comprehensive analysis due to the disparity of the predictions supplied by the detectors. This way, different configurations of noise and brightness have been tuned in conjunction with a set of 5,000 images to form an exhaustive set of experiments. From a quantitative point of view, the experimental results conclude that, in general, the insertion of noise and/or a reduction of the brightness of the image has a negative incidence on the performance of the detector methods. However, there are situations where adding a small quantity of noise and/or reducing the illumination conditions may be beneficial to detect objects that are not detected in the raw input image.

This analysis might be helpful when designing systems composed of any object detector. In particular, the knowledge leveraged by this work could be most beneficial for systems dealing with environments featuring the presence of noise and/or low-illumination conditions.

Author Contributions: Conceptualization, J.A.R.-R. and J.A.Á.-R.; methodology, E.L.-R.; software, J.A.R.-R. and J.A.Á.-R.; validation, J.A.R.-R.; formal analysis, E.L.-R. and M.A.M.-C.; investigation, J.A.R.-R. and M.A.M.-C.; resources, M.A.M.-C.; data curation, J.A.R.-R. and J.A.Á.-R.; writing—original draft preparation, J.A.R.-R. and J.A.Á.-R.; writing—review and editing, M.A.M.-C. and E.L.-R.; visualization,

E.L.-R.; supervision, M.A.M.-C.; project administration, E.L.-R.; funding acquisition, M.A.M.-C. and E.L.-R. All authors have read and agreed to the published version of the manuscript.

Funding: This work was partially supported by the Autonomous Government of Andalusia (Spain) under grant UMA20-FEDERJA-108, the Ministry of Science and Innovation of Spain grant PID2022-136764OA-I00, and the University of Málaga (Spain) under grants B4-2022, B1-2019_01, B1-2019_02, B1-2021_20 and B1-2022_14.

Data Availability Statement: Publicly available datasets were analyzed in this study. This data can be found here: [20] (http://images.cocodataset.org/zips/val2017.zip, accessed on 11 December 2023), (http://images.cocodataset.org/annotations/annotations_trainval2017.zip, accessed on 11 December 2023). The source code developed for this work is also available (https://github.com/icai-uma/The-impact-of-noise-and-brightness-on-object-detection-methods, accessed on 11 December 2023).

Acknowledgments: The authors thankfully acknowledge the computer resources, technical expertise and assistance provided by the SCBI (Supercomputing and Bioinformatics) center of the University of Malaga. They also gratefully acknowledge the support of NVIDIA Corporation with the donation of a RTX A6000 GPU with 48 Gb. The authors thankfully acknowledge the grant of the Universidad de Málaga and the Instituto de Investigación Biomédica de Málaga y Plataforma en Nanomedicina-IBIMA Plataforma BIONAND.

Conflicts of Interest: The authors declare no conflicts of interest.

References

1. Martin-Gonthier, P.; Magnan, P. RTS noise impact in CMOS image sensors readout circuit. In Proceedings of the 2009 16th IEEE International Conference on Electronics, Circuits and Systems-(ICECS 2009), Yasmine Hammamet, Tunisia, 13–16 December 2009; pp. 928–931.
2. Hemanth, D.J.; Estrela, V.V. *Deep Learning for Image Processing Applications*; IOS Press: Amsterdam, The Netherlands, 2017; Volume 31.
3. Górriz, J.M.; Ramírez, J.; Ortíz, A.; Martinez-Murcia, F.J.; Segovia, F.; Suckling, J.; Leming, M.; Zhang, Y.D.; Álvarez-Sánchez, J.R.; Bologna, G.; et al. Artificial intelligence within the interplay between natural and artificial computation: Advances in data science, trends and applications. *Neurocomputing* **2020**, *410*, 237–270. [CrossRef]
4. Jiang, P.; Ergu, D.; Liu, F.; Cai, Y.; Ma, B. A Review of Yolo algorithm developments. *Procedia Comput. Sci.* **2022**, *199*, 1066–1073. [CrossRef]
5. Liu, B.; Zhao, W.; Sun, Q. Study of object detection based on Faster R-CNN. In Proceedings of the 2017 Chinese Automation Congress (CAC), Jinan, China, 20–22 October 2017; pp. 6233–6236.
6. Mohammed Abd-Alsalam Selami, A.; Freidoon Fadhil, A. A study of the effects of gaussian noise on image features. *Kirkuk Univ. J.-Sci. Stud.* **2016**, *11*, 152–169. [CrossRef]
7. Rodríguez-Rodríguez, J.A.; Molina-Cabello, M.A.; Benítez-Rochel, R.; López-Rubio, E. The effect of noise and brightness on convolutional deep neural networks. In Proceedings of the International Conference on Pattern Recognition, Virtual, 10–11 January 2021; Springer: Cham, Switzerland, 2021; pp. 639–654.
8. Wu, Z.; Moemeni, A.; Castle-Green, S.; Caleb-Solly, P. Robustness of Deep Learning Methods for Occluded Object Detection—A Study Introducing a Novel Occlusion Dataset. In Proceedings of the 2023 International Joint Conference on Neural Networks (IJCNN), Gold Coast, Australia, 18–23 June 2023; pp. 1–10.
9. Zhang, L.; Zuo, W. Image Restoration: From Sparse and Low-Rank Priors to Deep Priors [Lecture Notes]. *IEEE Signal Process. Mag.* **2017**, *34*, 172–179. [CrossRef]
10. Zha, Z.; Yuan, X.; Wen, B.; Zhou, J.; Zhang, J.; Zhu, C. From rank estimation to rank approximation: Rank residual constraint for image restoration. *IEEE Trans. Image Process.* **2019**, *29*, 3254–3269. [CrossRef] [PubMed]
11. Xu, J.; Zhang, L.; Zhang, D. External Prior Guided Internal Prior Learning for Real-World Noisy Image Denoising. *IEEE Trans. Image Process.* **2018**, *27*, 2996–3010. [CrossRef] [PubMed]
12. López-Rubio, F.J.; Lopez-Rubio, E.; Molina-Cabello, M.A.; Luque-Baena, R.M.; Palomo, E.J.; Dominguez, E. The effect of noise on foreground detection algorithms. *Artif. Intell. Rev.* **2018**, *49*, 407–438. [CrossRef]
13. Rodríguez-Rodríguez, J.A.; Molina-Cabello, M.A.; Benítez-Rochel, R.; López-Rubio, E. The impact of linear motion blur on the object recognition efficiency of deep convolutional neural networks. In Proceedings of the International Conference on Pattern Recognition, Virtual, 10–11 January 2021; Springer: Cham, Switzerland, 2021; pp. 611–622.
14. *EMVA Standard 1288*; Standard for Characterization of Image Sensors and Cameras. European Machine Vision Association: Barcelona, Spain, 2010. Available online: https://www.emva.org/standards-technology/emva-1288/ (accessed on 11 December 2023).
15. Du, J. Understanding of object detection based on CNN family and YOLO. *J. Phys. Conf. Ser.* **2018**, *1004*, 012029. [CrossRef]
16. Abbas, S.M.; Singh, S.N. Region-based object detection and classification using faster R-CNN. In Proceedings of the 2018 4th International Conference on Computational Intelligence & Communication Technology (CICT), Ghaziabad, India, 9–10 February 2018; pp. 1–6.

17. Jocher, G.; Chaurasia, A.; Stoken, A.; Borovec, J.; Kwon, Y.; Michael, K.; Fang, J.; Yifu, Z.; Wong, C.; Montes, D.; et al. Ultralytics/yolov5: v7. 0-yolov5 sota realtime instance segmentation. *Zenodo* **2022**. [CrossRef]
18. Jocher, G.; Chaurasia, A.; Qiu, J. YOLOv8 by Ultralytics. 2023. Available online: https://github.com/ultralytics/ultralytics (accessed on 11 December 2023).
19. Ren, S.; He, K.; Girshick, R.; Sun, J. Faster r-cnn: Towards real-time object detection with region proposal networks. In Proceedings of the Advances in Neural Information Processing Systems 28: Annual Conference on Neural Information Processing Systems 2015, Montreal, QC, Canada, 7–12 December 2015.
20. Lin, T.Y.; Maire, M.; Belongie, S.; Hays, J.; Perona, P.; Ramanan, D.; Dollár, P.; Zitnick, C.L. Microsoft coco: Common objects in context. In Proceedings of the Computer Vision–ECCV 2014: 13th European Conference, Zurich, Switzerland, 6–12 September 2014; Proceedings, Part V 13; Springer: Cham, Switzerland, 2014; pp. 740–755.
21. *High Accuracy Star Tracker CMOS Active Pixel Image Sensor*; NOIH25SM1000S Datasheet; ONSemiconductor: Phoenix, AZ, USA, 2009.
22. *4" Color CMOS QSXGA (5 Megapixel) Image Sensor with OmniBSI Technology*; OV5640 Datasheet; OmniVision: Santa Clara, CA, USA, 2010.
23. ams-OSRAM AG Miniature CMOS Image Sensor. NanEye Datasheet. 2018. Available online: https://ams.com/naneye (accessed on 11 December 2023).
24. ams-OSRAM AG CMOS Machine Vision Image Sensor. CMV50000 Datasheet. 2019. Available online: https://ams.com/cmv50000 (accessed on 11 December 2023).
25. Padilla, R.; Netto, S.L.; Da Silva, E.A. A survey on performance metrics for object-detection algorithms. In Proceedings of the 2020 International Conference on Systems, Signals and Image Processing (IWSSIP), Niterói, Brazil, 1–3 July 2020; pp. 237–242.

Disclaimer/Publisher's Note: The statements, opinions and data contained in all publications are solely those of the individual author(s) and contributor(s) and not of MDPI and/or the editor(s). MDPI and/or the editor(s) disclaim responsibility for any injury to people or property resulting from any ideas, methods, instructions or products referred to in the content.

Article

Simple Conditional Spatial Query Mask Deformable Detection Transformer: A Detection Approach for Multi-Style Strokes of Chinese Characters

Tian Zhou [1,2], Wu Xie [1,2,*], Huimin Zhang [3,4,*] and Yong Fan [5]

1. Guangxi Key Laboratory of Image and Graphic Intelligent Processing, Guilin University of Electronic Technology, Guilin 541004, China; guetzt163@163.com
2. School of Computer Science and Information Security, Guilin University of Electronic Technology, Guilin 541004, China
3. Key Laboratory of Education Blockchain and Intelligent Technology, Ministry of Education, Guangxi Normal University, Guilin 541004, China
4. Guangxi Key Lab of Multi-Source Information Mining and Security, Guangxi Normal University, Guilin 541004, China
5. School of Mechanical and Electrical Engineering, Guilin University of Electronic Technology, Guilin 541004, China; fanysmee@126.com
* Correspondence: guetxie126@126.com (W.X.); zhang2023gxnu@126.com (H.Z.)

Abstract: In the Chinese character writing task performed by robotic arms, the stroke category and position information should be extracted through object detection. Detection algorithms based on predefined anchor frames have difficulty resolving the differences among the many different styles of Chinese character strokes. Deformable detection transformer (deformable DETR) algorithms without predefined anchor frames result in some invalid sampling points with no contribution to the feature update of the current reference point due to the random sampling of sampling points in the deformable attention module. These processes cause a reduction in the speed of the vector learning stroke features in the detection head. In view of this problem, a new detection method for multi-style strokes of Chinese characters, called the simple conditional spatial query mask deformable DETR (SCSQ-MDD), is proposed in this paper. Firstly, a mask prediction layer is jointly determined using the shallow feature map of the Chinese character image and the query vector of the transformer encoder, which is used to filter the points with actual contributions and resample the points without contributions to address the randomness of the correlation calculation among the reference points. Secondly, by separating the content query and spatial query of the transformer decoder, the dependence of the prediction task on the content embedding is relaxed. Finally, the detection model without predefined anchor frames based on the SCSQ-MDD is constructed. Experiments are conducted using a multi-style Chinese character stroke dataset to evaluate the performance of the SCSQ-MDD. The mean average precision (mAP) value is improved by 3.8% and the mean average recall (mAR) value is improved by 1.1% compared with the deformable DETR in the testing stage, illustrating the effectiveness of the proposed method.

Keywords: object detection; Chinese character stroke; transformer; deformable DETR; SCSQ-MDD

1. Introduction

With the rapid integration of artificial intelligence into the mechanical industry, the use of industrial robotic arms has increased [1–3]. However, at present, there are few robotic arms that can write good Chinese calligraphy characters on a flat surface. On the one hand, this is because the robotic arm does not have the same precise control of each stroke trajectory as the human arm. On the other hand, it is because the extent of processing of Chinese characters is not fine enough. Currently, there is no appropriate algorithm that can be used to perfectly predict the trajectory point of each stroke in every

Chinese character, so the input parameters of the robotic arm cannot enable it to make a complete trajectory movement. The three elements of Chinese images include stroke, position, and sequence, i.e., each stroke of a Chinese character, the corresponding position of each stroke in the image, and the sequence of each stroke. Chao et al. [4] proposed the use of a corner point detection technique to decompose Chinese characters into a set of strokes, subsequently using the operator's gestures to recognize the decomposed strokes as the robot's writing trajectory to complete the robot's Chinese character writing task. Wang et al. [5] proposed the use of a full convolutional network to extract the stroke skeleton and intersection region and track the whole stroke extraction process based on the pixels in the non-intersection region, finally using the tree search method to match the candidate strokes with the standard strokes to obtain the correct strokes. Although these methods have been successfully used to extract the strokes of Chinese characters and obtain the trajectory of the robot arm by processing the strokes, they are based on manual predefined rules to split the strokes, making them not truly unsupervised stroke extraction methods. Each stroke of a Chinese character and the specific position of each stroke are acquired in our work using object detection techniques in the field of image recognition.

Most previous object detection approaches aimed at improving the generation of proposal boxes and optimizing the filtering of proposal boxes by generating a series of sample candidate boxes using two-stage methods to classify the samples with convolutional neural networks (CNNs). These methods focused on improving detection accuracy and positioning precision, but the models' detection speeds slowed down due to the use of two-stage detection methods [6–11] to map the candidate boxes to the corresponding area of the feature maps after generating them. The generation of sample candidate boxes was removed, and the problem of the localization of target boxes was directly transformed into a regression prediction problem using one-stage methods. These methods focused on addressing the problem of slow detection speeds, but they are inferior to the two-stage methods as far as detection accuracy and positioning precision are concerned. Redmon et al. [12] proposed the You Only Look Once (YOLO) model to predict two bounding boxes and multiple category scores for each grid cell on the feature map and continuously update the values of the bounding boxes and category scores through a loss function. The average precision (AP) value on the Pascal VOC 2007 test dataset reached 63.4%, yet YOLO has the limitation of poor detection on small targets in groups. Therefore, Redmon et al. [13] proposed YOLOv3, which integrated low-level and high-level features by adding a feature pyramid network (FPN) structure and predicted three different feature layers, enabling the model to detect objects of different scales. The AP value on the COCO dataset reached 33.0%, but YOLOv3 has limitations such as imbalanced positive and negative samples and sensitivity to grid boundary values. Bochkovskiy et al. [14] proposed YOLOv4 based on this problem to redesign the sample matching criterion to reduce the impact of the positive and negative sample imbalance. YOLOv4 eliminated the grid sensitivity problem through the design of an activation function, and the AP value on the COCO dataset reached 41.2%. But YOLOv4 still has the problem of poor matching of manually designed anchor boxes on different tasks. Due to the excessive number of candidate boxes generated during the prediction process of one-stage methods [15–20], non-maximum suppression (NMS) processes are required to filter out a large number of candidate boxes, which not only reduces the inference speed but also fails to achieve truly end-to-end prediction, as shown in [21,22], due to the incorporation of a supervisory mechanism.

With transformer methods achieving good results in the natural language processing (NLP) field, researchers have also attempted to introduce transformers into the computer vision (CV) field. Bello et al. [23] proposed adopting self-attention as an alternative approach to convolutional neural networks for discriminating visual tasks to address the limitation of convolutional blocks only being calculated with local neighborhoods, resulting in a lack of global information. Because the traditional CNN model structure can only be utilized to model local information, it is difficult to model long-period information. The attention model has a strong periodic modeling ability, so self-attention can make up for the deficiency of CNNs in ultra-long-period modeling. To compensate for the lack of spatial position information in the transformer, Shaw et al. [24] proposed combining relative position encoding and a self-attention mechanism for modeling position information in images. Furthermore, Ramachandran et al. [25] proposed the use of only attention and relative position encoding instead of the convolutional module in a deep residual network (Resnet) to achieve an image model with full attention. Dosovitskiy et al. [26] presented a vision transformer (ViT) and converted images into token sequences that can be received by the transformer encoder through patch embedding, allowing the operation processes of multi-head self-attention to be performed on the image feature map. Meanwhile, Carion et al. [27] proposed a detection transformer (DETR) that flattens the feature map obtained from the images through the backbone. The feature map is converted into a token sequence, allowing it to be processed by the transformer encoder. The memory vector obtained from the encoder and the 100 object queries obtained from the decoder, after self-attention updating, are subjected to a cross-attention operation. Finally, the classification and regression values of the 100 queries are predicted. Using this method, many manual design components, such as the generation of sample candidate boxes and NMS processing, are effectively eliminated. Wu et al. [28] proposed improved relative position encoding (iRPE) and combined relative position encoding and absolute position encoding in a DETR, resulting in a 1.3% increase in the AP value compared with only using absolute position encoding. Chen et al. [29] proposed a group DETR by employing multiple groups of object queries and performing one-to-one label assignments for each group to support grouped one-to-many assignments, addressing the limitations of DETR, which relies on one-to-one assignments and lacks the ability to utilize multiple positive object queries. Bar et al. [30] presented an unsupervised pretraining method with region priors for object detection, known as DETReg, to pretrain the entire DETR detection network by extracting the proposal box and predicting the self-supervised image coding of regions through an object localization task and an object embedding task during pretraining. The corresponding feature embedding with the self-supervised image coding embedding is aligned to achieve the goal of pretraining the whole DETR detection network.

However, since every point needs to be calculated with all other points in the attention computing module of the DETR, the convergence is slow, and the image resolution is limited. Li et al. [31] attributed the slow convergence of the DETR to the discreteness of the Hungarian matching algorithm and the randomness of model training, leading to ground-truth (GT) box matching becoming a dynamic and unstable process. The DN-DETR (DeNoising DETR) was proposed to reconstruct the GT box by feeding the GT box with noise into the transformer decoder and training the model. Since this process does not require Hungarian matching, the difficulty of binary graph matching is effectively reduced, and the convergence speed is accelerated. Zhang et al. [32] attributed the slow convergence speed of the DETR to the complexity of matching object queries with target object features in different feature embedding spaces. They proposed the SAM-DETR (semantic-aligned matching DETR), wherein object queries are projected into the same embedding space as encoded image features, and then semantic alignment matching is performed, thereby improving detection accuracy and speeding up convergence. Gao et al. [33] argued that the reason for the slow convergence of the DETR is that the object query vector of the DETR needs to interact with the global features of the image so the decoder needs a long training time for the object query to accurately locate the object. The SMCA (Spatially Modulated Co-Attention) mechanism was proposed to improve the convergence speed of the DETR by introducing the Gaussian distribution model of objects into the common attention mechanism and adjusting the search range of each object query vector in the common attention mechanism within a certain distance near the object center.

Kitaev et al. [34] analyzed the traditional transformer and emphasized that the distribution of long sequences is almost always sparse, which indicates that a feature point in a sequence is usually highly correlated with only a few other points. Therefore, only the connections between a subset of points and the current point need to be focused on during the computation of the attention module, which avoids the need for correlation calculations with all points when calculating attention in the transformer, thereby reducing the overall model computation. Zhu et al. [35] proposed the deformable DETR by incorporating deformable attention into the DETR, which requires calculating the connections of each sampling point and its surrounding key points, thereby addressing the problem of the excessive attention computation of DETR. Meanwhile, the multiscale feature map concatenating is used in deformable DETR to solve the problem of slow accuracy in detecting small objects in DETR. Experiments showed that the deformable DETR improved convergence speed and accuracy compared to the ordinary DETR; however, the deformable attention module should focus more on key sampling points with important features when sampling key points is not considered in the deformable DETR. Meng et al. [36] proposed the conditional DETR and also analyzed the reason for the slow convergence speed of the DETR. They found that cross-attention highly relies on content embedding to locate the position of the prediction box, so the demand for high-quality content embedding increases along with the training difficulty. Therefore, the decoupling of the content query and spatial query was proposed, and a learnable conditional spatial query module was introduced to enable the model to learn conditional spatial queries from the decoder embedding. This allows each cross-attention head to focus on different areas, narrowing the spatial range for object classification and prediction box regression in different regional positions.

The goal of the method proposed in this paper is to discard all hyperparameters associated with the anchor frame, enabling the use of the deformable DETR without a predefined anchor frame. The sampling points that do not contribute features during the deformable attention module sampling are resampled so that these points provide feature contributions to the reference points. It is found through experiments that setting the offset of certain sampling points to 0, i.e., discarding some sampling points, leads to a small performance improvement. The reason for this phenomenon is that some points in the random sampling process are repeatedly sampled. When these duplicate sampling points are removed, the computational efficiency of the model is superior to the model with duplicate sampling points, making it easier to converge. Our proposed mask deformable DETR is an improved end-to-end stroke detection method via a deformable DETR with a sampling region prediction mechanism. Figure 1 shows the difference between the deformable attention with the addition of the mask mechanism and the original deformable attention. A mask mechanism is introduced in our method to predict which sampling points are the likely regions of interest for the current reference point and to recalculate and adjust the sampling points in non-important areas. By removing the attention calculation for invalid sampling points, the computational efficiency of the deformable attention module is increased. The convergence speed of the model is accelerated and preferable performance is achieved in a short period of time.

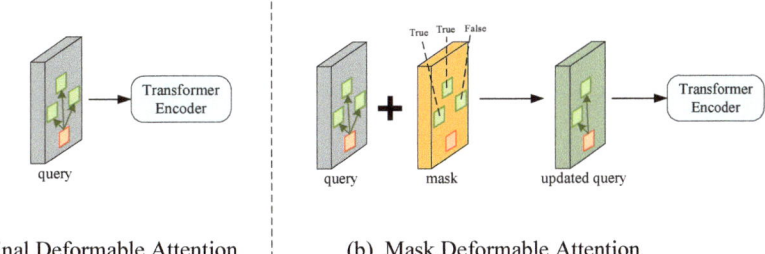

(a) Original Deformable Attention   (b) Mask Deformable Attention

Figure 1. (**a**) The simple process of the deformable attention used in existing methods. (**b**) The simple process of the proposed deformable attention with a mask mechanism. The proposed mask mechanism determines whether a sampling point has a contribution value to the current reference point by predicting the mask layer and discards the sampling points that do not have a contribution value. The gray block represents the query vector, while the yellow block represents the mask vector corresponding to the query vector, and the green block represents query vector after filtering and resampling. The red box represents the current reference point and the green boxes represent the random sampling points.

The most primitive feature map extracted from the backbone and the updated query vector of each encoder layer are jointly utilized to sample, concatenate, and fuse by the deformable DETR based on the mask mechanism. This is done to predict which sampling points in the query vector are candidate points with contribution values for the current reference point. Resampling is used to assign values to candidate points without contribution. Moreover, the mask prediction layer can be simply embedded into the encoder layer of the transformer without the need to modify complex logical structures. Excellent performance improvements are achieved while reducing computational costs. The main contributions of this study can be summarized as follows:

(1) A multiscale deformable attention module based on a mask mechanism is proposed to improve computational efficiency and speed up the convergence of the model by predicting the key sampling points around each reference point in the query vector. In addition, the points that contribute features to the current reference point are filtered out, whereas points that do not contribute features are resampled.

(2) A simple conditional spatial query structure is introduced. By processing the content query vector and the spatial query vector and performing simple linear fusion, the separation of the content query and spatial query is accomplished without introducing additional parametric quantities. The model can be used to focus on not only the content embedding but also the spatial embedding when performing cross-attention calculations. The dependence of the prediction task on content embedding is relaxed, and the training process is simplified.

(3) A splitting feedforward network (SFN) structure is proposed to perform split and cross-fusion calculations on the output vectors from the transformer decoder. To the best of our knowledge, this is the first work to apply the simple conditional spatial query mask deformable DETR (SCSQ-MMD) with an SFN module in the field of deformable DETR. Then, classification and regression predictions are performed in the SFN to enhance the focus on different features for classification and regression tasks.

In short, for the Chinese character writing task performed by robotic arms, an accurate and efficient algorithm is needed to support the detection of Chinese character strokes, especially the implementation of a complete end-to-end stroke detection method. In this paper, the deformable DETR model is improved by enhancing stroke detection accuracy through the above three novel contributions. Experimental results show that the stroke detection method proposed in this paper is superior to the traditional deformable DETR detection method, which can assist robotic arms in completing the Chinese character writing process.

2. Related Works

The network model in this study is improved using the deformable DETR and further extended through the use of the conditional DETR. The deformable DETR is introduced briefly in Section 2.1, and the idea of the conditional DETR is introduced in Section 2.2.

2.1. Deformable DETR and Multiscale Deformable Attention Mechanism

The deformable DETR [35] incorporates a multiscale deformable attention mechanism based on the DETR. First, a query vector is obtained by concatenating the input feature maps of multiple scales, which is fed into the encoder. Each reference point in this vector directly predicts k random offsets around the current point. Second, these k offsets are mapped to the query vector for sampling, and then the final value obtained by linearly interpolating the features of these k points is used to update the features of the current reference point.

The self-attention mechanism obtains the weight coefficients for each value by calculating the correlation between each query and the other keys in this vector. The weight coefficients and the corresponding value are then weighted and summed to obtain the final attention value. In this way, the connections between each point in the vector and the other points can be obtained by the attention module, and the interdependent features in the vector can be captured.

In the multi-head attention mechanism, the self-attention module is used to calculate for each head, without sharing parameters between each head. The final result is obtained by concatenating and fusing the results of the self-attention computed by multiple heads. The formula for calculating the multi-head attention mechanism is as follows:

$$\text{MultiHeadAttn}(z_q, x) = \sum_{m=1}^{M} W_m [\sum_{k \in \Omega} A_{mqk} \cdot W'_m x_k] \quad (1)$$

where q indexes a query element with the representation feature z_q, k indexes a key element with the representation feature x_k, and m indexes the attention head. W_m and W'_m are the trainable weights. A_{mqk} represents the attention weights of the k-th point in the m-th attention head.

The multiscale deformable attention mechanism is based on the common multi-head attention mechanism and adds sampling offsets to each attention head of each scale. The mechanism involves sampling the key of the local position in the global position for each query to obtain the value of the corresponding local position. Finally, the local attention weight and the local value are calculated to reduce the computation of attention, thereby accelerating the convergence speed of the model. The formula is as follows:

$$\text{MSDeformAttn}(z_q, p_q, x) = \sum_{m=1}^{M} W_m [\sum_{k=1}^{K} A_{mqk} \cdot W'_m x(p_q + \Delta p_{mqk})] \quad (2)$$

where m indexes the attention head, k indexes the sampled key, K is the total number of sampled keys ($k \ll HW$), and Δp_{mqk} and A_{mqk} are the sampling offsets and attention weights of the k-th sampling point in the m-th attention head, respectively.

2.2. Conditional DETR and Conditional Spatial Query Module

The reasons for the slow convergence of the DETR were analyzed in [36]. The spatial query only utilizes the common attention weight information and not the specific image information. The content query has to match both the spatial keys and content keys, meaning there is no way for it to learn good features in a short time. The attention weights for the cross-attention mechanism in the DETR are calculated based on the dot product between the query and the key. The formula is as follows:

$$(c_q + p_q)^T \cdot (c_k + p_k) = c_q^T c_k + c_q^T p_k + p_q^T c_k + p_q^T p_k \quad (3)$$

where c_q is the content query, c_k is the content key, p_q is the spatial query, and p_k is the spatial key.

By forcing the separation of content queries and spatial queries in the conditional cross-attention mechanism, content queries and spatial queries can focus on content attention weights and spatial attention weights, respectively. The content attention weights and spatial attention weights are derived from the content dot product and the spatial dot product, respectively. The formula is as follows:

$$c_q^T c_k + p_q^T p_k \quad (4)$$

A learnable conditional spatial query strategy is introduced in the conditional DETR to learn the conditional spatial query vectors from decoder embeddings for decoding multi-head cross-attention. Specifically, the conditional space query p_q is obtained by dot-producting the sine and cosine encoding results p_s of the reference point s with the linear mapping result T of the embedding f output by the decoder at the previous layer.

$$p_q = T \cdot p_s = \text{FFN}(f) \cdot (\text{sinusoidal}(\text{sigmoid}(s))) \tag{5}$$

The input query vector of the cross-attention module is obtained by concatenating the conditional spatial query p_q and the encoding c_q obtained by the self-attention module.

In this approach, the high dependence on content embedding is reduced by separating the spatial queries and content queries, allowing them to focus on spatial attention weights and content attention weights, respectively.

3. Methods

3.1. Mask Deformable Attention in the Multiscale Deformable Attention Module

The core of the deformable DETR model with the mask attention mechanism is the multiscale deformable attention module with the mask mechanism. Figure 2 shows the complete structure of this deformable attention module. The feature map obtained by the backbone contains the most accurate foreground location information from the original image, reflecting the location of the object and the size of the region in the original image. In order to use the mask to accurately predict whether each sampling point in the query vector contributes features to other reference points, the foreground position information from the feature map of the original image is needed. The same operation used in the original deformable DETR is adopted to concatenate the multiscale feature map into a query vector, which is then fed into the multiscale deformable attention module for computation. The generation process of the mask prediction layer and the filtering process of the sampling points are as follows: (1) The upper-layer feature map with the least missing information undergoes convolution to obtain object region position information, which is then fused with the features processed by the channel mapper. (2) The feature maps from several other levels are sampled to obtain information from the feature map of each level corresponding to the object region position. (3) The position information of object regions from multiple levels is fused and concatenated to generate a mask prediction layer. This mask layer is used to predict whether k key sampling points of each reference point z_q, obtained from the query vector through linear mapping, have contributed features to the current reference point. k is the number of sampled points. Since the query vector updated by each encoder layer contains the latest information of the current reference point, the mask prediction layer needs to be updated by each encoder layer to ensure that the mask always learns the crucial predicted features.

The deformable multi-head attention module with the mask mechanism is used not only to sample the key of the local position in the global position for each query vector but also to filter the local sampling points according to the value predicted by the mask. Sampling points without feature contributions are resampled to obtain new keys, and the updated local attention weights are then multiplied by local values. This process reduces the computation in the attention module. Sampling points without feature contributions are filtered out to avoid useless computation on points without contributions. The convergence speed of the model is further accelerated, and higher detection accuracy is achieved in a short period of time. The calculation formula for the deformable attention module with the mask mechanism is as follows:

$$\text{MMSDeformAttn}(z_q, p_q, x) = \sum_{m=1}^{M} W_m [\sum_{k=1}^{K} A_{mqk} \cdot W'_m x(p_q + mask_{mqk} \cdot \text{func}(\Delta p_{mqk}))] \tag{6}$$

where m indexes the attention head, k indexes the sampled keys, K is the total number of sampled keys ($k \ll HW$), and Δp_{mqk} and A_{mqk} are the sampling offsets and attention weights of the k-th sampling point in the m-th attention head, respectively. $mask_{mqk}$ is the mask value corresponding to the k-th sampling point in the m-th attention head ($mask_{mqk} \in$ [True, False]), which determines whether the current point needs to be resampled for calculation. The symbol func is the resampling function with two strategies: non-sampling and value decay.

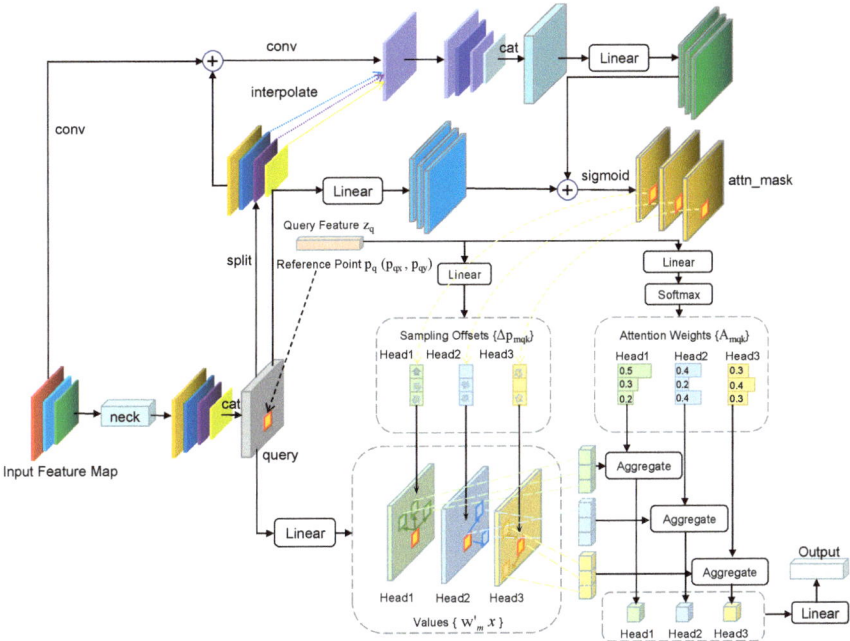

Figure 2. The proposed deformable attention module based on the mask mechanism. The purpose of the proposed mask mechanism is to generate a mask prediction layer that predicts which random sampling points have not contributed to the current reference point and discards these points. The random sampling process is accomplished by adding random integer offsets to the current reference point to obtain the exact position of the sampling points. Blocks of different colors represent different vectors, and blocks of different sizes represent different sized feature maps. The dotted lines represent the correspondences between the different blocks.

The difference between the proposed mask deformable attention mechanism and the attention weights in the original DETR is that although the original DETR can adjust the importance of different contributing points through attention weights, it computes the query vectors of each point with the key vectors of all other points. Although the importance of different contributing points can be adjusted using this approach, there is no way to reduce the number of computations. Even when the attention weight of a point is 0, the query, key, and value of that point still participate in the self-attention calculation process. In the mask deformable attention mechanism, each query vector is computed only with the key and value of the random sampling set of points around it. A resampling strategy is adopted in this mechanism to address the problem of non-contributing sampling points caused by random sampling, which replaces non-contributing points with other points close to the current reference point. This makes the model carry out a more effective attention computation process, thereby accelerating its convergence speed.

The resampling strategy of the F function is shown in Figure 3, with the direction-invariant and value-nonlinear decay strategy on the left, and the non-sampling strategy on the right. The non-sampling strategy sets the sampling offset value of the k-th sampling point in the m-th attention head to 0, indicating that the point corresponding to the predicted offset does not have a contribution value and is directly discarded. The strategy of direction-invariant and value-nonlinear decay is formulated as follows:

$$F(\Delta p_x, \Delta p_y) = (\frac{\Delta p_x}{\Delta p_x} \times \sqrt{|\Delta p_x|}, \frac{\Delta p_y}{\Delta p_y} \times \sqrt{|\Delta p_y|}) \qquad (7)$$

where Δp_x and Δp_y are the predicted x offset and y offset, respectively. $|\Delta p_x|$ and $|\Delta p_y|$ represent the absolute values of Δp_x and Δp_y respectively. When a certain point is predicted to be a non-critical point, the points of all regions with increasing values in this direction are also non-critical points. So, the offset needs to be updated using a constant direction and nonlinear decay of value. Specifically, the random sampling process involves adding an integer-valued random offset to the current reference point to obtain the specific position of the sampling point. If a point at a certain location is predicted to be a non-contributing point by the mask layer, the value of the random offset nonlinearly shrinks in the current direction to become closer to the position of the reference point using the linear decay strategy in Equation (7). With this strategy, non-contributing points can be replaced with other points around the reference point, thus leading to more efficient attention computations by the model.

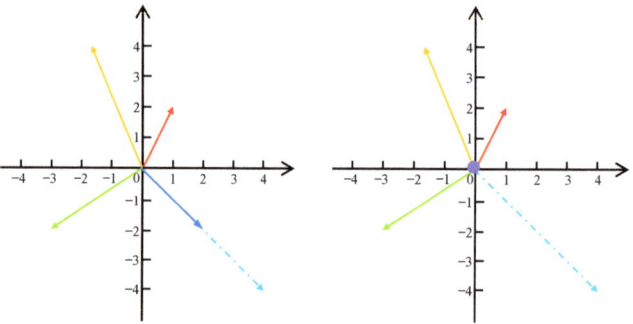

Figure 3. The resampling strategy of sampling points in the deformable attention module based on the mask mechanism. The offset in the same direction for a sample point that does not contribute to the current reference point is decreased, and this invalid reference point is replaced with another sample point in that direction that is closer to the reference point. Arrows of different colors represent different sampling points, and the coordinates where the arrows are located represent the offsets from the current reference point. Dashed arrows represent the original sampling point, solid arrows in that direction represent a reassignment of the sampling offset, and the dot represents discarding the current offset.

3.2. The Simple Conditional Spatial Query Strategy

The conditional DETR can be used to speed up the convergence of the model and improve detection accuracy. But, owing to the multiscale concatenating mode used in the deformable DETR, the size of the query vector of the input encoder is too long. The direct incorporation of the conditional spatial query module not only significantly increases the number of computations but also compromises the generality of the deformable attention module. Specifically, too many linear mappings are used in the conditional spatial query module in the original conditional DETR, which increases the number of computations and contrasts with the original intention of the design of this study—to reduce the number of computations. Moreover, the implementation of the original conditional DETR conflicts with that of the deformable DETR, which means the conditional spatial query module of the conditional DETR cannot be used directly in the proposed mask deformable DETR. In this study, the experiments prove that the detection performance of the deformable DETR is instead reduced by using the complex sub-module structure.

Therefore, the separation operation of the content queries and spatial queries in the self-attention module is discarded through the proposed simple conditional spatial query (SCSQ) strategy in this study. The strategy of the original conditional spatial query in the cross-attention module is simplified. In order to reduce the number of additional parameters introduced and preserve the generality of the deformable attention module, some of the linear mapping processes are omitted in the decoder in the simple conditional spatial query. After the query vector and the conditional spatial query vector are concatenated, feature fusion is performed through a linear layer to restore the dimensionality of the feature vector to its original length. These modification processes allow the application of the deformable DETR using the mask mechanism to the simple conditional spatial query module in the decoder. The separation of content queries and spatial queries is achieved by introducing a small number of additional parameters. Therefore, the model can focus on content embeddings and spatial embeddings separately when cross-attention computation is performed. Thus, the dependence of the prediction task on the content embeddings is relaxed, and the training processes are simplified.

The conditional cross-attention is formed by connecting the content query, the output of the self-attention in the decoder, and the spatial query. The keys consist of the content keys and spatial keys. The formula for calculating the conditional cross-attention is as follows:

$$\text{CondCrossAttn}(k, k_{pos}, q, q_{pos}, v) = \text{proj}([\text{SelfAttn}(q, k, v, q_{pos}), \text{CondSpatial}(q, q_{pos})]) \cdot \text{proj}([k, k_{pos}])^\text{T} \cdot v \qquad (8)$$

where k and v are the memory vectors output by the encoder, k_{pos} is the 2D spatial position information input by the encoder, q is the query input by the decoder, q_{pos} is the 2D spatial position information corresponding to the query input of the decoder, and $[,]$ is the concatenation operation. The proj symbol is the simple linear mapping function. The SelfAttn function is a common self-attention mechanism calculation process. The formula for SelfAttn is as follows:

$$\text{SelfAttn}(q, k, v, q_{pos}) = (q + q_{pos}) \cdot (k + k_{pos}) \cdot v \qquad (9)$$

where q_{pos} is the spatial embedding of the q vector, as shown in Formula (8). CondSpatial is the calculation process of the conditional spatial query, and the formula is as follows:

$$\text{CondSpatial}(q, q_{pos}) = \text{FFN}(q) \cdot \text{P}_\text{s}(q_{pos}) \qquad (10)$$

where FFN is the multiple linear mapping layers, and P_s is the projection process of position encoding. The formula for P_s is as follows:

$$\text{P}_\text{s}(q_{pos}) = \text{sinusoidal}(\text{sigmoid}(q_{pos})) \qquad (11)$$

where sinusoidal represents a sine and cosine positional encoding function.

3.3. SFN Structure and Cross-Fused Module

The traditional feedforward network (FFN) structure consists of a stack of multiple linear layers. The query vector is obtained through cross-attention computation using the memory obtained by the encoder after self-attention and the object queries of the decoder. This query vector is used to predict both the classification task and the regression task. Since the focuses of the classification task and regression task are different, the features they focus on should also differ. The classification task should focus more on stroke category information in the query vector, whereas the regression task should focus more on stroke position information. Therefore, a channel-splitting FFN structure is proposed in this study, called the splitting feedforward network (SFN) structure, which is shown in Figure 4. This structure initially splits the query vector output by the decoder into two different vectors, which are each used to predict different tasks. The problem caused by simple splitting is that the former and latter parts of features only focus on their own prediction tasks, resulting in the loss of correlations between the two parts. Considering that features important for classification may be potentially useful for the regression task, and features important for regression may be equally useful for the classification task, an alternative approach is considered. This approach involves cross-computing the former and latter parts of features, so a cross-fused module is proposed in this paper. This module facilitates the interaction between the corresponding weights of the features in the first part of the SFN and the features in the second part of the SFN. Similarly, it allows the corresponding weights of the features in the second part of the SFN to influence the first part of the SFN. In this way, both independence and correlation can be simultaneously emphasized by the former and latter parts of features. The resulting output features for the former part of the SFN are used for predicting the classification task, whereas the resulting output features for the latter part are used for predicting the regression task, thereby enhancing the independence of different prediction tasks. Specifically, the classification features are utilized to obtain a weight matrix through the linear layer, which is then applied to the regression task. The regression features are utilized to influence the classification task through a weight matrix obtained from the linear layer. In this way, the correlation between the classification task and the regression task can be strengthened. The *class* vector for the classification task and the *bounding box* vector for the regression task are obtained using the following equations:

$$Vector_{cls} = \text{FFN}(\sigma(\text{Linear}(reg)) \cdot cls) \quad (12)$$

$$Vector_{bbox} = \text{FFN}(\sigma(\text{Linear}(cls)) \cdot reg) \quad (13)$$

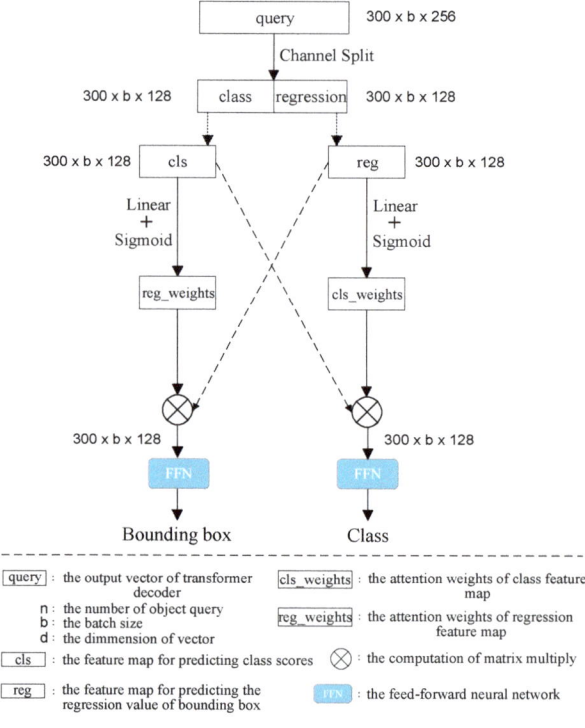

Figure 4. The structure of the proposed SFN module. The corresponding weight matrices are obtained from the vectors after linear and sigmoid computations. The weight matrices of the classification features are used as the weight coefficients of the regression task, and the weight matrices of the regression features are used as the weight coefficients of the classification task.

The difference between the SFN module proposed in this paper and the decoupled detector head in YOLO is that the decoupled detector head in YOLO ensures that the classification and regression tasks focus more on their features of interest by introducing classification and regression branches, allowing the detector head to converge faster and reducing latency while maintaining accuracy. However, the number of channels is reduced by slicing the channel of the query vector in the SFN module proposed in this paper to halve the computation of the detection head. At the same time, there is an intrinsic connection between the classification task and the regression task, e.g., the feature regions that are concerned with the classification features are also concerned with the regression task, and therefore cross-weights are used to strengthen the connection between the classification task and the regression task.

3.4. SCSQ-MDD Pipeline

The overall structure of the SCSQ-MDD (simple conditional spatial query mask deformable DETR) is shown in Figure 5. The overall process of the model can be described as follows: (1) The input image is fed into the Resnet feature extraction network to obtain three feature maps of different scales, and the number of channels of feature maps of the three scales is then unified by the channel mapper, followed by a convolution operation to obtain the feature map of the fourth scale. (2) The feature maps of four scales are concatenated to obtain the feature vector (encoder embeddings), which contains image information from four different scales. The absolute position encoding of encoder embeddings is obtained using the sine and cosine position encoding methods. The query, key, value, unprocessed three-layer feature maps, and absolute position encoding obtained from encoder embeddings are input into the encoder. The mask layer is predicted in each encoder layer using the query and the unprocessed three-layer feature maps. In the deformable attention module, the mask layer specifies which sampling offsets need to be updated. The updated query vector is obtained in each encoder layer, and the model predicts a new mask layer based on the updated query vector, ensuring that the latest features can always be learned by the mask. The query, updated after six encoder layers, serves as a memory vector and undergoes cross-attention calculations with the object queries of the decoder. (3) By initializing the object queries and their corresponding positional encoding, the object queries, positional encoding, and memory vector obtained from the encoder can be input into the decoder. Initially, the calculations for the query positions and object queries are performed using a simple conditional spatial query to obtain the vectors (query embedding) of the conditional spatial query in each layer of the decoder. Then, calculations of the query, key, and value from the object queries are carried out with self-attentions to obtain the updated query vectors. The query vectors and the conditional spatial query vectors are concatenated and linearly fused to obtain new queries, which are vectors calculated and fused separately after isolating the content query and spatial query. At the same time, the memory vector serves as the key and value in the encoder used for cross-attention calculations with the new query, and the updated query is obtained through the deformable attention module. (4) After six decoder layer updates, the final query vector is obtained, which is then segmented and cross-fused. The first part of the features is used for predicting classification tasks, whereas the latter part is used for predicting regression tasks.

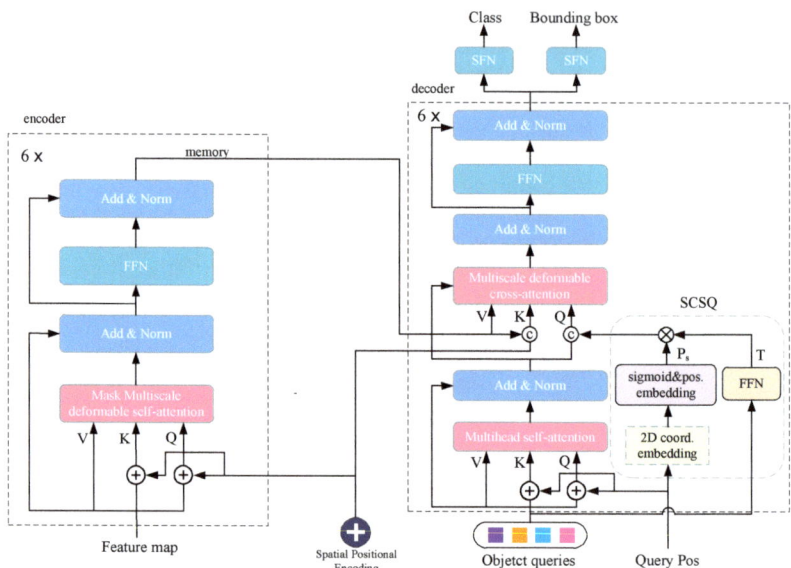

Figure 5. The overall network structure of the proposed SCSQ-MDD (simple conditional spatial query mask deformable DETR). The mask deformable attention mechanism proposed in this paper is incorporated into the improved multiscale deformable self-attention module. The purpose of the SCSQ is to obtain a spatial query vector, which is spliced and fused with the content query vector and used for cross-attention computations).

3.5. Chinese Stroke Detection Method Based on the SCSQ-MDD

Figure 6 shows the overall flow of the Chinese character stroke detection method based on the SCSQ-MDD. The inputs are images of Chinese characters with five different stroke styles: "SimKai", "SimHei", "SimSun", "MSYH", and "Deng". First, the sample data are processed using data augmentation methods, such as random flip, resize, and random crop, to improve the model's detection generalization ability. Second, the sample data are fed into the Resnet network to extract features from both the Chinese character images and the Chinese character strokes. Third, since the original Chinese character images contain direct connections between strokes, the original three-layer feature maps extracted by Resnet and the current query vector are used to jointly predict the mask layer. Then, this mask layer is used to filter valid reference points in the query vector, discarding invalid reference points or resampling them as valid reference points using Equation (7). The attention is calculated only for the valid reference points using Equation (6), reducing the attention calculation process for the invalid reference points. This accelerates the learning of stroke feature information in the Chinese character image by the query vector. Fourth, the query vector of the decoder is obtained by performing simple conditional spatial query calculations using the object queries of the decoder and the 2D positional embedding vector. By separating the content query and the spatial query, the representation features of the detection boxes for the strokes can be learned faster by the query vector. Cross-attention calculations are performed between the memory vector updated by the encoder and the query vector of the decoder. The components of each query vector focus on the feature information of a stroke in the Chinese character image. The query vector obtained after six updates contains both the stroke feature information and the stroke position information. Finally, separate classification and regression predictions on query vectors are performed through the SFN module. The stroke detection results are visualized based on the category scores and the parameters of the regression detection box.

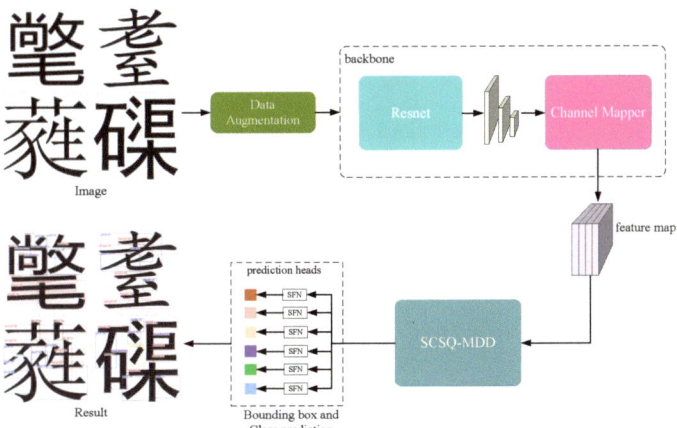

Figure 6. The overall flow of the Chinese character stroke detection network. Each Chinese character image consists of one style of stroke, and there are a total of 5 stroke styles of Chinese character images as input. The purpose of the backbone is to obtain the basic features of the image. The proposed SCSQ-MDD method is used to generate a high-level representation containing information about the location of the strokes and the category of the strokes of the Chinese characters, which is ultimately used in the head detector.

3.6. Application of the SCSQ-MDD to Robotic Arms

The method proposed in this study can be applied to robotic arms for Chinese character writing tasks. Figure 7 shows the process of applying the SCSQ-MDD stroke detection method to robotic arms. First, the images of Chinese characters captured by the camera were recognized, and standard Chinese character images were generated based on the recognition results. Second, the SCSQ-MDD stroke detection method was used to detect all the strokes of the standard Chinese characters. Then, the reduction rules of the strokes were defined, and the pixel points of each stroke were reduced using the detected stroke categories and stroke positions. Finally, a set of pixel points was passed as a parameter to the listening program of the robotic arms, and the operating system of the robotic arms was used to complete the writing task of Chinese characters.

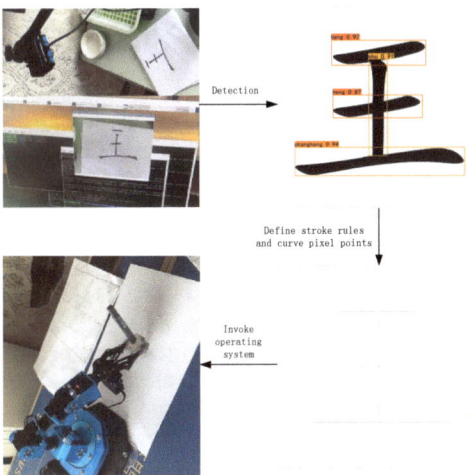

Figure 7. The process of applying the SCSQ-MDD method to robotic arms. A Chinese character consists

of multiple strokes, and the position and category information of all the strokes are obtained by stroke detection.

4. Experiments and Results

Since there is no publicly available Chinese character stroke dataset, the experimental data used in this paper were from a self-labeled standard Chinese character stroke dataset containing 1200 standard Chinese character images with five different styles and 52 stroke categories. Details of the 52 stroke categories are shown in Table 1. This dataset was divided into three sub-datasets for different tasks, where the training sets were used for model training, the validation sets were used to evaluate the detection metrics, and the test sets were used to test the model. The Pytorch framework was used to deploy the entire SCSQ-MDD model. In total, 800 standard Chinese character stroke images were used for model training, 200 standard Chinese character stroke images were used for model validation, and 200 standard Chinese character stroke images were utilized to test the model. The training platform used was a Quadro RTX 5000 graphics card with a batch size of 2. After 200 epochs of training, a total of more than 80,000 iterations were performed.

Table 1. The 52 stroke categories for standard Chinese character images based on Chinese Pinyin.

dian	fandian	duanheng	heng	changheng
shu	zuoxieshu	youxieshu	pie	shupie
fanpie	na	ti	piedian	shuti
hengzheti	wangou	shugou	shuwangou	xiegou
wogou	henggou	hengzhegou	hengzhexiegou	tizhegou
hengzhewangou	hengzuozhewangou	hengpiewangoungzhegou	hengpiewanwan	hengzhezhezhegou
hengzuozhezhezhegou	shuzhezhegou	shuwan	hengzhewan	hengzhe
hengzuozhe	xieshuzhe	shuzhe	shutizhe	piezhe
banpiezhe	hengpie	hengxiaopie	tixiaopie	banhengpie
hengna	hengzhezhepie	shuzhepie	hengxiegou	shuzhezhe
hengzhezhe	hengzhezhezhe			

4.1. Implementation Details

The proposed SCSQ-MDD method used Resnet50/Resnet101 as the backbone to extract the basic features of Chinese character images. Six layers of transformer encoder layers were used on the transformer encoder side, and a mask multiscale deformable attention module with eight heads, embedding dimensions of 256, and four sample points was applied at each layer. Six layers of transformer decoder layers were used on the transformer decoder side, and a multi-head attention module with eight heads and embedding dimensions of 256 was applied at each layer. The classification loss used the FocalLoss function, with gamma set to 2.0 and alpha set to 0.25. The bounding box regression loss used the L1Loss function. The intersection over union (IOU) loss used the GIOULoss function. The Adam optimizer with a learning rate of 0.0002 was used.

4.2. Comparison between the SCSQ-MDD and the Deformable DETR

Figure 8 shows the trends in the loss function during network training for the deformable DETR and the SCSQ-MDD. The loss curves of both the deformable DETR and the SCSQ-MDD converged at the 50th epoch, with the loss curve of the SCSQ-MDD converging slightly faster than the loss curve of the ordinary deformable DETR.

Figure 9 shows the trends in the accuracies of the deformable DETR and SCSQ-MDD networks during training/validation. It can be seen that the accuracy curve of the SCSQ-MDD lies slightly above the accuracy curve of the deformable DETR, demonstrating that the detection accuracy of the SCSQ-MDD method proposed in this paper is superior to that of the deformable DETR.

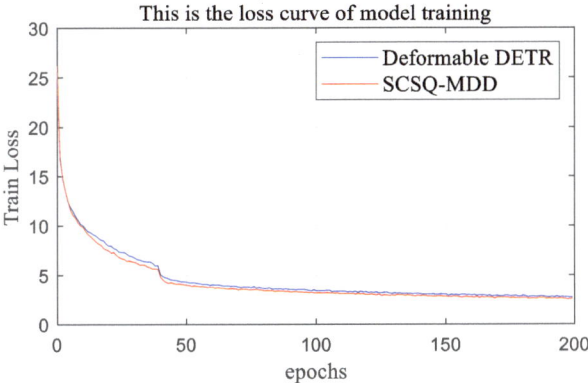

Figure 8. Trends in the loss function for the deformable DETR and the SCSQ-MDD during network training. The training loss of the proposed SCSQ-MDD decreased faster compared to the deformable DETR.

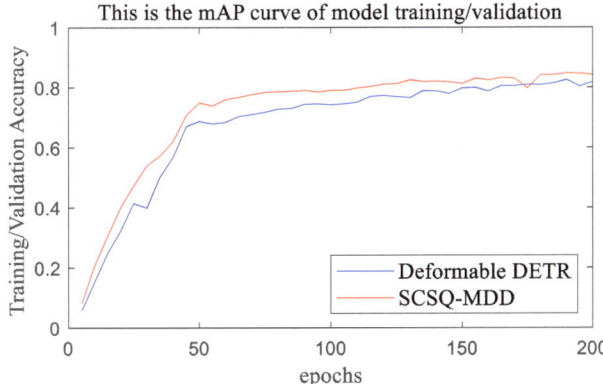

Figure 9. Trends in the accuracies of the deformable DETR and the SCSQ-MDD during the training/validation stage. The proposed SCSQ-MDD method performed significantly better compared to the deformable DETR method in terms of detection accuracy.

Table 2 shows the results of the comparison experiments using the deformable DETR and the SCSQ-MDD. It was found that our model resulted in substantial improvements in accuracy with only a slight increase in the number of parameters. Compared with the deformable DETR, the use of the mask deformable DETR increased the number of parameters by 1M, FLOPS by 41G, AP by 1.5%, AP50 by 2.0%, and AR by 0.8%. The use of the SCSQ-MDD increased the number of parameters by 6M, FLOPS by 41G, AP by 3.8%, AP50 by 3.1%, and AR by 1.1%. The FLOPS of the SCSQ-MDD exhibited almost no improvement compared to the mask deformable DETR, with increases in the AP of 2.3%, AP50 of 1.1%, and AR of 0.3%.

Table 2. The comparative detection results of the deformable DETR and the SCSQ-MDD (simple conditional spatial query mask deformable DETR) on the test set. The metrics used to evaluate detection accuracy include the AP, AR, params, FLOPS, and FPS.

Method	Epochs	AP	AP_{50}	AP_{75}	AP_M	AP_L	AR	Params	FLOPS	FPS
Deformable DETR [35]	150	79.8	90.5	89.1	71.8	79.8	90.6	40M	144G	5
Mask Deformable DETR	150	81.3	92.5	91.7	71.7	81.3	91.4	41M	185G	4
SCSQ-MDD	150	83.6	93.6	93.0	71.7	81.7	91.7	46M	185G	4

4.3. Comparison between the SCSQ Mask Deformable DETR and Mainstream Detection Methods

Table 3 shows the comparison results between the proposed SCSQ-MDD and mainstream detection methods. The AP of the method proposed in this paper was 88.1%, the AP50 was 95.6%, the AP75 was 95.5%, and the AR was 93.5%. Compared with Faster RCNN, the proposed method improved the AP by 9.6%, the AP50 by 0.5%, the AP75 by 1.6%, and the AR by 0.9%. Compared with ATSS, the proposed method improved the AP by 16.3%, the AP50 by 5.8%, the AP75 by 12.4%, and the AR by 14.5%. Compared with YOLOv5 with 1500 training epochs, the AP of our method with 500 epochs increased by 4% and the AP50 increased by 1.5%.

Table 3. The comparative detection results of the SCSQ-MDD and mainstream methods on the test set. The metrics used to evaluate detection accuracy include the AP, AP_{50}, AP_{75}, and AR. A check mark ($\sqrt{}$) indicates whether to enable the SFN module.

Method	Backbone	SFN	AP	AP_{50}	AP_{75}	AP_M	AP_L	AR
Faster RCNN [8]	ResNet50		78.5	96.4	94.4	68.9	78.5	82.6
ATSS [37]	ResNet50		71.8	89.8	83.1	79.6	71.7	79.0
YOLOv5	ResNet50		84.1	94.1	–	–	–	–
YOLOv5	ResNet50	$\sqrt{}$	84.9	94.9	–	–	–	–
YOLOv7 [17]	ResNet50		87.9	95.0	–	–	–	–
SCSQ-MDD	ResNet50	$\sqrt{}$	88.1	95.6	95.5	72.1	88.2	92.9
SCSQ-MDD	ResNet101	$\sqrt{}$	88.6	96.9	96.0	75.1	88.6	93.5

4.4. Ablation Study on Improved Mask Deformation Attention

Detailed ablation experiments were conducted for the SCSQ-MDD network structure design and the embedding of the mask deformable attention, SCSQ, and SFN modules. The performance of the network model was evaluated by comparing the prediction accuracy during the testing phase. The designs of the SCSQ, mask deformable attention, and SFN modules were explored for their usefulness in training the Chinese character stroke detection network, with a check mark ($\sqrt{}$) indicating whether the specific technique or module was used.

Table 4 shows the ablation results of various options of the proposed deformable attention module based on the mask mechanism. Using the mask mechanism to resample the offset of deformable attention effectively improved detection accuracy, with a 1.5% increase in the AP. Adding the simple conditional spatial query module further improved the AP value by 1.1%. Using the proposed SFN module in this paper to split channels further improved the AP by 1.2%. Overall, using both the simple conditional spatial query module and the SFN module improved the AP value in this experiment. It can be seen that when either the SCSQ module or the SFN module was added alone, the SFN module led to more improvements than the SCSQ module.

Table 4. Ablation results of the SCSQ-MDD network structure on the test set. "Mask Deformable Attn." refers to the deformable attention module based on the mask mechanism and "Simple Conditional Spatial Query (SCSQ)" refers to the simple conditional spatial query strategy. A check mark ($\sqrt{}$) indicates whether to enable the specified module.

Deformable DETR	Mask Deformable Attn	SCSQ	SFN	AP	AP50	AP75	AR
$\sqrt{}$				79.8	90.5	89.1	90.6
$\sqrt{}$	$\sqrt{}$			81.3	92.5	91.7	91.4
$\sqrt{}$	$\sqrt{}$	$\sqrt{}$		82.4	92.4	92.0	91.3
$\sqrt{}$	$\sqrt{}$		$\sqrt{}$	82.5	93.1	92.6	90.7
$\sqrt{}$		$\sqrt{}$	$\sqrt{}$	81.5	91.1	89.7	92.1
$\sqrt{}$	$\sqrt{}$	$\sqrt{}$	$\sqrt{}$	83.6	93.6	93.3	91.7

In this paper, a mask mechanism is used to filter the sampled reference points and resample the invalid reference points to reduce the randomness of reference points during the feature updating processes. This enables the model to converge faster compared to not using a mask mechanism. Therefore, better detection performance can be achieved by a model with a mask mechanism in the same cycle compared to a model without it. Moreover, a simple conditional spatial query strategy is used in this paper to separate the content query and spatial query, reducing the dependence of the prediction task on the content embedding and accelerating the convergence of our model compared to models that do not use a simple conditional spatial query. Finally, a channel-splitting FFN network is adopted in this paper as the prediction head. The classification and regression tasks can focus on their respective features and interconnections, thereby improving the model's detection accuracy rate. However, since a mask prediction branch, as well as a resampling strategy, are introduced in this study, and a conditional spatial query strategy is used to compute the conditional spatial query vector, additional computations are needed, and the runtime of the model is longer compared to the original deformable DETR.

5. Conclusions

In this study, a deformable DETR method based on a mask mechanism with a simple conditional spatial query for detecting Chinese character strokes is proposed. This method is utilized to address the problem of random sampling in the deformable attention module in the original deformable DETR, to further accelerate convergence speed, and to improve accuracy. The mask mechanism designed in this study can be used to effectively reduce the uncertainty of deformable attention in sampling, thus reducing unnecessary computational costs. The simple conditional spatial query module is added to significantly improve the detection performance of the model with only a small increase in the number of parameters. Moreover, for the transformer task, the final query vector output of the decoder is split to specify the specific predictions for different tasks, which can be used to slightly improve the model's performance without any increase in the computational cost and number of parameters.

This method provides a new solution for Chinese character stroke detection tasks with an improved detection paradigm. Moreover, as a method capable of handling Chinese character strokes, this method can accomplish the Chinese character writing task using robotic arms at the stroke level. Meanwhile, in addition to detecting strokes, a library for stroke order also needs to be built. The rules are established for each stroke of each Chinese character in order to complete the process of Chinese character reduction.

Although we have completed the task of detecting the strokes of standard Chinese characters, the task of detecting the strokes of handwritten Chinese characters is still difficult due to their irregularity and stylistic heterogeneity. In the future, we will focus on stroke detection and the restoration of handwritten Chinese characters. Meanwhile, this work has important implications for early education in Chinese character calligraphy, the dissemination of multi-font Chinese character graphics on social networks, and writing using industrial robotic arms.

Author Contributions: Conceptualization, T.Z., W.X. and Y.F.; methodology, T.Z.; software, T.Z.; validation, T.Z.; investigation, T.Z., W.X. and Y.F.; resources, W.X.; data curation, T.Z.; writing—original draft preparation, T.Z.; writing—review and editing, T.Z., W.X. and H.Z.; visualization, T.Z.; supervision, W.X. and H.Z.; project administration, T.Z.; funding acquisition, W.X. All authors have read and agreed to the published version of the manuscript.

Funding: This research was supported by the National Natural Science Foundation of China under Grant No. 61966008 and Grant No. 62103114.

Institutional Review Board Statement: Not applicable.

Informed Consent Statement: Not applicable.

Data Availability Statement: Data are contained within the article.

Conflicts of Interest: The authors declare no conflicts of interest.

References

1. Ma, C.H.; Lu, C.L.; Shih, H.C. Vision-Based Jigsaw Puzzle Solving with a Robotic Arm. *Sensors* **2023**, *23*, 6913. [CrossRef]
2. Xia, X.; Li, T.; Sang, S.; Cheng, Y.; Ma, H.; Zhang, Q.; Yang, K. Path Planning for Obstacle Avoidance of Robot Arm Based on Improved Potential Field Method. *Sensors* **2023**, *23*, 3754. [CrossRef]
3. Zhang, Z.; Wang, Z.; Zhou, Z.; Li, H.; Zhang, Q.; Zhou, Y.; Li, X.; Liu, W. Omnidirectional Continuous Movement Method of Dual-Arm Robot in a Space Station. *Sensors* **2023**, *23*, 5025. [CrossRef]
4. Chao, F.; Huang, Y.; Lin, C.M.; Yang, L.; Hu, H.; Zhou, C. Use of Automatic Chinese Character Decomposition and Human Gestures for Chinese Calligraphy Robots. *IEEE Trans.-Hum.-Mach. Syst.* **2019**, *49*, 47–58. [CrossRef]
5. Wang, T.Q.; Jiang, X.; Liu, C.L. Query Pixel Guided Stroke Extraction with Model-Based Matching for Offline Handwritten Chinese Characters. *Pattern Recognit.* **2022**, *123*, 108416. [CrossRef]
6. Girshick, R.B.; Donahue, J.; Darrell, T.; Malik, J. Rich Feature Hierarchies for Accurate Object Detection and Semantic Segmentation. In Proceedings of the 2014 IEEE Conference on Computer Vision and Pattern Recognition, Portland, OR, USA, 23–28 June 2013; pp. 580–587.
7. He, K.; Zhang, X.; Ren, S.; Sun, J. Spatial Pyramid Pooling in Deep Convolutional Networks for Visual Recognition. *IEEE Trans. Pattern Anal. Mach. Intell.* **2014**, *37*, 1904–1916. [CrossRef] [PubMed]
8. Ren, S.; He, K.; Girshick, R.B.; Sun, J. Faster R-CNN: Towards Real-Time Object Detection with Region Proposal Networks. *IEEE Trans. Pattern Anal. Mach. Intell.* **2015**, *39*, 1137–1149. [CrossRef]
9. Girshick, R.B. Fast R-CNN. In Proceedings of the 2015 IEEE International Conference on Computer Vision (ICCV), Santiago, Chile, 7–13 December 2015; pp. 1440–1448.
10. Dai, J.; Li, Y.; He, K.; Sun, J. R-FCN: Object Detection via Region-based Fully Convolutional Networks. *arXiv* **2016**, arXiv:1605.06409.
11. Cai, Z.; Vasconcelos, N. Cascade R-CNN: Delving Into High Quality Object Detection. In Proceedings of the 2018 IEEE/CVF Conference on Computer Vision and Pattern Recognition, Salt Lake City, UT, USA, 18–23 June 2017; pp. 6154–6162.
12. Redmon, J.; Divvala, S.K.; Girshick, R.B.; Farhadi, A. You Only Look Once: Unified, Real-Time Object Detection. In Proceedings of the 2016 IEEE Conference on Computer Vision and Pattern Recognition (CVPR), Las Vegas, NV, USA, 27–30 June 2016; pp. 779–788.
13. Redmon, J.; Farhadi, A. YOLOv3: An Incremental Improvement. *arXiv* **2018**, arXiv:1804.02767.
14. Bochkovskiy, A.; Wang, C.Y.; Liao, H.Y.M. YOLOv4: Optimal Speed and Accuracy of Object Detection. *arXiv* **2020**, arXiv:2004.10934.
15. Redmon, J.; Farhadi, A. YOLO9000: Better, Faster, Stronger. In Proceedings of the 2017 IEEE Conference on Computer Vision and Pattern Recognition (CVPR), Honolulu, HI, USA, 21–26 July 2017; pp. 6517–6525.
16. Li, C.; Li, L.; Jiang, H.; Weng, K.; Geng, Y.; Li, L.; Ke, Z.; Li, Q.; Cheng, M.; Nie, W.; et al. YOLOv6: A Single-Stage Object Detection Framework for Industrial Applications. *arXiv* **2022**, arXiv:2209.02976.
17. Wang, C.Y.; Bochkovskiy, A.; Liao, H.Y.M. YOLOv7: Trainable bag-of-freebies sets new state-of-the-art for real-time object detectors. *arXiv* **2022**, arXiv:2207.02696.
18. Liu, W.; Anguelov, D.; Erhan, D.; Szegedy, C.; Reed, S.E.; Fu, C.Y.; Berg, A.C. SSD: Single Shot MultiBox Detector. *arXiv* **2015**, arXiv:1512.02325.
19. Lin, T.Y.; Goyal, P.; Girshick, R.B.; He, K.; Dollár, P. Focal Loss for Dense Object Detection. *IEEE Trans. Pattern Anal. Mach. Intell.* **2017**, *42*, 318–327. [CrossRef] [PubMed]
20. Ge, Z.; Liu, S.; Wang, F.; Li, Z.; Sun, J. YOLOX: Exceeding YOLO Series in 2021. *arXiv* **2021**, arXiv:2107.08430.
21. Xie, T.; Zhang, Z.; Tian, J.; Ma, L. Focal DETR: Target-Aware Token Design for Transformer-Based Object Detection. *Sensors* **2022**, *22*, 8686. [CrossRef] [PubMed]
22. Li, S.; Sultonov, F.; Tursunboev, J.; Park, J.H.; Yun, S.; Kang, J.M. Ghostformer: A GhostNet-Based Two-Stage Transformer for Small Object Detection. *Sensors* **2022**, *22*, 6939. [CrossRef]
23. Bello, I.; Zoph, B.; Vaswani, A.; Shlens, J.; Le, Q.V. Attention Augmented Convolutional Networks. In Proceedings of the 2019 IEEE/CVF International Conference on Computer Vision (ICCV), Seoul, Republic of Korea, 27 October–2 November 2019; pp. 3285–3294.
24. Shaw, P.; Uszkoreit, J.; Vaswani, A. Self-Attention with Relative Position Representations. *arXiv* **2018**, arXiv:1803.02155.
25. Ramachandran, P.; Parmar, N.; Vaswani, A.; Bello, I.; Levskaya, A.; Shlens, J. Stand-Alone Self-Attention in Vision Models. *arXiv* **2019**, arXiv:1906.05909.
26. Dosovitskiy, A.; Beyer, L.; Kolesnikov, A.; Weissenborn, D.; Zhai, X.; Unterthiner, T.; Dehghani, M.; Minderer, M.; Heigold, G.; Gelly, S.; et al. An Image is Worth 16x16 Words: Transformers for Image Recognition at Scale. *arXiv* **2020**, arXiv:2010.11929.
27. Carion, N.; Massa, F.; Synnaeve, G.; Usunier, N.; Kirillov, A.; Zagoruyko, S. End-to-End Object Detection with Transformers. *arXiv* **2020**, arXiv:2005.12872.
28. Wu, K.; Peng, H.; Chen, M.; Fu, J.; Chao, H. Rethinking and Improving Relative Position Encoding for Vision Transformer. In Proceedings of the 2021 IEEE/CVF International Conference on Computer Vision (ICCV), Montreal, BC, Canada, 11–17 October 2021; pp. 10013–10021.
29. Chen, Q.; Chen, X.; Zeng, G.; Wang, J. Group DETR: Fast Training Convergence with Decoupled One-to-Many Label Assignment. *arXiv* **2022**, arXiv:2207.13085.

30. Bar, A.; Wang, X.; Kantorov, V.; Reed, C.; Herzig, R.; Chechik, G.; Rohrbach, A.; Darrell, T.; Globerson, A. DETReg: Unsupervised Pretraining with Region Priors for Object Detection. In Proceedings of the 2022 IEEE/CVF Conference on Computer Vision and Pattern Recognition (CVPR), New Orleans, LA, USA, 18–24 June 2022; pp. 14585–14595.
31. Li, F.; Zhang, H.; guang Liu, S.; Guo, J.; shuan Ni, L.M.; Zhang, L. DN-DETR: Accelerate DETR Training by Introducing Query DeNoising. In Proceedings of the 2022 IEEE/CVF Conference on Computer Vision and Pattern Recognition (CVPR), New Orleans, LA, USA, 18–24 June 2022; pp. 13609–13617.
32. Zhang, G.; Luo, Z.; Yu, Y.; Cui, K.; Lu, S. Accelerating DETR Convergence via Semantic-Aligned Matching. In Proceedings of the 2022 IEEE/CVF Conference on Computer Vision and Pattern Recognition (CVPR), New Orleans, LA, USA, 18–24 June 2022; pp. 939–948.
33. Gao, P.; Zheng, M.; Wang, X.; Dai, J.; Li, H. Fast Convergence of DETR with Spatially Modulated Co-Attention. In Proceedings of the 2021 IEEE/CVF International Conference on Computer Vision (ICCV), Montreal, BC, Canada, 11–17 October 2021; pp. 3601–3610.
34. Kitaev, N.; Kaiser, L.; Levskaya, A. Reformer: The Efficient Transformer. *arXiv* **2020**, arXiv:2001.04451.
35. Zhu, X.; Su, W.; Lu, L.; Li, B.; Wang, X.; Dai, J. Deformable DETR: Deformable Transformers for End-to-End Object Detection. *arXiv* **2020**, arXiv:2010.04159.
36. Meng, D.; Chen, X.; Fan, Z.; Zeng, G.; Li, H.; Yuan, Y.; Sun, L.; Wang, J. Conditional DETR for Fast Training Convergence. In Proceedings of the 2021 IEEE/CVF International Conference on Computer Vision (ICCV), Montreal, BC, Canada, 11–17 October 2021; pp. 3631–3640.
37. Zhang, S.; Chi, C.; Yao, Y.; Lei, Z.; Li, S.Z. Bridging the Gap Between Anchor-Based and Anchor-Free Detection via Adaptive Training Sample Selection. In Proceedings of the 2020 IEEE/CVF Conference on Computer Vision and Pattern Recognition (CVPR), Seattle, WA, USA, 13–19 June 2020; pp. 9756–9765.

Disclaimer/Publisher's Note: The statements, opinions and data contained in all publications are solely those of the individual author(s) and contributor(s) and not of MDPI and/or the editor(s). MDPI and/or the editor(s) disclaim responsibility for any injury to people or property resulting from any ideas, methods, instructions or products referred to in the content.

Article

Enhanced Knowledge Distillation for Advanced Recognition of Chinese Herbal Medicine

Lu Zheng [1,2], Wenhan Long [1], Junchao Yi [1,3], Lu Liu [4] and Ke Xu [1,3,*]

[1] College of Computer Science, South-Central Minzu University, Wuhan 430074, China
[2] Key Laboratory of Information Physics Integration and Intelligent Computing of National Ethnic Affairs Commission, Wuhan 430074, China
[3] Hubei Provincial Engineering Research Center of Agricultural Blockchain and Intelligent Management, Wuhan 430074, China
[4] School of Computing and Mathematical Sciences, University of Leicester, Leicester LE1 7RH, UK
* Correspondence: xuke@scuec.edu.cn

Abstract: The identification and classification of traditional Chinese herbal medicines demand significant time and expertise. We propose the dual-teacher supervised decay (DTSD) approach, an enhancement for Chinese herbal medicine recognition utilizing a refined knowledge distillation model. The DTSD method refines output soft labels, adapts attenuation parameters, and employs a dynamic combination loss in the teacher model. Implemented on the lightweight MobileNet_v3 network, the methodology is deployed successfully in a mobile application. Experimental results reveal that incorporating the exponential warmup learning rate reduction strategy during training optimizes the knowledge distillation model, achieving an average classification accuracy of 98.60% for 10 types of Chinese herbal medicine images. The model boasts an average detection time of 0.0172 s per image, with a compressed size of 10 MB. Comparative experiments demonstrate the superior performance of our refined model over DenseNet121, ResNet50_vd, Xception65, and EfficientNetB1. This refined model not only introduces an approach to Chinese herbal medicine image recognition but also provides a practical solution for lightweight models in mobile applications.

Keywords: Chinese herbal medicine; knowledge distillation; dual-teacher supervision; adaptive attenuation; portable application

Citation: Zheng, L.; Long, W.; Yi, J.; Liu, L.; Xu, K. Enhanced Knowledge Distillation for Advanced Recognition of Chinese Herbal Medicine. *Sensors* 2024, 24, 1559. https://doi.org/10.3390/s24051559

Academic Editor: Stefano Berretti

Received: 13 December 2023
Revised: 8 February 2024
Accepted: 24 February 2024
Published: 28 February 2024

Copyright: © 2024 by the authors. Licensee MDPI, Basel, Switzerland. This article is an open access article distributed under the terms and conditions of the Creative Commons Attribution (CC BY) license (https://creativecommons.org/licenses/by/4.0/).

1. Introduction

Chinese herbal medicine stands as a distinctive therapeutic approach within traditional Chinese medicine (TCM), offering a diverse range of remedies for various ailments. However, the expansive array of Chinese herbal medicines used across different regions has given rise to a concerning trend: the proliferation of counterfeit and substandard substitutes on the market. This poses significant risks as ordinary consumers, lacking in-depth knowledge, often inadvertently consume these falsified products. The complexity of Chinese herbal medicine compounds this issue, making it challenging for laypersons to accurately identify genuine herbs. As a consequence, mistaken ingestion remains a frequent occurrence among consumers. Presently, the identification and classification of these herbs heavily rely on individuals with specialized expertise in this field. To address these challenges, the integration of deep learning technology into the recognition and classification of Chinese herbal medicine becomes imperative. The remarkable advancements in image recognition offered by deep learning present a promising solution. This integration holds immense potential in revolutionizing traditional Chinese medicine (TCM) by providing a systematic approach to identifying and authenticating herbal medicines. The application of deep learning in research aimed at recognizing and categorizing Chinese herbal medicine marks a crucial step forward in preserving and advancing the legacy of TCM. Its incorporation promises to empower both practitioners and consumers by enhancing the authentication and classification of these invaluable remedies.

In recent studies, deep learning technologies have been harnessed for advancing Chinese herbal medicine identification. Huang et al. [1] proposed a Chinese herb image classification method based on AlexNet. Through meticulous data augmentation and parameter fine-tuning, they achieved an impressive classification accuracy of 87.5% after 300 epochs. Gao et al. [2] introduced a recognition approach for natural grassland plant species using Inception_V3 with Tensorflow, achieving a peak accuracy of 89.41% in the model's validation dataset's Top1 error. Zhang et al. [3] contributed by classifying 17 types of Chinese herbs through a VGG network, attaining an outstanding average recognition accuracy of 96% in the validation dataset. Their model was further deployed on mobile devices, demonstrating practical application. Wang et al. [4] proposed an image recognition method for Chinese herbal plants based on the AlexNet network. Utilizing a deep coding and decoding network, they successfully trained the model to classify 15 types of Chinese herbal images, achieving an impressive average classification accuracy of 99.38%, but the large model parameter amount led to large training and inference computation, requiring more memory and computing resources, which was not suitable for mobile deployment. Hu et al. [5] introduced a dual-channel U-shaped convolutional neural network with feature calibration. They generated a training model for single-view fritillaria image data, surpassing the classification results of traditional machine learning methods. Moreover, by incorporating multi-view fritillaria images and employing a three-dimensional convolutional neural network, they developed a more precise fritillaria classification model. While these scholars have conducted profound research in Chinese herbal medicine image recognition, there is still room for improvement in model accuracy. An improved model has to not only maintain high classification accuracy but also focus on reducing the model parameter amount to improve performance. Moreover, concerns arise regarding the redundancy of model parameters, hindering deployment on mobile devices due to inadequate detection speed.

To address the aforementioned challenges, this paper proposes an approach called dual-teacher supervised decay (DTSD) for adaptive-decay knowledge distillation, which aims to enhance the performance of the standard model. By enhancing the output soft label, adapting decay parameters, and dynamically combining loss functions from the teacher model, DTSD is employed in the MobileNet_v3_Small network to enable accurate predictions despite its smaller size. Consequently, the accuracy of the MobileNet_v3_Small network is elevated to match that of more complex networks. The proposed model is then integrated into an intelligent Chinese herbal medicine recognition system for mobile devices, facilitating the efficient recognition and classification of Chinese herbal medicine.

The main contributions of our work include:

1. Proposing a dual-teacher supervised model to reorganize the predictive distribution of dual teachers to achieve more accurate and robust soft labeling, which improves the performance of the model.
2. Dynamically adjusting the temperature parameter T and the weight distribution value λ between the teacher model and the real label to gradually reduce the influence of the teacher model in the training process, so that the student model can more flexibly balance the complexity and the model's generalization ability in the training process.
3. Adopting JS scatter with symmetry to replace the cross-entropy loss of the predicted values of the soft label and the student model to better capture the similarity between the distributions and prompt the student model to better inherit the knowledge of the tutor model.
4. A lightweight MobileNet_v3 network-based herbal medicine recognition system is implemented, and by applying our proposed DTSD method to the MobileNet_v3_Small network, we improve the accuracy and robustness of herbal medicine recognition while maintaining a small model size.

The rest of this paper is organized as follows: The preliminaries and dataset collection are presented in the first half of Section 2. Also in that section, the dataset augmentation strategies are described. Then, we introduce the knowledge distillation model with dual-teacher supervised decay in the second half of Section 2. Comparative experiment results and analysis are given in Section 3 to verify the effectiveness of the proposed methods, followed by a short conclusion in Section 4.

2. Materials and Methods

2.1. Dataset Collection

The dataset utilized in this study consists of images depicting 10 distinct types of Chinese herbal medicine. These images were gathered from the Internet using a Web crawling technique and subsequently underwent a meticulous process of curation and filtration. A Web crawler first needs to determine the initial URL to be crawled and then builds a queue of URLs by parsing links on the page. The crawler accesses the web pages one by one according to the URLs in the queue. The page is requested from the server via an HTTP request, and then the HTML data returned by the server are downloaded locally. The downloaded pages are usually in HTML format, and the crawler needs to parse the HTML to extract useful information. The parsed data were the herbal images. This process resulted in a total of 1000 images of Chinese herbal medicines being compiled. The dataset, as illustrated in Figure 1, encompassed 10 specific types of Chinese herbal medicines, namely *Radix paeoniaealba*, *Radix stemoonae*, *Fructus aurantia tablets*, *Polygon atum*, *turmeric*, *Pollen typhae*, *Cnidium monnieri*, *motherwort*, *Chinese wolfberry*, and *curcuma*. By cropping and compression, each category comprised 100 images, all of which sized at 320 pixels by 320 pixels and possessing a resolution of 96 dots per inch. During the training of the teacher network, the dataset was partitioned into a 7:2:1 ratio, with 700 images (70%) allocated to the training dataset, 200 images (20%) to the validation dataset, and 100 images (10%) to the test dataset. The dataset was divided into datasets according to the 7:2:1 ratio for each category. The authors labeled various types of herbs.

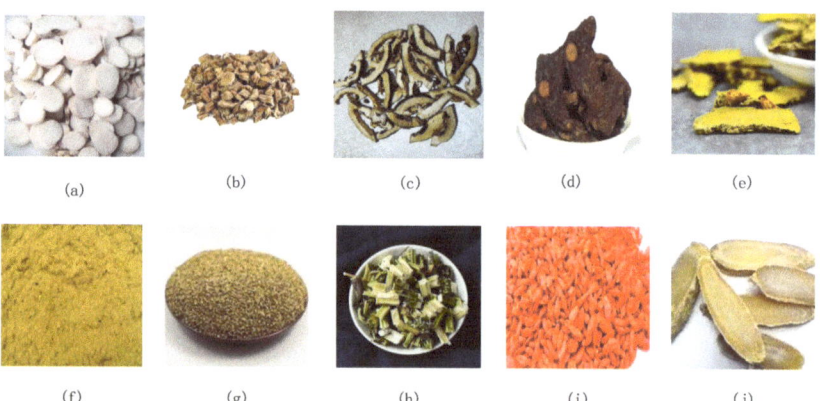

Figure 1. Sample of the Chinese herbal medicine dataset: (**a**) Radix paeoniae; (**b**) Radix stemonae; (**c**) Fructus aurantia tablets; (**d**) Polygonatum; (**e**) turmeric; (**f**) Pollen typhae; (**g**) Cnidium monnieri; (**h**) motherwort; (**i**) Chinese wolfberry; (**j**) curcuma.

2.2. Data Enhancement

The performance and recognition ability of the model during training are influenced by the generalization and quantity of data. When the available data are limited, overfitting becomes a more prominent issue in deep learning models. To address this challenge and improve the model's generalization capabilities, data augmentation techniques are employed prior to training. These techniques enable the generation of more diverse data representations, as depicted in Figure 2. In order to preserve the original data features, the dataset was expanded to a size of 10,000 through generalization under simulated real conditions. The augmentation strategies [6] primarily included an image transformation class and an image cropping class.

Figure 2. Data augmentation: (**a**) original image; (**b**) random augmentation strategy; (**c**) cutout strategy; (**d**) random erasing strategy; (**e**) hide-and-seek strategy.

2.2.1. Image Transformation Class

For the rand augmentation strategy, specific probability distributions were set for each sub-strategy. The transformations included rotation (±30 degrees), flipping (50% probability of a horizontal flip), cropping (up to 20% crop of the original image size), contrast adjustment (±10%), and other sub-strategies. Each sub-strategy was randomly applied with a uniform probability distribution, reducing the need for manual selection. Moreover, all sub-strategies were also applied with equal probabilities, resulting in multiple augmentation sub-strategies being concurrently applied to a single image through probability combination. This approach allowed for the adjustment of image brightness, contrast, saturation, and hue simultaneously, simulating variations in shooting angles and actual lighting conditions. The probability of applying each sub-strategy was set at 10%, ensuring a balanced augmentation without overpowering the original image characteristics. By incorporating random factors to mimic real-world lighting differences, the parameters aligned more closely with reality, reducing the impact of image angle and lighting variations, and enhancing the model's robustness.

2.2.2. Image Cropping

In the cropping class, cut out, random erasing, and hide-and-seek strategies were employed, each with a distinct probability of application: 15% for cut out, 10% for random erasing, and 5% for hide and seek. The main objective of these strategies is to imitate classification scenarios where the subject is partially occluded in real-world situations. The size of the cropped area ranged from 10% to 20% of the original image, randomly chosen for each application. This helped prevent the model from becoming overly sensitive to salient regions of the image, thus avoiding overfitting.

2.3. Teacher Model: ResNet_vd

ResNet, introduced by Kaiming [7], aimed primarily to reduce the computational expense during network training and address issues related to diminishing or amplifying gradients leading to performance degradation with increasing network depth. This architecture employs stacked nonlinear layers to accommodate skip connections, thereby establishing an identity mapping. This ensures that deeper layers perform as effectively as shallower networks [8].

The ResNet_vd model, an enhancement of ResNet by Tong et al., introduces various versions of residual modules [9]. Experiments conducted by He et al. demonstrate that ResNet_vd achieves significantly higher accuracy compared to other structural variations, leading to an approximate 0.85% increase in the top-1 accuracy on ImageNet. The network's structure is illustrated in Figure 3.

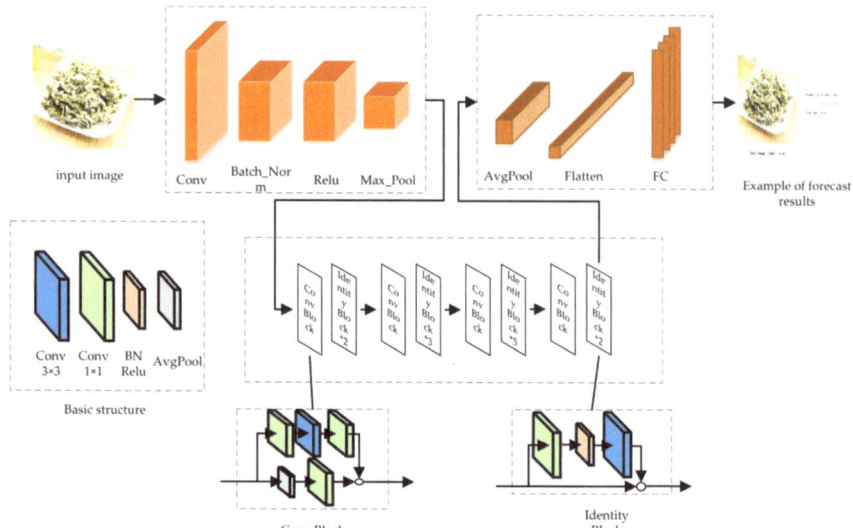

Figure 3. Resnet_vd model structure diagram.

2.4. Teacher Model: DenseNet

DenseNet, proposed by Gao [10] et al., emerged subsequent to an analysis of ResNets, highway networks, FractalNets, and other models. These authors highlighted a crucial attribute shared by these models: the construction of shortcuts between preceding and succeeding network layers, ensuring an identity mapping between them. Leveraging this characteristic, they developed an enhanced connection mode: each layer receives the feature maps from all preceding layers as input, as depicted in Figure 4. Research indicates that this connection method notably enhances the transfer of features and gradients within the network.

Compared to the residual network (ResNet), DenseNet achieves equivalent accuracy on the ImageNet dataset while utilizing less than half the number of parameters and computational resources [11]. Simultaneously, it demonstrates robust resistance to overfitting and displays strong generalization performance [12].

2.5. Student Model: MobileNet_v3_Small

In the realm of deep convolutional network models, achieving high accuracy often comes at the cost of increased model size and slower prediction speeds due to the incorporation of various techniques. The choice of MobileNet_v3 as the student model in our study over other lightweight models was motivated by its unique balance of efficiency and performance. Compared to other lightweight architectures, MobileNet_v3 offers an optimal trade-off between accuracy and speed, crucial for real-time applications on embedded devices. This balance is achieved through its advanced architectural innovations that reduce computational demand without significant loss in accuracy [13,14].

MobileNet_v3 represents the next evolutionary step: a lightweight network that amalgamates the essence of MobileNet_v1 and MobileNet_v2 while introducing enhancements. It was selected for its superior efficiency in processing speed and reduced parameter count, critical for deployment in resource-constrained environments. This iteration revolves around four core blocks: (1) a depthwise-separable convolution; (2) an inverted residual structure with linear bottleneck; (3) a lightweight attention block; (4) the utilization of h-swish as an activation function, replacing the conventional swish [15–18]. The architecture of MobileNet_v3, depicted in Figure 5, demonstrates these key components.

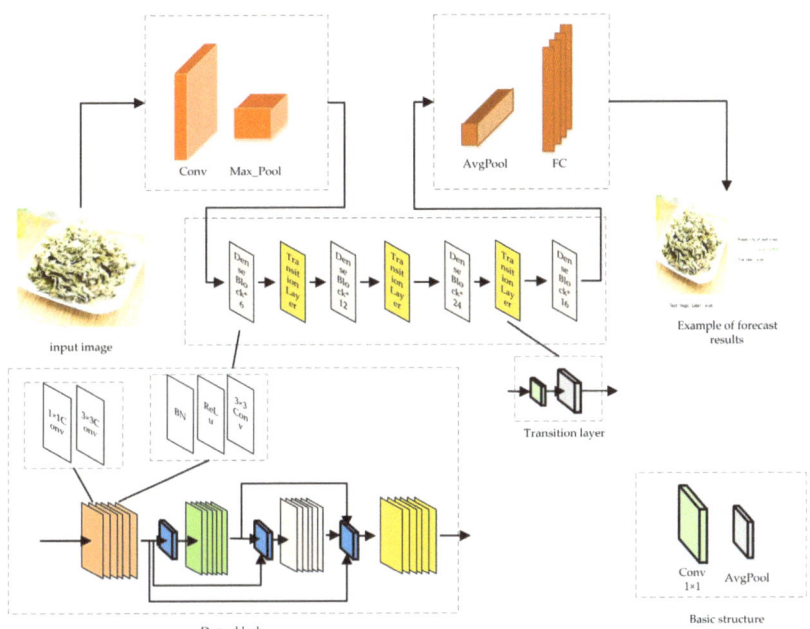

Figure 4. Densenet121 model structure diagram.

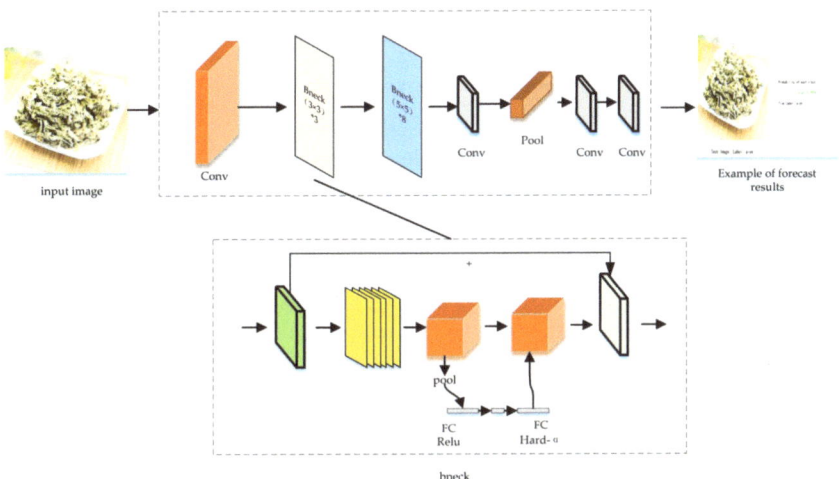

Figure 5. MobileNet_v3 model structure diagram.

2.6. Knowledge Distillation Model with Dual-Teacher Supervised Decay

In deep neural networks, the presence of a large number of parameters leads to redundancy. Knowledge distillation, as proposed by Hinton et al. [19], emerges as a technique to address this issue by compressing the model and reducing the parameter count [20–24]. The fundamental idea behind knowledge distillation involves incorporating soft labels associated with the teacher network into the total loss. This integration guides the training of the student network, facilitating knowledge transfer. The improvement in the performance of the student network is achieved while keeping the number of parameters constant. The resulting performance metrics closely align with those of the larger model. The detailed process is illustrated in Figure 6, and it unfolds as follows:

1. The teacher network initially trains on hard targets. Once the model is trained, just before the network performs softmax normalization on the output, each term is divided by a fixed temperature, T. This process yields the soft targets used to guide the learning of the student network.
2. During the training of the student network, the loss value employed for updating parameter weights during backpropagation is divided into two components. One part represents the cross-entropy loss computed on the true labels of the training dataset. The other part corresponds to the loss calculated on the soft output of the teacher network. Ultimately, these two losses are weighted and combined to generate the overall loss, which is then applied to the training of the student network model [25–28].

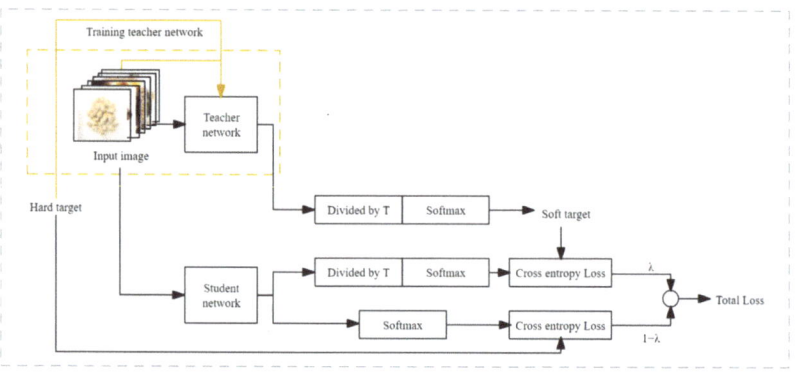

Figure 6. Flow chart of knowledge distillation.

The standard knowledge distillation process has shown its innovative aspects, but it still faces the following issues, leading to a decrease in the accuracy of the student model compared to the general model in certain training scenarios:

1. Quality of guidance from the teacher model: Sometimes, the complex model might not predict perfectly. This is like a chef giving slightly incorrect cooking instructions to an apprentice. When these predictions, or guidance, are enhanced to make them more detailed for the student model (akin to increasing the "temperature" to make the lessons more intense), it can introduce errors or "noise". This may lead the student model to learn incorrectly, like an apprentice learning flawed cooking techniques.
2. Adjusting the intensity of teaching (temperature): In past research, the intensity or detail in the teacher's guidance was often set at a fixed level, usually moderate. But it is now understood that this should vary throughout the training, much like adjusting teaching methods for students as they progress. The "temperature", or level of detail and complexity in the teacher's guidance, needs to be adaptable, increasing or decreasing at different stages of the student model's learning.
3. Balancing real data vs. teacher's predictions (loss weighting λ): In traditional teaching methods, the balance between real-world data (hard labels) and the teacher's predictions (soft labels) is constant. However, it is more effective if this balance changes over time. As the student model learns, the emphasis should gradually shift from what the teacher model predicts to what is actually observed in real-world data, allowing the student model to become more adept at handling real situations independently.

Based on these issues, an improved model of adaptive-decay knowledge distillation with dual-teacher supervision is proposed, with specific improvements as follows:

1. The combination of soft labels: In standard knowledge distillation, we expand the teacher model from a single teacher to dual teachers to obtain multiple prediction

distributions. To maintain the accuracy of the prediction distributions while acquiring more dark knowledge, we recombine the prediction distributions of the two teacher models. This is done by taking the maximum value of the predictions from the two teacher models in each dimension as the category classification result for that dimension, thereby obtaining a soft label with greater accuracy and richer dark knowledge. The formula is shown as Formula (1). Here, p and q represent the predicted labels given by the two teacher models, respectively. Through this formula, we generate a new probability distribution composed of the maximum values from two different probability distributions in each dimension.

$$Max\_out(p,q) = [Max(p\_1,q\_1), Max(p\_2,q\_2), \ldots Max(p\_n,q\_n)] \quad (1)$$

2. Selection of T: By analyzing the distillation distribution of the model output probability for different T cases, as shown in Figure 7, the different types are 10 classifications for herbal recognition, and the standard output is the blue solid line. When the value of T is smaller than 1 (the red dotted line), the gap between the true prediction value and the dark knowledge is enlarged, that is, the proportion of the true prediction increases. When the value of T is greater than 1 (green dotted line), the total prediction distribution is smoother, which means the proportion of dark knowledge is increased. Therefore, in the early stage of training, T is set to a value smaller than 1, so that the student model can quickly find the basic proper parameters in the early stage. With the deepening of training and the expansion of the proportion of dark knowledge, the student model with high accuracy further learns the dark knowledge part of the correct prediction distribution given by the teacher model, so as to improve its accuracy. Thus, the value of T is set to the value of the function that grows with the training epochs. As illustrated in Formula (2) x is the training metric; through this function, the temperature T changes with the x in an S-shaped curve and is defined as the deepening of the experiment (step/epochs), increasing in an S-shaped curve, and the main value range is [0–3], so that the student model can learn different degrees of dark knowledge in different epochs. This is shown in Figure 8. The student model in the early stage as a low weight; as the model training process continues to rise, the relationship is well reflected as a sigmoid function, that is, an "S" curve. We have adjusted the parameters of the sigmoid function so as to be more in line with the whole training process of the model.

$$T\_function(x) = 3 \times Sigmoid(10 \times (x - 0.5)) \quad (2)$$

3. Selection of λ: In knowledge distillation, when the student model is at distinct training phases, the combined weights of the teacher model and the true label are likewise diverse. In the early stage of training, transfer learning and real labels are mainly mixed for learning and fitting, which guarantees that the high accuracy based on the pretrained model can be acquired in the whole model training. Nevertheless, with the deepening of the training epochs, since the student model has reached a successful convergence situation through self-study, the accuracy cannot be further improved. Therefore, by increasing the proportion of the teacher model on and on, the student model learns the dark knowledge distribution from the teacher model prediction distribution, thereby improving the model performance. The change in λ is shown in Figure 8 and Formula (3), where x indicates the training times. By this function, λ decreases in an S-curve with the deepening of the training process.

$$\lambda\_function(x) = 1 - Sigmoid(10 \times (x - 0.5)) \quad (3)$$

4. Calculation of the loss:
    (a) In the loss calculation of the soft label and the student model, the Jensen–Shannon divergence with symmetry is utilized to replace the cross-entropy

loss as the similarity measure metrics of two prediction distributions, as shown in Formula (5).

(b) In view of the one-hot characteristic of the hard label, the loss between the hard label and the student model is still computed via the cross-entropy loss, as shown in Formula (6).

(c) The total loss is derived from Formulas (7)–(12).

$$KL(P,Q) = \sum p(x) \log \frac{p(x)}{q(x)} \tag{4}$$

$$JS(P_1, P_2) = \frac{1}{2} KL(P_1, \frac{P_1 + P_2}{2}) + \frac{1}{2} KL(P_2, \frac{P_1 + P_2}{2}) \tag{5}$$

$$CE(Lable, Predict) = -\sum_{j=1}^{N} Label_j \cdot \log(Predict_j) \tag{6}$$

$$T = T\_function(step/epochs) \tag{7}$$

$$\lambda = \lambda\_funciton(step/epochs) \tag{8}$$

$$Out_{teacher} = Max\_out(Out_{teacher1}, Out_{teacher2}) \tag{9}$$

$$L_{soft} = JS(Out_{teacher}/T, Out_{student}/T) \tag{10}$$

$$L_{hard} = CE(Hard_{label}, Out_{student}) \tag{11}$$

$$Loss = (1 - \lambda) * L_{soft} + \lambda * L_{hard} \tag{12}$$

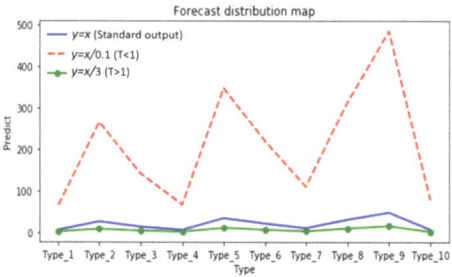

Figure 7. Distillation distribution of model output probabilities for different T cases.

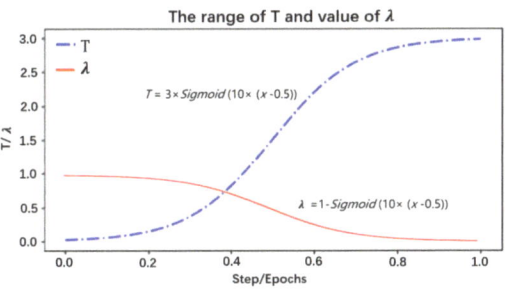

Figure 8. The distribution of T/λ.

In Formula (4), KL represents the relative entropy formula, P and Q represent two probability distributions, respectively, and $p(x)$ and $q(x)$ are the specific probabilities in a certain dimension. In Formula (5), JS is the loss function for calculating the two probability distributions, P_1 and P_2 are two distinct probability distributions, and the function returns the JS divergence loss between these two distributions. In Formula (6), CE is the cross-entropy function of two probability distributions, Label and Predict are usually the prediction probability distributions given by the true label and the student model, respectively. In Formulas (7)–(12), the above formulae are called. T represents the temperature metric during distillation, λ represents the combination weight between two different losses, where the quotient of step (current training number) and epochs (total training number) is used as the training progress index. Outx reflects the probability distribution given by model x. L represents the two losses computed by different calculation methods, and eventually they are merged by the λ weight to synthesize the final loss.

The improved dual-teacher supervised decay (DTSD) has two teacher models selected as the complicated and high-accuracy models ResNet50_vd and DenseNet121, respectively. The student model chosen for this paper was MobileNet_v3_Small, which is known for its lightweight design. The model's structure is depicted in Figure 9.

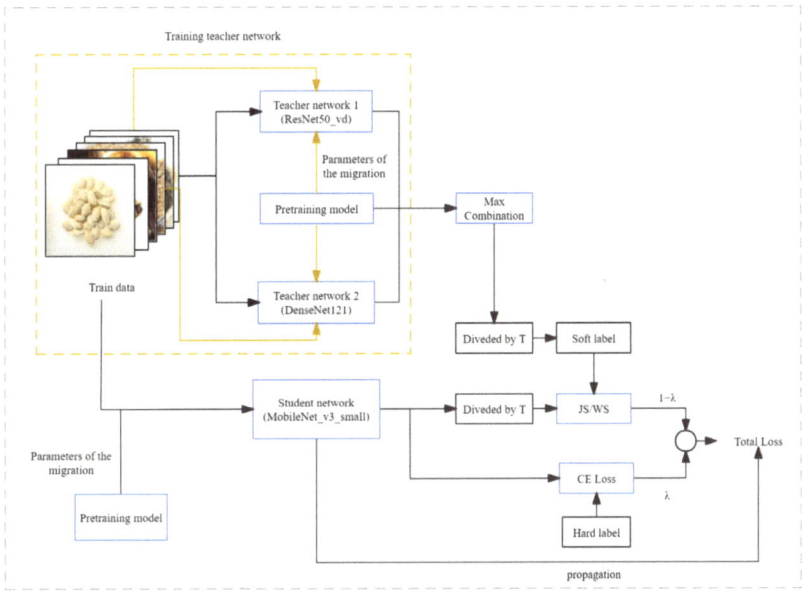

Figure 9. Flow chart of DTSD distillation.

3. Results and Discussion

3.1. Experimental Setting

The primary experimental setup for this paper consisted of: (1) A desktop computer operating on Windows 10, equipped with an Intel Xeon E5-2630 v4 processor at 2.2 GHz, 64 GB of RAM, a 1.5 TB mechanical hard drive, and an NVIDIA Tesla P4 graphics card with 32 GB of video memory. This setup leveraged GPU acceleration for computations and was configured with the Paddle deep learning framework within a programming language environment. (2) Baidu's open platform, Ai Studio, featuring an NVIDIA Tesla V100 graphics card, utilizing GPU computing power and acceleration, with the Paddle deep learning framework also implemented in a programming language environment.

3.2. Experimental Design

First, we pretrained the ResNet50_vd, DenseNet121, and MobileNet_v3_Small network models on the public dataset ImageNet2012 [29], mainly fine-tuning the models to verify that pretraining could improve the model performance. Subsequently, we trained the student model MobileNet_v3_Small under the guidance of the dual teacher models ResNet50_vd and DenseNet121, to validate the DTSD model and perform optimal parameter tuning. Furthermore, to verify the trained MobileNet_v3_Small_DTSD, we compared it with the similar MobileNet_v3_Small and other classic models in comparative experiments. To minimize the randomness of training, the data presented in the table are average values obtained from multiple measurements. Moreover, the dataset used was an augmented dataset of 10 types of Chinese herbal medicines.

3.3. Experiments on Boosting Training with Pretrained Models

To verify the impact of the pretrained models on the training process and the final model performance, comparative experiments were conducted for the two teacher models, ResNet50_vd and DenseNet121, as well as the student model MobileNet_v3_Small, using pretrained models. All the mentioned models were first pretrained on the public dataset ImageNet2012, followed by transfer learning on a Chinese herbal medicine dataset.

Taking MobileNet_v3_Small as an example, the comparison of pretrained models is illustrated in Figure 10. The loss values of the models using pretraining converged more quickly than those of the normally trained models (as shown in Figure 10a), reaching a desirable convergence state in the early stages of training. Consequently, under the same number of training epochs, models trained with pretraining exhibited higher performance. Furthermore, the accuracy (Acc) of the models using pretrained models maintained a higher precision compared to those without pretraining, achieving high performance from the early stages of training (as depicted in Figure 10b).

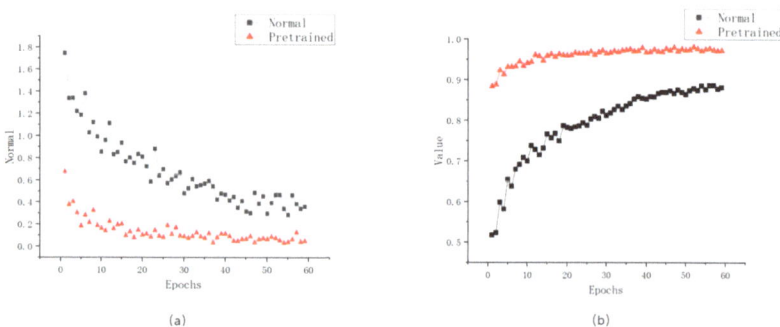

Figure 10. Comparison of the effects of pretrained models on model training: (**a**) loss value figure; (**b**) accurate figure.

As shown in Table 1, the final accuracies of the three models that underwent pretraining through transfer learning all showed improvements, with increases of 12.40%, 7.35%, and 9.5%, respectively. Therefore, it can be concluded that pretrained models played a significant role in enhancing the performance of the models. Moreover, with the demonstration of the superiority of the DTSD technique, all subsequent experiments utilized transfer learning for pretraining.

3.4. Improved Model Verification

The above experiments provide evidence that the use of a pretrained model contributes to an improvement in the performance of the model. To further verify the effect of the DTSD model on MobileNet_v3_Small, the high-precision teacher models ResNet50_vd (98.9%) and DenseNet121 (98.7%) were trained on the 10 Chinese herbal medicine augmented datasets in advance.

Table 1. Accuracy of the models with of pretraining.

Network	Pretraining Model	Accuracy (%)
ResNet50_vd	×	86.10
	√	98.50
DenseNet121	×	91.05
	√	98.40
MobileNet_v3_Small	×	88.35
	√	97.85

"×" indicates that the model has not been pre-trained. "√" indicates that the model has been pre-trained.

In order to obtain optimal performance for MobileNet_v3_Small_DTSD, three learning rate decline strategies were implemented while applying the improved knowledge distillation DTSD. These strategies were exponential warmup, piecewise and cosine. When the other experimental parameters were the same, the DTSD distillation model using the exponential warmup learning rate decline strategy had the highest accuracy of 98.60%. Consequently, as shown in Table 2, it can be inferred that the learning rate decline strategy of exponential warmup conferred advantages to the DTSD training model.

Table 2. Comparison of parameter combination results of MobileNet_v3_Small_DTSD.

Student Model	Teacher Model	Acc Of Teacher Model (%)	Learning Rate Decline Strategy	Accuracy (%)
MobileNet_v3_Small	ResNet50_vd	98.90	Piecewise	97.80
	DensNet121	98.70		
MobileNet_v3_Small	ReNet50_vd	98.90	Cosine	98.15
	DenseNet121	98.70		
MobileNet_v3_Small	ResNet50_vd	98.90	Exponential warmup	98.60
	DenseNet121	98.70		

The parameter and metric changes of the whole training process of MobileNet_v3_Small_DTSD are shown in Figure 11, where variables 11a, 11b, and 11c denote the changes in total loss, soft loss and hard loss, respectively. The total loss is the overall loss of the model during the training process, which is usually a combination of multiple loss functions. When training a neural network, there are usually multiple tasks or multiple loss metrics, and each loss function corresponds to one task or one metric. The total loss is the weighted sum or average of these loss functions and is used to measure the performance of the entire model. Soft loss is a technique used in training, mainly to help the model learn better. Soft loss is usually achieved by introducing some extra penalty or regularization terms in the loss function, which can help the model generalize better to new data and avoid overfitting. The introduction of a soft loss can help to adjust the learning direction of the model to better fit the training data. Hard loss usually refers to the loss calculated in the inference stage of the model, which is the performance of the model on the test data. Hard loss is the difference between the true label and the predicted label of the model, which is used to measure the prediction accuracy or performance of the model. During training, the model usually adjusts its own parameters by optimizing the total loss to minimize the hard loss. It is evident from the figure that the model achieved satisfactory convergence during the initial phase of training. In Figure 11d, the λ value (lambda) is usually used in regularization terms (e.g., L1 regularization, L2 regularization) to balance the model's fitting effect with the effect of the regularization terms. The λ-value change curve shows the effect of different λ values on the model's performance during the training process. λ decays in an inverted S-shape, which means that the student model mainly performs distribution fitting with the hard label through transfer learning in the early stage, and gradually shifts the learning center to the soft label of the teacher model with the deepening of the experiment. The T value (temperature) is usually used in temperature-regulated soft label methods to smooth the label distribution to improve the training of the model. The T-value variation curve shows the effect of different T values on the performance of the model during the training process. However, the temperature T depicted in Figure 11e exhibits a contrasting situation. During the initial phase, the correct label part of the teacher model is mainly enlarged, so that the student model can quickly fit. In the subsequent phase, the proportion of dark knowledge is gradually expanded, because the accuracy of the student model can be further enhanced. The accuracy change curve shows how the accuracy of the model changes during the training process. The accuracy on the training set and the validation set are usually treated as two separate parts of the curve. This curve can be used to observe whether the model is overfitting or underfitting. Figure 11f reflects the evaluation accuracy of MobileNet_v3_Small_DTSD. Once the model quickly reaches good performance in the early stage, it mainly focuses on acquiring the dark knowledge from the teacher model to achieve higher accuracy in the subsequent stage.

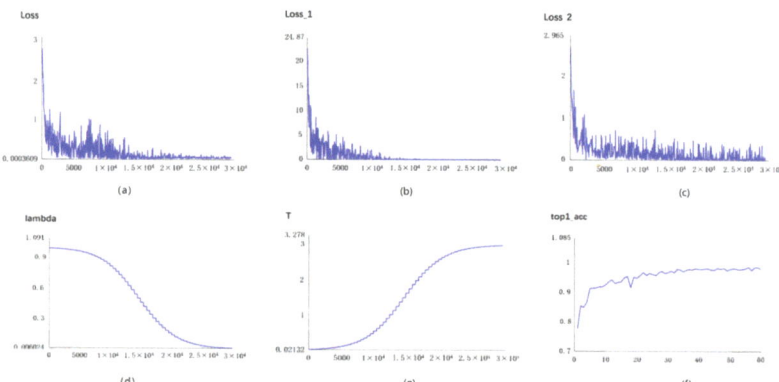

Figure 11. Metrics of mobileNet_v3_Small_DTSD changes in training process: (**a**) total loss; (**b**) soft loss; (**c**) hard loss; (**d**) λ value change; (**e**) T value change; (**f**) accuracy change curve.

3.5. Comparative Experiments with Similar Models

To prove the superiority of the best MobileNet_v3_Small_DTSD model, which was achieved by the adjustment of the learning rate decline strategy, a comparative analysis was conducted with MobileNet_v3_Small using various techniques. The comparison models were mainly as follows: (1) MobileNet_v3_Small without a transfer learning pretrained model; (2) MobileNet_v3_Small_Pre with pretrained models; (3) MobileNet_v3_Small_SSLD trained with SSLD (semi-supervised label knowledge distillation) technique.

According to the data presented in Table 3, it can be observed that the improved DTSD technique achieved an accuracy of 98.60% under identical training parameters. This accuracy was notably 11.15% higher than that of the original training model. Furthermore, the improved DTSD technique outperformed the SSLD technique, which also incorporates knowledge distillation, by 1.50%. Furthermore, the DTSD technique surpasses the PRE model, which uses transfer learning, by 1.35%.

Table 3. Comparison results with MobileNet_v3_Small.

Learning Rate Decline Strategy	Batch Size	Accuracy
MobileNet_v3_Small	16	86.45
MobileNet_v3_Small_PRE	16	97.25
MobileNet_v3_Small_SSLD	16	97.10
MobileNet_v3_Small_DTSD	16	98.60

As shown in Figure 12, the normally trained model in Figure 12a performed poorly in terms of both aspects (loss and accuracy), while the improved DTSD model maintained the same convergence state as the other two models PRE and SSLD in terms of loss. In Figure 12b, the circular green line (using the DTSD technique) is compared with the square red line (using transfer learning). During the initial stage of training, the performance of the DTSD model is inferior to that of the transfer learning model. The ongoing decay of λ causes the student model to shift the training focus to the soft label of the teacher model, and the constant growth of temperature T expands the proportion of dark knowledge, which makes the student model learn more parameters. These two factors make an obvious intersection point appear in the two graphs. Furthermore, DTSD successfully achieved a leadership position after coming from behind. The confusion matrix diagram in Figure 12c is the performance of MobileNet_v3_Small_DTSD on the test dataset under the training parameters. Eventually, the model performed well. Therefore, the utilization of the DTSD technique inside the same model should help to break through the limitation of the model accuracy.

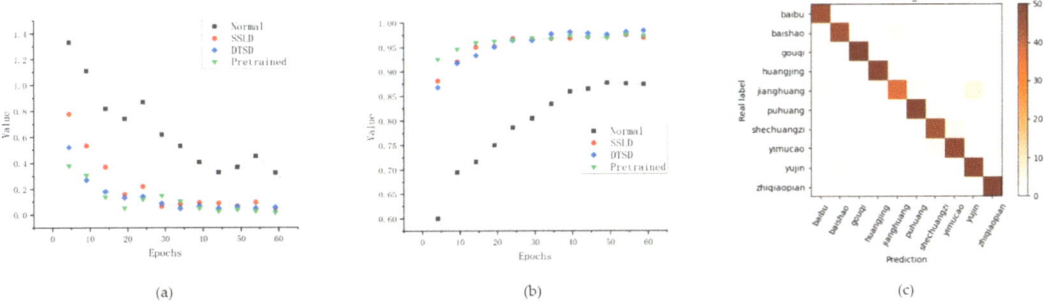

Figure 12. Performance comparison of MobileNet series: (a) loss; (b) Accuracy; (c) Confusion matrix.

3.6. Experimental Comparisons with Other Models

To further verify the performance of the DTSD technique, it was compared with other mainstream models including EfficientNetB1, Xception65, ResNet50_vd, DenseNet121, and others. The main evaluation criteria encompassed accuracy, model volume, and prediction cost, where the prediction cost was the average time of predicting 500 test herbal images. DenseNet121 is a model in the DenseNet family. Dense connections in DenseNet help alleviate gradient sparsity, making the model easier to train and improving its generalization ability. Dense connections allow features to be passed through shorter paths, increasing the efficiency of information transfer and reducing information loss. ResNet50_vd is relatively deep, with 50 layers, which allows the model to learn higher-level abstract features and improve the ability to capture and represent complex patterns. Xception65 utilizes a structure of depth-separable convolution, which divides standard convolution into two steps: deep convolution and point-by-point convolution. This structure helps to reduce the number of parameters and improve the computational efficiency of the model.

EfficientNetB1 employs a compound coefficient (composite coefficient) approach to design an efficient model structure by scaling the depth, width, and resolution of the network in a balanced way. This structure can achieve better performance with certain computational resources.

The training parameter values for the experiments in this paper are set as shown in Table 4, Batch size is 16, Basic learning rate is 0.0037, Learning rate decline strategy used is ExponentialWarmup, Epoches is 60, and pre-training of the model is performed.

Table 4. Setting of training parameters.

Parameter	Batch Size	Basic Learning Rate	Learning Rate Decline Strategy	Epochs	Pretraining Model
Parameter value	16	0.0037	Warmup	60	√

"√" indicates that the model has been pre-trained.

As shown in Table 5, when comparing different models horizontally, it is evident that MobileNet_v3_Small, which uses DTSD technology for knowledge distillation, achieved an accuracy of 98.60%, ranking second in terms of accuracy. However, its model size (10 MB) and prediction cost (0.0172 s) were optimal among the five models. The feasibility of the DTSD technique was thus proved.

Table 5. Comparison of results between MobileNet_v3_Small_DTSD and other mainstream models.

Network	Accuracy (%)	Model Volume (MB)	Prediction Time (s)
DenseNet121	97.10	29.30	0.0186
ResNet50_vd	97.35	90.90	0.0237
Xception65	97.65	131	0.0198
EfficientNetB1	98.95	27.5	0.0233
MobileNet_v3_Small_DTSD	98.60	10	0.0172

3.7. Application

To validate the practical application of the model, we developed a mobile recognition app based on it. As shown in Figure 13, the app's main features include an input image, a historical search, and text search capabilities. After selecting the image search function, users can upload relevant images of Chinese herbs. The app processes these images through cropping and utilizes the corresponding model and a backend database of Chinese herbs to return information about the herb. The example result, shown in Figure 14, indicates that the tested image has been classified as "aiye" with a matching true label. The probability of correct classification is 99.76%. Practical validation confirmed that the MobileNet_v3_DTSD model, trained using the dual-teacher adaptive-decay approach based on improved knowledge distillation and data augmentation, maintained its lightweight and rapid processing characteristics while also exhibiting robustness and high accuracy. Future efforts will focus on further optimizing these improvements to enhance the model's performance in real-life applications.

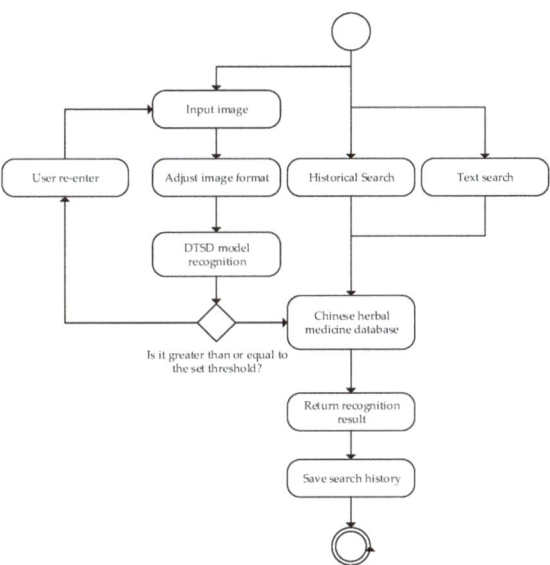

Figure 13. Process design of our Chinese herbal medicine identification system.

Figure 14. The system outputs the result of the test image.

4. Conclusions

In this paper, to realize lightweight Chinese herbal medicine image recognition on a mobile terminal, a Chinese herbal medicine recognition model with dual-teacher supervised decay based on knowledge distillation was proposed. By improving the single-teacher model to a dual-teacher model, the output soft label, adaptive decay parameters, and dynamic combination loss of the teacher model, it was applied to the lightweight model network MobileNet_v3, and finally deployed into a mobile application. The experimental results indicated that the mean classification accuracy of a set of 10 Chinese herbal medicine images was 98.60%. Moreover, the average time taken to identify a single image was 0.0172 s, and the model size was 10 MB. Upon successful deployment of the application, it demonstrated the capability to fulfill the speed and accuracy requirements of real-life scenarios, hence offering valuable technical reference for mobile phone applications. Despite the remarkable results achieved in this paper in lightweight herbal image recognition, there are still some shortcomings that need further consideration and improvement. The robustness of the model in dealing with real-world challenges such as complex scenes and lighting changes still needs to be improved to ensure high accuracy in a variety of environments. The next step is to expand and diversify the dataset to improve the model's adaptability to different herbal species and environmental conditions. Secondly, techniques such as adversarial training should be introduced to enhance the robustness of the model against noise and interference.

Author Contributions: Conceptualization, L.Z.; methodology, W.L. and L.Z.; software, J.Y.; validation, K.X.; supervision, L.L.; data curation, W.L. and J.Y.; formal analysis, L.Z.; investigation, W.L. and L.Z.; resources, L.L. and K.X.; writing, W.L. and L.Z.; review and editing, L.Z.; visualization, W.L. and L.Z. All authors have read and agreed to the published version of the manuscript.

Funding: This work is funded by Special Project on Regional Collaborative Innovation in Xinjiang Uygur Autonomous Region (Science and Technology Aid Program) under grant number 2022E02035, Hubei Provincial Administration of Traditional Chinese Medicine Research Project on Traditional Chinese Medicine under grant number ZY2023M064, Wuhan knowledge innovation special Dawn project under grant number 2023010201020465, and National innovation and entrepreneurship training program for college students under grant number 202310524017.

Institutional Review Board Statement: Not applicable.

Informed Consent Statement: Not applicable.

Data Availability Statement: The data presented in this study are available on request from the corresponding author.

Conflicts of Interest: The authors declare no conflict of interest.

References

1. Huang, F.; Yu, L.; Shen, T. Research and implementation of Chinese herbal medicine plant image classification based on alexnet deep learning model. *J. Qilu Univ. Technol.* **2020**, *34*, 44–49.
2. Gao, H.; Gao, X.; Feng, Q. Natural grassland plant species identification method based on deep learning. *Grassl. Sci.* **2020**, *37*, 1931–1939.
3. Zhang, W.; Zhang, Q.; Pan, J. Classification and recognition of Chinese herbal medicine based on deep learning. *Smart Health* **2020**, *6*, 1–4+13.
4. Wang, Y.; Sun, W.; Zhou, X. Research on Chinese herbal medicine plant image recognition method based on deep learning. *Inf. Tradit. Chin. Med.* **2020**, *37*, 21–25.
5. Hu, K. Research and Implementation of Fritillaria Classification Algorithm Based on Deep Learning. Master's Thesis, Chengdu University, Chengdu, China, 2020.
6. Cubuk, E.D.; Zoph, B.; Shlens, J.; Le, Q.V. Randaugment: Practical automated data augmentation with a reduced search space. In Proceedings of the IEEE/CVF Conference on Computer Vision and Pattern Recognition Workshops, Seattle, WA, USA, 14–19 June 2020; pp. 702–703.
7. He, K.; Zhang, X.; Ren, S. Deep residual learning for image recognition. In Proceedings of the IEEE Conference on Computer Vision and Pattern Recognition, Las Vegas, NV, USA, 26 June–1 July 2016; pp. 770–778.
8. Alzubaidi, L.; Zhang, J.; Humaidi, A.J.; Al-Dujaili, A.; Duan, Y.; Al-Shamma, O.; Santamaría, J.; Fadhel, M.A.; Al-Amidie, M.; Farhan, L. Review of deep learning: Concepts, CNN architectures, challenges, applications, future directions. *J. Big Data* **2021**, *8*, 53. [CrossRef]
9. He, T.; Zhang, Z.; Zhang, H. Bag of Tricks for Image Classification with Convolutional Neural Networks. In Proceedings of the 2019 IEEE/CVF Conference on Computer Vision and Pattern Recognition (CVPR), Long Beach, CA, USA, 16–20 June 2019.
10. Huang, G.; Geoff, P.; Laurens, V.; Kilian, W. Convolutional Networks with Dense Connectivity. *IEEE Trans. Pattern Anal. Mach. Intell.* **2019**, *44*, 8704–8716. [CrossRef]
11. Guo, X.; Fan, T.; Shu, X. Tomato leaf diseases recognition based on improved multi-scale AlexNet. *Trans. Chin. Soc. Agric. Eng.* **2019**, *35*, 162–169.
12. Fang, Z.; Ren, J.; Marshall, S.; Zhao, H.; Wang, S.; Li, X. Topological optimization of the DenseNet with pretrained-weights inheritance and genetic channel selection. *Pattern Recognit.* **2021**, *109*, 107608. [CrossRef]
13. Koonce, B. MobileNetV3. In *Convolutional Neural Networks with Swift for Tensorflow*; Apress: Berkeley, CA, USA, 2021; Volume 1, pp. 125–144. [CrossRef]
14. Rosebrock, A. *Deep Learning for Computer Vision with Python-Starter Bundle*; PyImageSearch: Baltimore, MD, USA, 2017; pp. 189–190.
15. Gao, A.; Geng, A.; Song, Y.; Ren, L.; Zhang, Y.; Han, X. Detection of maize leaf diseases using improved MobileNet V3-small. *Int. J. Agric. Biol. Eng.* **2023**, *16*, 225–232. [CrossRef]
16. Chen, W.; Tong, J.; He, R. An easy method for identifying 315 categories of commonly-used Chinese herbal medicines based on automated image recognition using AutoML platforms. *Inform. Med. Unlocked* **2021**, *25*, 100607. [CrossRef]

17. Chen, J.; Wang, W.; Zhang, D. Attention embedded lightweight network for maize disease recognition. *Plant Pathol.* **2021**, *70*, 630–642. [CrossRef]
18. Bi, S.; Gao, F.; Chen, J. Detection method of citrus based on deep convolution neural network. *Trans. Chin. Soc. Agric. Mach.* **2019**, *50*, 181–186.
19. Hinton, G.; Vinyals, O.; Dean, J. Distilling the knowledge in a neural network. *arXiv* **2015**, arXiv:1503.02531.
20. Zhuang, F.; Luo, P.; He, Q. Survey on transfer learning research. *J. Softw.* **2015**, *26*, 26–39.
21. Li, Y.; Hao, Z.; Lei, H. Survey of convolutional neural network. *J. Comput. Appl.* **2016**, *36*, 2508–2515, 2565.
22. Ma, J.; Du, K.; Zheng, F.; Zhang, L.; Sun, Z. Disease recognition system for greenhouse cucumbers based on deep convolutional neural network. *Trans. Chin. Soc. Agric. Eng. (Trans. CSAE)* **2018**, *34*, 186–192.
23. Jabir, B.; Falih, N. Deep learning-based decision support system for weeds detection in wheat fields. *Int. J. Electr. Comput. Eng.* **2022**, *12*, 816. [CrossRef]
24. Khan, S.; Rahmani, H.; Shah, S.A. A guide to convolutional neural networks for computer vision. *Synth. Lect. Comput. Vis.* **2018**, *8*, 1–207.
25. Li, Z.; Guo, R.; Li, M. A review of computer vision technologies for plant phenotyping. *Comput. Electron. Agric.* **2020**, *176*, 105672. [CrossRef]
26. Li, K.; Zou, C.; Bu, S. Multi-modal feature fusion for geographic image annotation. *Pattern Recognit.* **2017**, *73*, 1–14. [CrossRef]
27. Kolhar, S.; Jagtap, J. Plant trait estimation and classification studies in plant phenotyping using machine vision—A review. *Inf. Process. Agric.* **2023**, *10*, 114–135. [CrossRef]
28. Cao, X.; Li, R.; Wen, L. Deep multiple feature fusion for hyperspectral image classification. *IEEE J. Sel. Top. Appl. Earth Obs. Remote Sens.* **2018**, *11*, 3880–3891. [CrossRef]
29. Deng, J.; Dong, W.; Socher, R.; Li, L.J.; Li, K.; Fei-Fei, L. Imagenet: A large-scale hierarchical image database. In Proceedings of the 2009 IEEE Conference on Computer Vision and Pattern Recognition, Miami, FL, USA, 20–25 June 2009; pp. 248–255.

Disclaimer/Publisher's Note: The statements, opinions and data contained in all publications are solely those of the individual author(s) and contributor(s) and not of MDPI and/or the editor(s). MDPI and/or the editor(s) disclaim responsibility for any injury to people or property resulting from any ideas, methods, instructions or products referred to in the content.

Article

Lizard Body Temperature Acquisition and Lizard Recognition Using Artificial Intelligence

Ana L. Afonso [1], Gil Lopes [2] and A. Fernando Ribeiro [3],*

[1] Mechanical Engineering Department, University of Minho, 4800-058 Guimarães, Portugal; ana.leonor.afonso@gmail.com
[2] LIACC, University of Maia, 4475-690 Maia, Portugal; alopes@umaia.pt
[3] Centro ALGORITMI, University of Minho, 4800-058 Guimarães, Portugal
* Correspondence: fernando@dei.uminho.pt

Abstract: The acquisition of the body temperature of animals kept in captivity in biology laboratories is crucial for several studies in the field of animal biology. Traditionally, the acquisition process was carried out manually, which does not guarantee much accuracy or consistency in the acquired data and was painful for the animal. The process was then switched to a semi-manual process using a thermal camera, but it still involved manually clicking on each part of the animal's body every 20 s of the video to obtain temperature values, making it a time-consuming, non-automatic, and difficult process. This project aims to automate this acquisition process through the automatic recognition of parts of a lizard's body, reading the temperature in these parts based on a video taken with two cameras simultaneously: an RGB camera and a thermal camera. The first camera detects the location of the lizard's various body parts using artificial intelligence techniques, and the second camera allows reading of the respective temperature of each part. Due to the lack of lizard datasets, either in the biology laboratory or online, a dataset had to be created from scratch, containing the identification of the lizard and six of its body parts. YOLOv5 was used to detect the lizard and its body parts in RGB images, achieving a precision of 90.00% and a recall of 98.80%. After initial calibration, the RGB and thermal camera images are properly localised, making it possible to know the lizard's position, even when the lizard is at the same temperature as its surrounding environment, through a coordinate conversion from the RGB image to the thermal image. The thermal image has a colour temperature scale with the respective maximum and minimum temperature values, which is used to read each pixel of the thermal image, thus allowing the correct temperature to be read in each part of the lizard.

Keywords: artificial intelligence; body temperature acquisition; computer vision; lizards; object detection; YOLO

Citation: Afonso, A.L.; Lopes, G.; Ribeiro, A.F. Lizard Body Temperature Acquisition and Lizard Recognition Using Artificial Intelligence. *Sensors* **2024**, *24*, 4135. https://doi.org/10.3390/s24134135

Academic Editors: Sylvain Girard and Lorenzo Scalise

Received: 17 April 2024
Revised: 17 June 2024
Accepted: 21 June 2024
Published: 26 June 2024

Copyright: © 2024 by the authors. Licensee MDPI, Basel, Switzerland. This article is an open access article distributed under the terms and conditions of the Creative Commons Attribution (CC BY) license (https://creativecommons.org/licenses/by/4.0/).

1. Introduction

Lizards are ectothermic animals, which means that they do not produce enough metabolic heat to maintain their body temperature, having to resort to the use of external heat sources. In biology laboratories, measuring body temperature in lizards can provide relevant information to biologists. Traditionally, the body temperature of lizards is measured using a contact thermometer. This method is extremely invasive and painful for the animal. Also, it is impossible to obtain the temperature from different lizard body parts.

Thus, a new method emerged that consisted of filming the lizard kept in captivity with a Forward-Looking Infrared Camera (FLIR), also known as a thermal camera. Later, using specialised software for this type of camera (in this case, FLIR Tools), temperatures of the body parts of the animal under study were obtained by manually clicking on each of the body parts in the video and recording their value. This process is carried out every 20 s of the video. The entire process does not occur in real time. There is also a possible loss of information regarding changes in the lizard's body temperature between each measurement process.

This new method proved to be advantageous for the animal, as it is not an invasive method. However, the entire procedure of obtaining temperature values in different parts of the body follows a time-consuming, difficult, monotonous, and not very rigorous method (for example, due to potential inaccuracies when manually clicking on parts of the lizard's body). Therefore, it is desirable to overcome the adversities presented by this new measurement method. To this end, a different approach is proposed. By using artificial intelligence and a combination of an RGB camera and a thermal camera, it is possible to detect the lizard and its body parts automatically and obtain the respective desired body temperature values quickly and coherently. This system can be applied to images of previously recorded videos. In both cases, the final values are automatically saved into a text file. In addition, a greater flow of data allows more detailed maintenance of the lizards, and the time that was spent by biologists in manually obtaining measurements can be used for other purposes.

The method presented in this paper provides biologists with a faster and non-intrusive way to measure the temperatures of lizards placed in a box in a controlled laboratory setting. In these controlled environments, different temperatures can be applied to various sections of a box, allowing researchers to monitor the temperature preferences of lizards as they choose where to move to get warmer. This capability is crucial for studying the behavioural responses of lizards to temperature changes, enabling detailed observations of their thermoregulation strategies.

The significance of this research lies in its contribution to more efficient and humane methods of monitoring lizard body temperatures, which are essential for understanding their behaviour and physiological needs. By automating the temperature acquisition process, our method reduces the stress and potential harm to the animals, providing a more ethical approach to studying their behaviour. Additionally, the insights gained from such studies can inform broader ecological research and conservation efforts, particularly in understanding how lizards might adapt to changing environmental conditions.

1.1. Artificial Intelligence

Artificial intelligence (AI) speeds up human tasks with a guaranteed level of precision and accuracy. With the emergence of new algorithms, the progress in computing power and storage, and the accessibility to a vast quantity of data, AI suffered notable breakthroughs and is already being applied to numerous fields, such as the field of biology.

Researchers are regularly confronted with complex and time-consuming problems. Thus, AI emerges to offer solutions to these problems and promote innovation in laboratories. Biological research and artificial intelligence are becoming increasingly related. Developing tools for the analysis and interpretation of vast amounts of data is one of the most significant uses of artificial intelligence in biology. AI is already present in a variety of biology research works, such as:

- Protein 3D structure prediction: AI helps predict the three-dimensional structure of proteins and subsequently understand their function, enabling the development of new specialised drugs [1].
- Drug development: AI helps speed up drug development [2].
- Conservation and wildlife tracking and monitoring: AI helps protect wildlife and natural resources and helps automate wildlife tracking and monitoring [3].

1.2. Machine Learning and Deep Learning

Machine learning (ML) is a subset of AI that aims to give a computer the ability to learn from experience, using data instead of being explicitly programmed. An ML model is the output generated after training the ML algorithm with data [4]. Supervised learning (SL) is one of the main ML approaches, where a set of labelled training data, sample data (input), and associated target responses (output) are provided to the algorithm for it to learn a function that maps an input to an output, and a predictive model is created [5]. This model is then used to make predictions on never-seen samples.

The SL algorithm needs to have the capability of generalising from training data to unseen samples. The model testing should not be carried out on the training data because it gives the false impression of success; instead, it should be carried out on new examples.

Overfitting and underfitting are two common problems in ML. Overfitting occurs when the model can predict correctly all the labels of the training data but does not generalise well to unseen data; in this case, the model has a low bias and a high variance (high complexity model) [6]. On the other hand, underfitting occurs when the model cannot generalise well to unseen data and makes mistakes trying to predict the labels of the training data; in this case, the model has a high bias and a low variance (low complexity model) [6]. Overfitting and underfitting can occur due to several reasons, such as an inadequate size and quality of the training dataset.

Bias represents how closely the average prediction is to the true value, and variance quantifies how much, on average, predictions vary for different sets of training data [7]. To obtain the ideal model, it is essential to find the optimal balance between bias and variance.

Deep learning (DL) is a subset of ML based on neural networks. Neural networks are inspired by the structure of the human brain and the way it works and consist of three types of layers: the input layer, the hidden layer, and the output layer. An Artificial Neural Network (ANN) is a type of neural network with one or two hidden layers. A Convolutional Neural Network (CNN) is a type of ANN.

CNNs were first introduced with the design of LeNet-5 by Yann LeCun et al. [8] The introduction of Graphics Processing Units (GPUs), faster Central Processing Units (CPUs), and the increasing amount of training data available have driven the development of new architectures, such as AlexNet [9], ZFNet [10], GoogLeNet [11], ResNet [12], VGGNet [13], and EfficientNet [14], as well as several object detectors based on CNNs, including the R-CNN (Region-Based Convolutional Neural Networks) family of two-stage detection networks and the one-stage detection networks SSD (Single Shot MultiBox Detector) [15], RetinaNet [16], EfficientDet [17], and the YOLO (You Only Look Once) family.

Hao et al. [18] proposed a lightweight detection algorithm based on the one-stage detection network SSD for sheep facial identification, achieving a mAP of 83.47% and a detection speed of 68.53 frames per second. Jia et al. [19] developed a marine organism object detection model also based on a one-stage detection network, the improved EfficientDet, obtaining a mAP of 91.67% and a processing speed of 37.5 frames per second.

Roy et al. [20] presented a comparative study between the one-stage detection networks RetinaNet, SSD, YOLOv3, and YOLOv4 and the two-stage detection networks Mask R-CNN and Faster R-CNN for wildlife detection. The findings indicated that YOLO variants outperformed the other networks, with the one-stage detection network YOLOv4 achieving the best performance (mAP of 91.29%). Hu et al. [21] conducted a study utilising Detectron2, RetinaNet, YOLOv4, and YOLOv5 models to determine the count of cattle in satellite images, with YOLOv5 achieving the best results, producing an average precision of 91.60% and a recall of 91.20%. Both studies by Roy et al. [20] and Hu et al. [21] demonstrate the effectiveness of the YOLO family in animal detection.

Jubayer et al. [22] found that the overall performance of YOLOv5 in detecting mould on food surfaces was superior to that of YOLOv4 and YOLOv3, achieving an average precision of 99.6%. Long et al. [23] developed a system for fish detection, where YOLOv5 also obtained the highest mAP value of 95.95%, superior to YOLOv3 and YOLOv4. Ahmad et al. [24] conducted a study comparing the performance of YOLO-Lite, YOLOv3, YOLOR, and YOLOv5 in identifying insect pests, with YOLOv5 emerging once more as the most successful, achieving an average precision of 98.3%.

1.3. Current Research Status

The automated detection of animals and the extraction of body temperature values play critical roles in various domains within animal studies.

Advances in deep learning have stimulated the growth of studies focused on the automatic detection of animals for various purposes, such as forest wildlife monitoring and conservation [25], agriculture and farming [26,27], and species identification and classification [28,29]. While most studies on automatic animal detection predominantly focus on mammals and birds, studies addressing reptiles, particularly lizards, are relatively scarce. Aota et al. [30] addressed this gap by developing a deep neural network-based system for detecting the invasive lizard species *Anolis carolinensis* in drone images. This study aims to contribute to an effective and efficient approach to conserving ecosystems, as this invasive species threatens the native insect population of the Ogasawara Islands in Japan.

The body temperature of an animal is a crucial indicator of its health and well-being. However, traditional methods for obtaining these values are challenging. Consequently, there has been a notable increase in studies dedicated to developing automated methods for temperature extraction in animals. A substantial portion of these studies focuses on obtaining temperature data to monitor and assess the health status of pigs and cows [31,32]. Conversely, there is a notable scarcity of studies concerning the automated extraction of body temperature in lizards.

This paper addresses this research gap by developing a system capable of automatically detecting lizards and their body parts using YOLOv5s, followed by the automatic and contactless extraction of temperature values from the detected parts. This system allows biologists to easily obtain valuable data on the body temperature of lizards to use in their research without causing pain or stress to the animal. Karameta et al. [33] obtained the body temperature of insular agamid lizards by inserting a type K thermocouple directly into the animal's cloaca to study how seasonality impacts the thermal biology of an island population of lizards, providing insights into their survival strategies and potential adaptations to future environmental changes. The use of a non-invasive (contactless) and automatic system to extract these temperature values, such as the one developed in this paper, would have been a huge advantage in this study.

Furthermore, the system developed in this article offers the potential to be adapted and adjusted to extract the body temperature of various species of lizards and other reptiles.

2. Methodologies

This work presents a system capable of detecting the entire lizard and six pre-defined parts of its body (snout, head, back, left leg, left palm, and tail) in an image or a video and then displaying and recording the temperature values in these regions. It consists of two main parts: the development of a model for detecting the lizard and its body parts and the acquisition of the temperature values of the detected parts. All algorithms were developed in the Python language and supported with the OpenCV library.

2.1. Detection of Lizard Body Parts

Detection of lizard body parts was developed using the YOLOv5 ML algorithm.

2.1.1. YOLOv5

Object detection is a task focused on localising and classifying objects present in images or videos.

YOLO (You Only Look Once) is a state-of-the-art, real-time object detection algorithm. The fifth version of YOLO (YOLOv5) was proposed in 2020 by the company Ultralytics and is the version selected to use in this project, taking into account the YOLOv5 detection accuracy and detection speed. It is important to note that at the time of the practical development of this paper, YOLOv5 was the current version in use; therefore, later versions were not considered.

The YOLOv5 architecture is composed of three parts: CSP-Darknet53 as the backbone, Spatial Pyramid Pooling Fusion (SPPF) and CSP-PAN (Path Aggregation Network) structures in the neck [34], and the same head as YOLOv3. CSP-Darknet53 is formed by applying a Cross Stage Partial Network (CSPNet) to Darknet-53. The amount of computation may be significantly decreased with CSPNet, and both the inference speed and accuracy can be improved [35]. In the neck, the SPPF is a faster variation of a Spatial Pyramid Pooling (SPP) block. Figure 1 shows the architecture diagram of YOLOv5s.

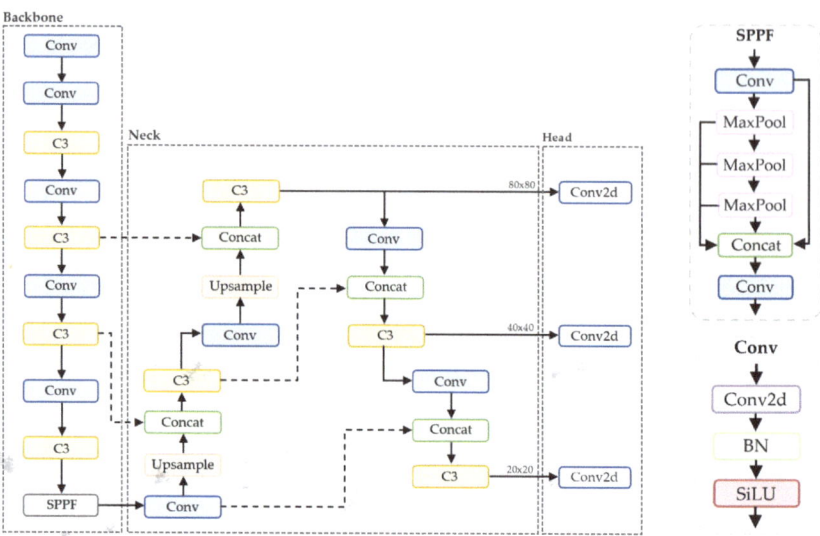

Figure 1. Architecture diagram of YOLOv5s.

Contrary to previous versions, YOLOv5 uses the PyTorch framework instead of the Darknet framework [36]. To reduce overfitting and improve the model's ability to generalise, YOLOv5 uses some data augmentation techniques, such as mosaic augmentation.

YOLOv5 is divided into five different model sizes: YOLOv5n (nano), YOLOv5s (small), YOLOv5m (medium), YOLOv5l (large), and YOLOv5x (extra-large). Larger models contain more parameters, need more memory to train, require larger and well-labelled datasets, and take longer to execute but will generally produce better results. On the other hand, smaller models are faster but may abdicate some accuracy.

To evaluate the performance of a certain object detection model, some metrics are used, such as intersection over union (IoU), confusion matrix, precision (P), recall (R), F1 score, average precision (AP), and mean average precision (mAP).

The intersection over union metric estimates how well a predicted bounding box matches the ground truth bounding box and is given by a ratio between the intersection area (area where the boxes overlap) and the union area (total area of both boxes) of the predicted bounding box with the ground truth bounding box.

A confusion matrix is a table in which the values predicted by the classifier are compared with the ground truth labels. This table is composed of four types of predictions: false positive (FP), false negative (FN), true positive (TP), and true negative (TN).

Precision counts the percentage of predicted positives that are actually positive and is calculated using Equation (1). Recall measures the percentage of positives correctly detected and is calculated using Equation (2). The F1 score combines precision and recall and ranges between 0 and 1. The F1 score is obtained using Equation (3).

$$\text{Precision} = \frac{\text{Correct Predictions}}{\text{Total Predictions}} = \frac{TP}{TP+FP} \qquad (1)$$

$$\text{Recall} = \frac{\text{Correct Predictions}}{\text{Total Ground Truth}} = \frac{TP}{TP+FN} \qquad (2)$$

$$\text{F1score} = 2 \times \frac{\text{Precision} \times \text{Recall}}{\text{Precision} + \text{Recall}} \qquad (3)$$

The area under the PR curve (AUC) gives the average precision (AP) and is calculated using Equation (4). The mean average precision (mAP) is obtained by taking the mean of the average precision obtained in every class, as shown in Equation (5).

$$AP = \int_{r=0}^{1} p(r)dr \qquad (4)$$

$$mAP = \frac{1}{N}\sum_{i=1}^{N} AP_i \qquad (5)$$

2.1.2. Selection of YOLOv5 Model Size

Initially, to choose the ideal YOLOv5 model size for the required application (detection of specific body parts of a lizard), training and inference were carried out for each one of the YOLOv5 model sizes under the same conditions.

An RGB dataset containing 10288 images was initially created from scratch to be later used in training. For training, 100 epochs and a batch size of 16 were used.

Tables 1 and 2 show the values obtained for precision, recall, mAP, training duration, number of parameters, GFLOPs (Giga Floating-point Operations Per Second), and inference time (time each model took to analyse a new image and make a prediction) using YOLOv5n, YOLOv5s, YOLOv5m, YOLOv5l, and YOLOv5x.

Table 1. Values obtained for precision, recall, mAP, training duration, number of parameters, GFLOPs, and inference time using YOLOv5n and YOLOv5s.

Metrics	YOLOv5n	YOLOv5s
Precision (%)	98.60	99.00
Recall (%)	97.60	98.40
mAP_0.5 (%)	97.90	98.40
mAP_0.5:0.95 (%)	71.50	74.30
Training Duration	2 h 18 min 22 s	3 h 22 min 34 s
Parameters (M)	1.8	7.0
GFLOPs	4.2	15.8
Inference Time (ms)	5.5	9.9

Table 2. Values obtained for precision, recall, mAP, training duration, number of parameters, GFLOPs, and inference time using YOLOv5m, YOLOv5l, and YOLOv5x.

Metrics	YOLOv5m	YOLOv5l	YOLOv5x
Precision (%)	99.10	99.10	99.10
Recall (%)	98.70	98.90	99.00
mAP_0.5 (%)	98.70	98.80	98.90
mAP_0.5:0.95 (%)	76.10	76.20	76.30
Training Duration	6 h 36 min 8 s	10 h 39 min 19 s	17 h 32 min 47 s
Parameters (M)	2.1	4.6	86.2
GFLOPs	47.9	107.7	203.9
Inference Time (ms)	13.1	25.0	47.8

To select the most suitable model for the application under analysis, the best balance between speed and accuracy was sought. Although YOLOv5n was the fastest and lightest model, its results were the lowest and, therefore, the model was disregarded (Table 1). The heaviest models, YOLOv5l and YOLOv5x, obtained the best results for the evaluation metrics; however, they took a long time to complete the training (more than 10 h) and presented a higher inference time, which is a major obstacle due to time limitations. Therefore, these models were also disregarded (Table 2). Finally, both YOLOv5s and YOLOv5m models obtained good results for the evaluation metrics. Since the difference between the values of the metrics obtained for each of these models was not very significant, the YOLOv5s model was chosen as it is lighter, leading to faster training and shorter inference time.

2.1.3. RGB Image Dataset

An RGB image dataset was created from scratch based on custom data. All filming took place in a controlled environment at CIBIO (Centre in Biodiversity and Genetic Resources), University of Porto, Portugal.

Firstly, a scenario was built consisting of a cardboard box, a lamp, a camera, and some black tape (Figure 2).

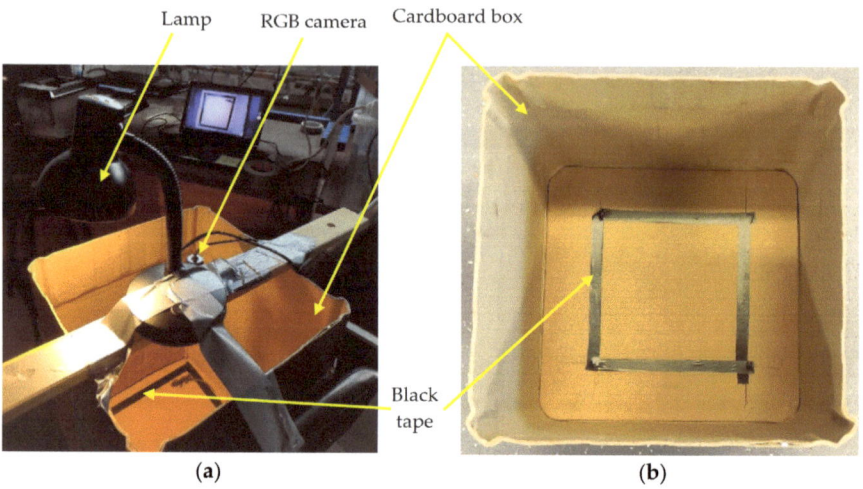

Figure 2. Scenario: (**a**) setup and (**b**) cardboard box used.

The lizard was placed inside the cardboard box, and the camera filmed its behaviour for about 10 min. In total, about 10 videos were collected using animals with different body sizes, colours, and patterns. All RGB images that compose the dataset were obtained from those videos, making a dataset of 4306 RGB images.

The image labelling was carried out using Roboflow. For each image, bounding boxes were drawn around each part of the lizard's body to be identified and labelled with the respective class. In total, seven classes were identified: "Lizard" (yellow bounding box in Figure 3), "Snout" (red bounding box in Figure 3), "Head" (cyan bounding box in Figure 3), "Dorsum" (blue bounding box in Figure 3), "Tail" (green bounding box in Figure 3), "Leg_L" (purple bounding box on the left hind leg in Figure 3), and "Palm_L" (orange bounding box on the left hind palm in Figure 3).

In Roboflow, inside the dataset, the images were split into three sets:

- "Training set": is used to train the model.
- "Validation set": is used during training to compute the validation mAP after each epoch. It is also used to evaluate the performance of the trained model.
- "Test set": is used to analyse the final performance of the model.

The "training set" contained 3014 RGB images (70%), the "validation set" contained 861 RGB images (20%), and the "test set" contained 431 RGB images (10%).

All images were resized to 640 × 640 as it is YOLOv5's default size, and some augmentation techniques were applied to the "training set" images to create new examples to use in the training of the model. The techniques used in the training images were modifications in saturation (between −10% and +10%), brightness (between −10% and +10%), exposure (between −10% and +10%), blur (up to 1 pixel), and noise (up to 1% of pixels). After augmentation, the dataset went from 4306 RGB images to 10,334 RGB images.

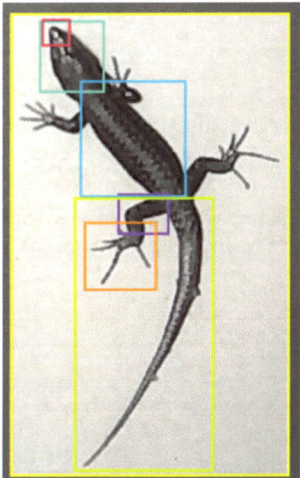

Figure 3. Example of a labelled dataset image in Roboflow.

2.1.4. Training and Inference

All training was carried out on Google Collaboratory, which runs in the cloud, and the NVIDIA Tesla T4 GPU (16 GB of memory) was used. Firstly, training was performed for the number of epochs and batch sizes represented in Table 3 to find the model with the best training results.

Table 3. Number of epochs and batch sizes used for training.

Batch	Epoch
16	100
	200
20	100
	200
32	100
	200
	300
	400
	500
64	100
	200
	300
	400

Secondly, inference was run on some images, and two thresholds were defined:
- Confidence threshold: Defines the minimum score the model considers the prediction to be correct; otherwise, it completely discards the prediction. This threshold was set to 0.50, meaning all predicted bounding boxes with a confidence score below 50% were discarded. This value was chosen based on a careful analysis of the results obtained using different threshold values.
- IoU threshold: Defines the minimum overlap between the predicted bounding box and the ground truth bounding box for the prediction to be considered correct. This threshold was set to 0.50 after a careful analysis of the results obtained using different threshold values.

The training and inference results are shown in Section 3.1.

2.2. Temperature Acquisition

After detecting the lizard's position, its temperature acquisition was then possible to be acquired as described next.

2.2.1. Thermal and RGB Image Acquisition

To obtain the thermal images used in this work, the FLIR T335 thermal camera was added to a scenario similar to the one described in Section 2.1.3. As shown in Figure 4, the thermal camera was positioned above the RGB camera with a certain horizontal offset to try to match the point of view of both cameras as much as possible.

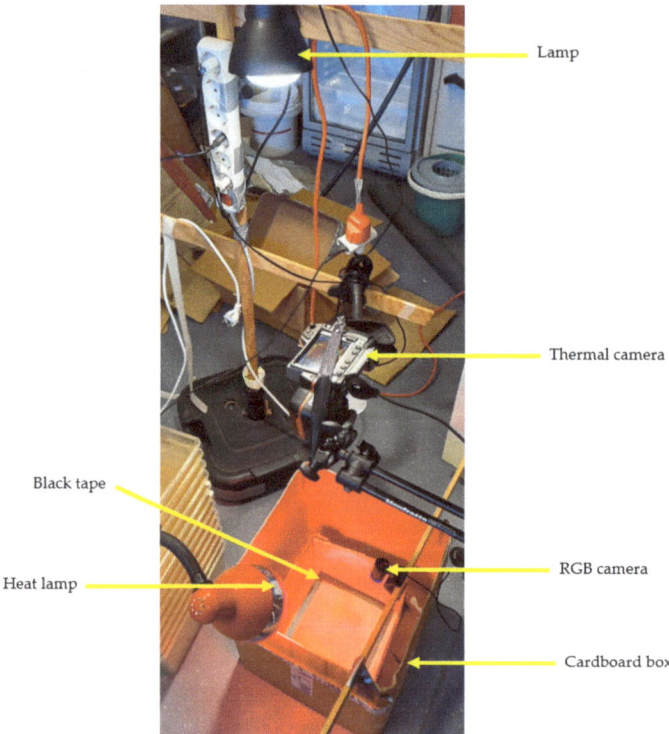

Figure 4. The scenario used to obtain thermal images and their associated RGB images.

The lizard was placed inside the cardboard box, and both cameras simultaneously filmed the animal's behaviour for a few minutes. The RGB camera and the thermal camera were placed side-by-side, as presented in Figure 5, and the videos were saved in the same way (screen recording). Some videos were recorded with the heat lamp on and others with the heat lamp off to observe more significant changes between videos in the animal's colour in the thermal images (change in the animal's body temperature). Using the RGB camera helps to determine the position of the lizard in the thermal camera, especially if the lizard is at the same temperature as the background.

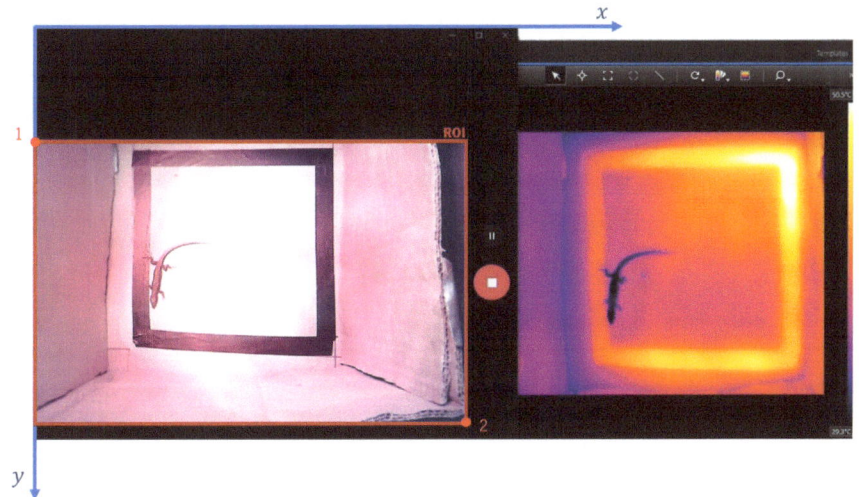

Figure 5. RGB camera output (**left side**) and thermal camera output (**right side**).

2.2.2. YOLOv5s Model Application

To detect the lizard and its six body parts, the model analysed in Section 3.1 was used. Since it was desirable to apply detection only to the RGB image, a region of interest (ROI) involving only the RGB image was created. The region of interest was defined using Equation (6).

$$ROI = image\ [y{:}\ y + height,\ x{:}\ x + width] \qquad (6)$$

where, from Figure 5's coordinate axes:

- "image" represents the input image, with the RGB and thermal images side-by-side;
- "x" is represented by the x-coordinate of point 1 in Figure 5;
- "y" is represented by the y-coordinate of point 1 in Figure 5;
- "y + height" is represented by the y-coordinate of point 2 in Figure 5;
- "x + width" is represented by the x-coordinate of point 2 in Figure 5.

Figure 6 shows the detection of the lizard and its six body parts in the defined region of interest.

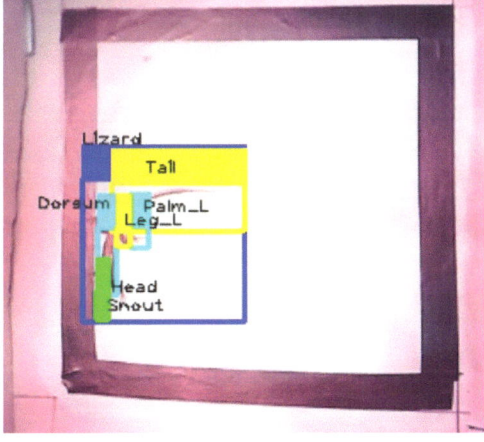

Figure 6. Detection of the lizard and its six body parts on the ROI (left).

The detections were only made for the RGB image and not for the thermal image, as the model was trained only with RGB images and not with thermal images. Using the model to detect the lizard and its body parts in thermal images would generate erroneous detections. Following this, the process is described with two examples: the whole lizard and its tail.

2.2.3. Bounding Box: Identified Class and Background

After the detection process, the bounding boxes generally involve the detected class and part of the background. To make the distinction between the background and the identified class clear, the following method was used, involving five sequential steps:

1. Creation of a black binary mask with the same dimensions as ROI.
2. In the black binary mask created in Step 1, all pixels within the region of each bounding box are set to white, as shown in the examples in Figure 7.

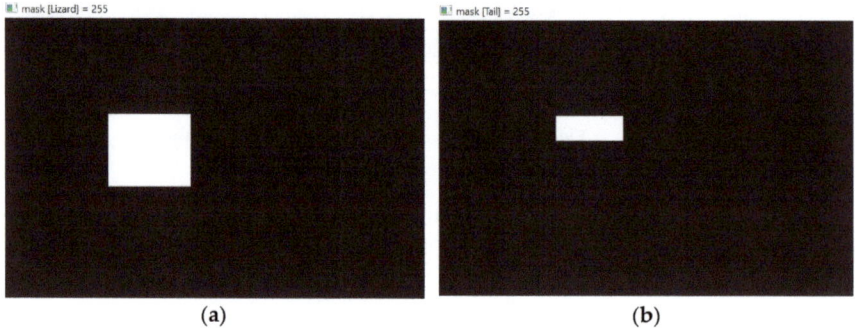

Figure 7. All pixels within the (**a**) "Lizard" and (**b**) "Tail" bounding boxes are white.

3. Application of a bitwise AND operation between the ROI and the binary mask from Step 2. This retains only the pixels that both have non-zero values (Figure 8), which are the pixels that fall within the bounding box.

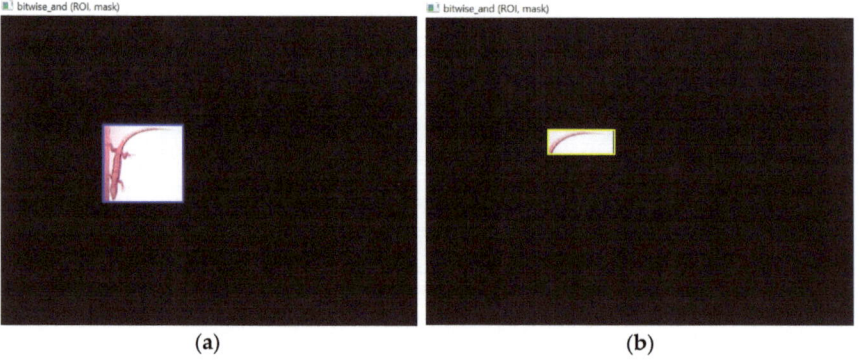

Figure 8. Isolation of the ROI defined by the (**a**) "Lizard" and (**b**) "Tail" bounding boxes.

4. Conversion to grayscale, as shown in Figure 9.

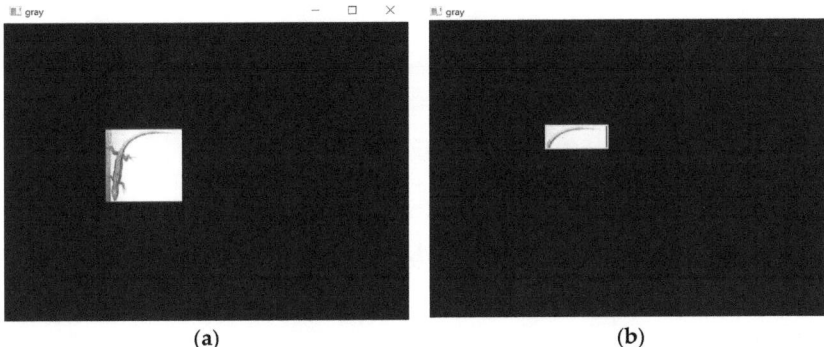

Figure 9. The (**a**) "Lizard" and (**b**) "Tail" bounding boxes are in grayscale.

5. Conversion from grayscale to binary using an inverse-binary threshold (Figure 10). This is user threshold-dependent since the threshold value must be chosen by the user.

Figure 10. The (**a**) "Lizard" and (**b**) "Tail" bounding boxes are in black and white (binary).

As demonstrated in Figure 10, inside each bounding box, the background pixels turned black, and the pixels of the class to be identified turned white. This allowed us to not only highlight the most important part within each bounding box (identified class) but also make it possible to distinguish between the lizard (white pixels) and the background (black pixels).

A single pixel was selected to represent each bounding box based on what was discussed and decided by the biologists. The main requirement was that in each bounding box, the pixel had to belong to the detected class, not the background. For this purpose, the pixel in the centre of each bounding box was initially considered (Figure 11).

However, as can be seen in Figure 11, not all pixels in the centre of the bounding boxes belong to the detected class, as some belong to the background. Undesirably, the pixel in the centre of the "Lizard" and "Tail" bounding boxes belonged to the background and not to the respective class.

To solve this problem, after using the method explained at the beginning of this section, a condition was created in which it was determined whether the central pixel in each bounding box was white (if center_pixel == 255) or not (else:). If it is determined that the pixel is white, that pixel would represent the bounding box; otherwise, it would search for the nearest white pixel to the central pixel (determined initially), and that would be the new pixel that should represent the bounding box. To find the coordinates of the nearest white pixel, a function called "nearest_white_pixel" was defined.

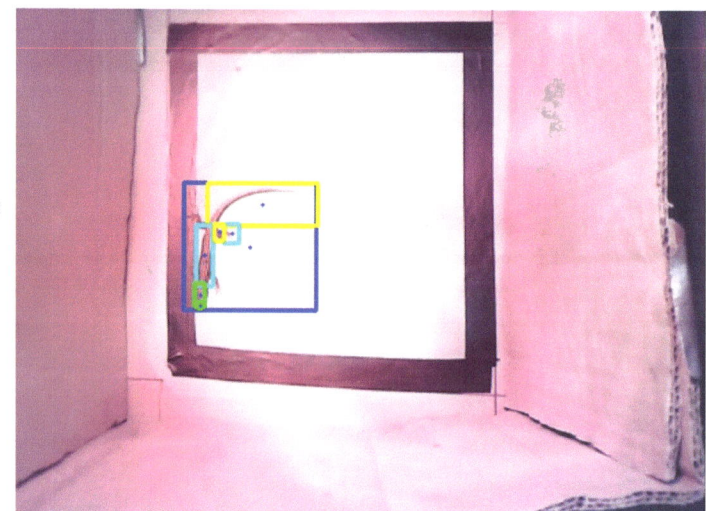

Figure 11. Bounding boxes and their respective central pixels are represented by a blue circle (in the ROI).

In Figure 12, the green circle represents the closest white pixel found in the bounding box, starting from the central black pixel. These are the new pixels considered for the thermal analysis.

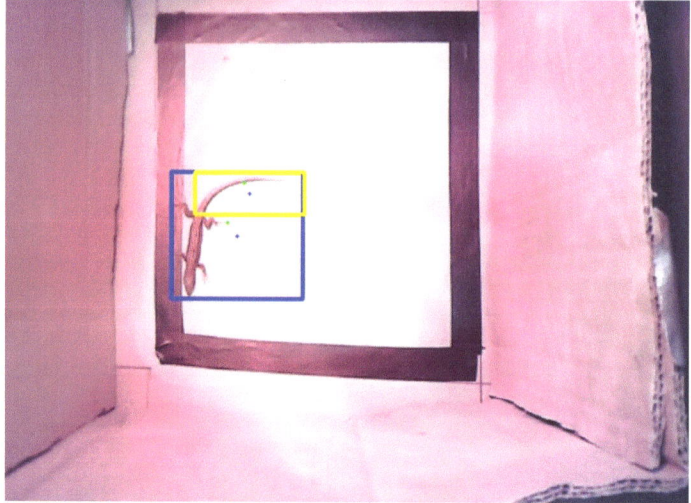

Figure 12. The initial pixel (centre) represents the "Lizard" (blue rectangle) and "Tail" (yellow rectangle) bounding boxes, marked with a blue dot. The final pixel representative of each bounding box is marked with a green dot.

2.2.4. Perspective Transformation and Temperature Detection

Perspective transformation is used to establish a relationship between pixels in the RGB image and corresponding pixels in the thermal image. In perspective transformation, a 3x3 transformation matrix is determined by four points in the RGB image and the corresponding four points in the thermal image.

Using Python OpenCV's library, Equation (7) calculates the transformation matrix.

$$matrix = cv2.getPerspectiveTransform(src, dst) \quad (7)$$

where:
- The "src" parameter represents the coordinates of the quadrilateral vertices in the source image (RGB image).
- The "dst" parameter represents the coordinates of the corresponding quadrilateral vertices in the destination image (thermal image).

The parameters "src" and "dst" are defined by the function shown in Equation (8).

$$np.array([[x_{min}, y_{min}], [x_{max}, y_{min}], [x_{max}, y_{max}], [x_{min}, y_{max}]], dtype = np.float32) \quad (8)$$

According to Equation (8) and Figure 13, for the "src" parameter, $[x_{min}, y_{min}]$ corresponds to the coordinates of point 1, $[x_{max}, y_{min}]$ corresponds to the coordinates of point 2, $[x_{max}, y_{max}]$ corresponds to the coordinates of point 3, and $[x_{min}, y_{max}]$ corresponds to the coordinates of point 4. For the "dst" parameter, $[x_{min}, y_{min}]$ corresponds to the coordinates of point 1′, $[x_{max}, y_{min}]$ corresponds to the coordinates of point 2′, $[x_{max}, y_{max}]$ corresponds to the coordinates of point 3′, and $[x_{min}, y_{max}]$ corresponds to the coordinates of point 4′.

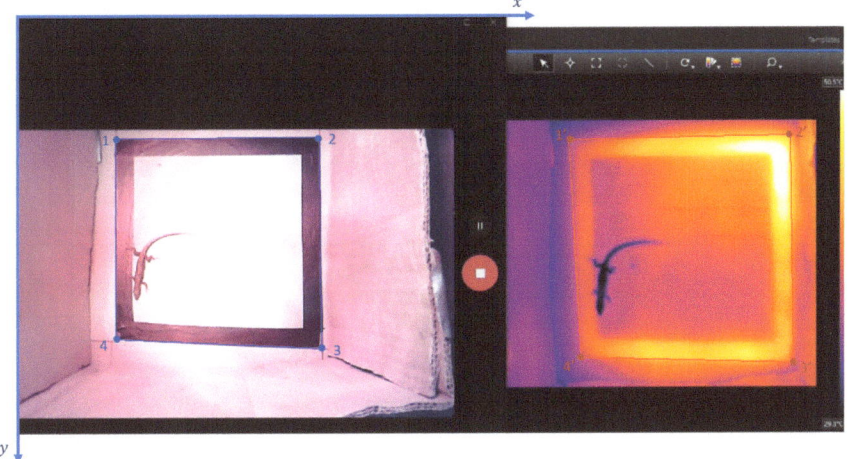

Figure 13. The "src" parameter is represented by the coordinates of points 1, 2, 3, and 4, and the "dst" parameter is represented by the coordinates of points 1′, 2′, 3′, and 4′.

This way, it is possible to transform any set of coordinates using the transformation matrix. In Equation (9), the transformation matrix described in Equation (7) is applied to the point represented as [centre_x, centre_y, 1].

$$transf\_coord = np.dot(matrix, np.array([centre\_x, centre\_y, 1], dtype = np.float32)) \quad (9)$$

Subsequently, the transformed coordinates obtained ("[x, y, w]") in Equation (9) are normalised using Equation (10). In this equation, "x" and "y" are divided by "w".

$$transf\_coord = transf\_coord[:2]/transf\_coord [2] \quad (10)$$

In Equation (11), the normalised transformed coordinates are assigned to "xn" and "yn".

$$xn, yn = tuple(transf\_coord) \quad (11)$$

In the Equations (12) and (13), "xn" and "yn" coordinates are rounded. "xn" and "yn" represent the final transformed coordinates in the thermal image corresponding to the original point in the RGB image (center_x, center_y).

$$xn = int(xn + 0.5) \tag{12}$$

$$yn = int(yn + 0.5) \tag{13}$$

After applying the equations mentioned above to each pixel marked in the RGB image (left side of Figure 14), it was possible to obtain the corresponding pixels in the thermal image (right side of Figure 14).

Figure 14. Bounding boxes and their representative pixels marked with blue dots (RGB image) and corresponding pixels marked with red dots in the thermal image.

For each pixel marked in the thermal image, the respective temperature value was obtained through its colouring. It is important to highlight that the colour temperature scale can vary between images.

To make this possible, a function was created that allows obtaining the temperature based on a given pixel colour (input) and a set of parameters (T_{max}, T_{min}, Y_{max}, Y_{min}, X_{med}). The temperature value is calculated using a linear interpolation, as shown in Equation (14), where "final" represents the row index.

$$T = T_{min} + \frac{(Y_{max} - final)}{(Y_{max} - Y_{min})} \cdot (T_{max} - T_{min}) \tag{14}$$

The maximum temperature (T_{max}), minimum temperature (T_{min}), maximum Y (Y_{max}), minimum Y (Y_{min}), and median X (X_{med}) values were defined based on the input image. Looking at the colour temperature scale present on the right side of Figure 15 (column of 10 pixels width represents the colour scale), it can be stated that the minimum temperature (T_{min}) is 29.3 °C, and the maximum temperature (T_{max}) is 50.5 °C. Also, the maximum Y (Y_{max}) value corresponds to the y-coordinate of the bottom corner of the bar (for T_{min}), the minimum Y (Y_{min}) value corresponds to the y-coordinate of the top corner of the bar (for T_{max}), and the median X (X_{med}) value corresponds to the position of the bar on the x-axis.

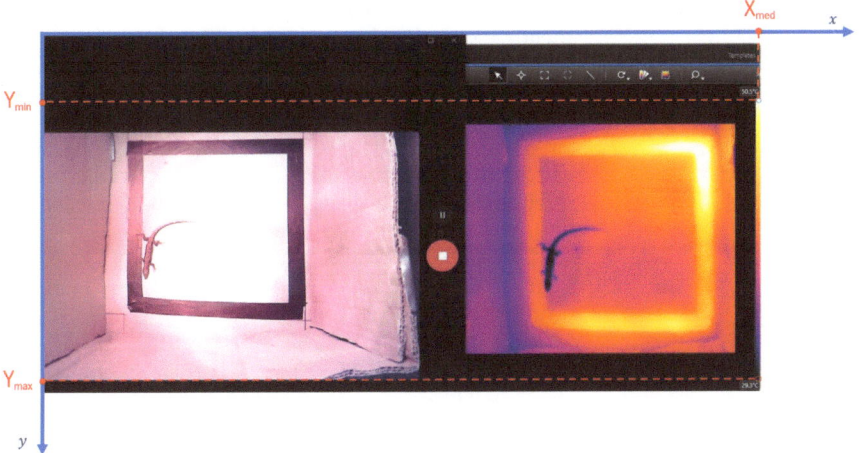

Figure 15. Annotation of the maximum Y (Y_{max}), minimum Y (Y_{min}), and median X (X_{med}) relative to the coordinate axis of the input image.

The temperature values obtained for each class were automatically stored in a text file together with the day, time of measurements, and the class name.

3. Results and Discussion

This section contains the results of the training and inferences carried out to obtain the best model for detecting the lizard and its body parts. It also demonstrates the system's potential in acquiring the temperature values of the parts detected by the model.

3.1. Detection of Lizard Body Parts: Training and Inference

The best training results were obtained using a batch size of 32 and 500 epochs for the neural network. This training took 15 h, 14 min, and 21 s.

Table 4 presents the values obtained for precision, recall, and mAP metrics for each of the seven classes. At the end, the average values of these metrics are shown.

Table 4. Values obtained for precision, recall, and mAP after training using a batch size of 32 and a number of epochs of 500.

Class	Precision (%)	Recall (%)	mAP_0.5 (%)	mAP_0.5:0.95 (%)
Lizard	99.80	99.90	99.40	92.20
Snout	95.60	95.00	93.70	42.70
Head	99.50	99.60	99.40	74.50
Dorsum	99.70	99.90	99.40	78.70
Tail	99.60	99.20	99.40	94.50
Leg_L	99.50	99.20	99.40	66.80
Palm_L	99.40	98.40	99.40	78.60
AVERAGE	99.00	98.80	98.60	75.40

By analysing Table 4, it is observed that the "Snout" class was the one that presented the lowest value in all the metrics. The reason for this may be due to the small size of this body part of the lizard in relation to the other parts, making its correct identification more complex.

Comparing the average values of mAP_0.5 and mAP_0.5:0.95 metrics, it is possible to perceive that the value of mAP_0.5:0.95 is significantly lower than mAP_0.5. This is common since, unlike mAP_0.5, mAP_0.5:0.95 evaluates the model over a wider range of IoU thresholds. The increasing of the IoU threshold results in stricter requirements, causing the mAP value to decrease; therefore, obtaining a high mAP_0.5:0.95 value can be challenging.

The average value obtained for mAP_0.5:0.95 (75.40%) can be considered good. Also, the average precision (99.00%), recall (98.80%), and mAP_0.5 (98.60%) values obtained are very good.

Figure 16 displays all the graphs of the average values obtained after training. Each graph represents the change in a certain value (y-axis) as the number of epochs increases during training (x-axis). As mentioned previously, the number of epochs used in training was 500, so the x-axis will go up to the value of 500. The four graphs on the right side of Figure 16 correspond to the previously mentioned metrics: precision ("metrics/precision"), recall ("metrics/recall"), mAP_0.5 ("metrics/mAP_0.5"), and mAP_0.5:0.95 ("metrics/mAP_0.5:0.95").

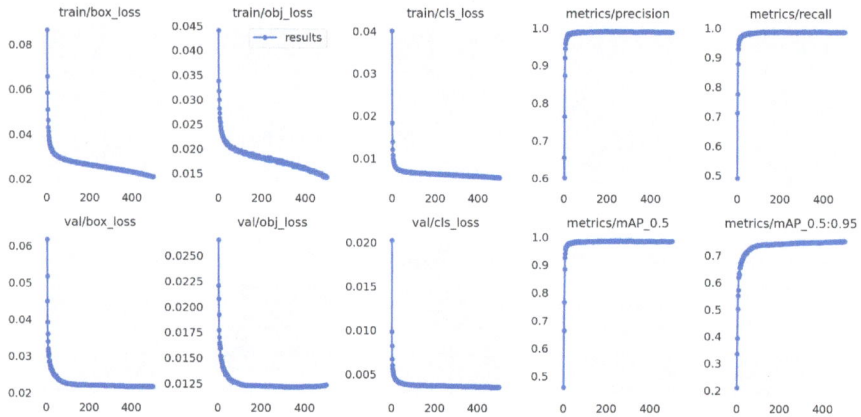

Figure 16. Resulting graphs after training using a batch size of 32 and 500 epochs.

Observing the behaviour of the "metrics/precision", "metrics/recall", and "metrics/mAP_0.5" graphs, it is possible to understand that they begin to stabilise after about 80 epochs. The "metrics/mAP_0.5:0.95" graph started to stabilise later, after about 400 epochs.

The stabilisation of a graph indicates that there will no longer be significant improvements in the measured value. Therefore, to avoid the occurrence of overfitting and the decrease in metric values, the training was considered completed for the number of epochs of 500.

The remaining six graphs on the left side of Figure 16 represent the training losses ("train/box_loss", "train/obj_loss", and train/cls_loss") and the validation losses ("val/box_loss", "val/obj_loss", and val/cls_loss"). Where "box_loss" is the box regression loss, "obj_loss" is the object loss, and "cls_loss" is the class loss. In these six graphs, it is possible to observe that, as desired, the loss values decreased as the number of epochs increased. Furthermore, a rapid decline was observed until around epoch 10.

By analysing the loss graphs in Figure 16, it can be concluded that overfitting did not occur.

The F1 score curve illustrates the F1 score across different thresholds, offering insights into the model's balance between false positives and false negatives. Figure 17 shows that the maximum F1 value is 0.99 when the confidence score is 0.601.

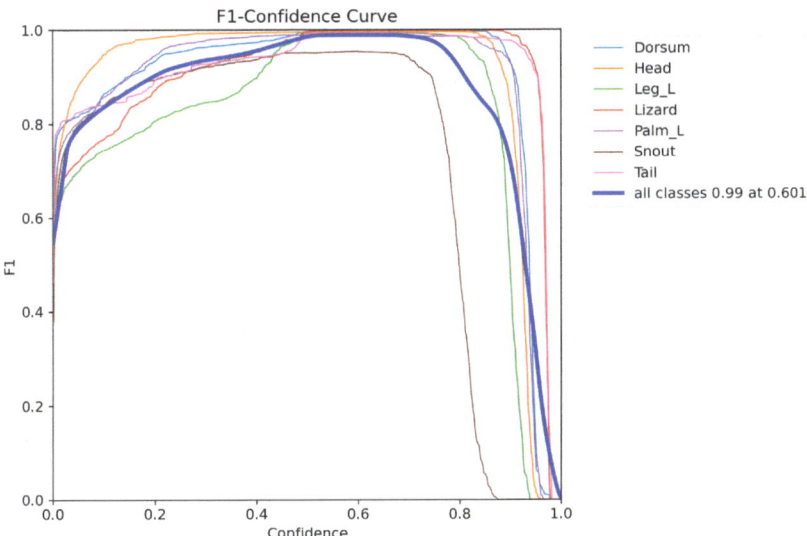

Figure 17. F1–confidence curve.

Figure 18 presents the precision–recall curve, where a larger area under the curve indicates better overall performance. The "Snout" class has a smaller area under the PR curve, indicating that the model has more difficulty in correctly detecting this class compared to the others. (mAP_0.5 of 0.937).

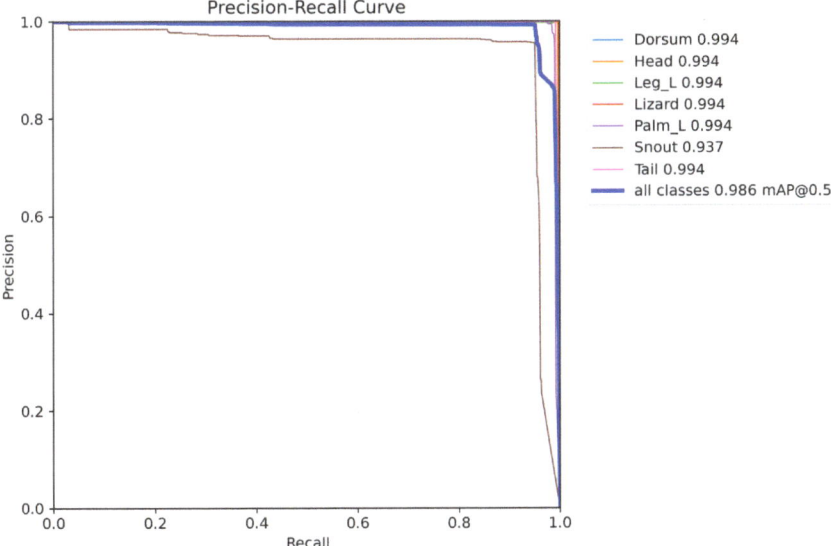

Figure 18. Precision–recall curve.

To evaluate how well the trained model generalises to unseen images, the inference was run on the images from the "test set". Figure 19 shows a sample image used in the inference. As expected, the predictions were acceptable. All classes were correctly indicated with confidence scores ranging from 78% to 97%.

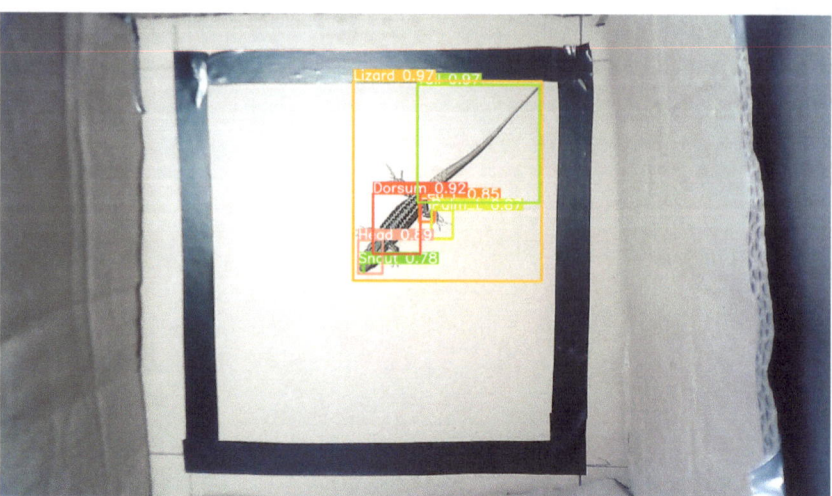

Figure 19. Example of an image from the "test set" with predictions.

Noise was added to Figure 19, as demonstrated in Figure 20, to analyse the model's performance on noisy images and variations in image quality. As shown in Figure 20, the model was able to correctly detect the lizard and its six body parts with confidence scores ranging from 77% to 96%. However, one more bounding box corresponding to the "Dorsum" class was incorrectly detected, with a confidence score of 60%.

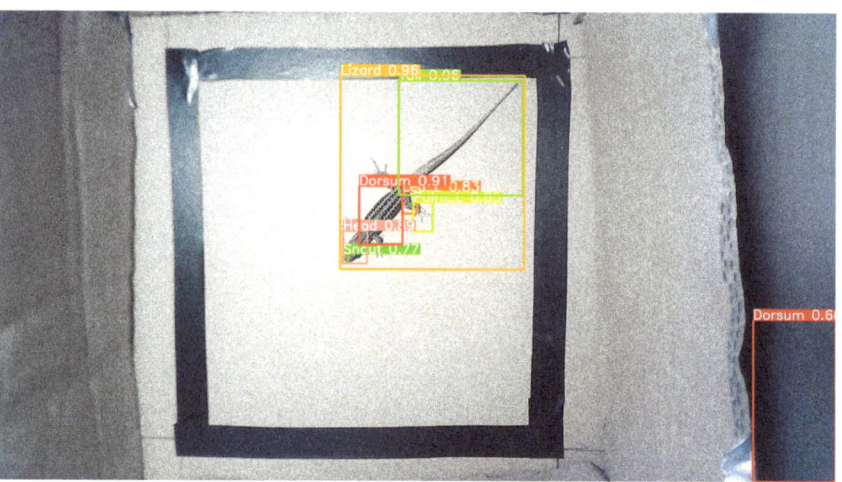

Figure 20. Example of an image from the "test set" with noise and predictions.

When the model is faced with cases for which it was not trained, it tends to show a decrease in the confidence score of the detected classes and may even generate false positives.

3.2. Temperature Acquisition

Applying the methodologies presented in Section 2.2, it was possible to successfully obtain the final temperature values in different images and videos. Figure 21 shows an example of the temperature values obtained for the lizard and its body parts in an image and a video.

```
                                              Temperature_Values_Video - Notepad
                                              Ficheiro Editar Formatar Ver Ajuda
                                              2024-03-30 15:36:45 --- Lizard -- 38.3ºC
                                              2024-03-30 15:36:45 --- Tail -- 40.0ºC
                                              2024-03-30 15:36:45 --- Dorsum -- 39.4ºC
                                              2024-03-30 15:36:45 --- Head -- 39.1ºC
                                              2024-03-30 15:36:45 --- Palm_L -- 40.7ºC
                                              2024-03-30 15:36:45 --- Leg_L -- 39.5ºC
  Temperature_Values - Notepad                2024-03-30 15:36:45 --- Snout -- 39.2ºC
  Ficheiro Editar Formatar Ver Ajuda          --------------------------------------
  2024-03-30 15:34:04 --- Lizard -- 37.4 ºC   2024-03-30 15:36:46 --- Lizard -- 38.3ºC
  2024-03-30 15:34:04 --- Tail   -- 36.4 ºC   2024-03-30 15:36:46 --- Tail -- 39.9ºC
  2024-03-30 15:34:04 --- Palm_L -- 37.0 ºC   2024-03-30 15:36:46 --- Dorsum -- 39.4ºC
  2024-03-30 15:34:04 --- Dorsum -- 29.6 ºC   2024-03-30 15:36:46 --- Head -- 39.2ºC
  2024-03-30 15:34:04 --- Head   -- 29.7 ºC   2024-03-30 15:36:46 --- Palm_L -- 40.7ºC
  2024-03-30 15:34:04 --- Leg_L  -- 31.4 ºC   2024-03-30 15:36:46 --- Leg_L -- 39.3ºC
  2024-03-30 15:34:04 --- Snout  -- 30.2 ºC   2024-03-30 15:36:46 --- Snout -- 39.1ºC
                                              --------------------------------------
                                              2024-03-30 15:36:47 --- Lizard -- 38.3ºC
                                              2024-03-30 15:36:47 --- Tail -- 39.9ºC
                                              2024-03-30 15:36:47 --- Dorsum -- 39.4ºC
                                              2024-03-30 15:36:47 --- Head -- 38.9ºC
                                              2024-03-30 15:36:47 --- Palm_L -- 40.7ºC
                                              2024-03-30 15:36:47 --- Leg_L -- 39.5ºC
                                              2024-03-30 15:36:47 --- Snout -- 39.1ºC
              (a)                                              (b)
```

Figure 21. Notepad with date, hour, class, and temperature values obtained: (**a**) from an image and (**b**) from a video.

The accuracy of temperature measurements is given by the thermal camera used; in this case, the FLIR T335 thermal camera, which has an accuracy of ± 2 °C of the reading.

3.3. Comparison with Other Studies for Automatic Detection and Temperature Extraction

In recent years, several studies have been carried out to develop methods for automatically detecting specific body parts of animals and extracting their body temperature values. These efforts were driven by the need to address the limitations and challenges associated with traditional manual temperature measurement techniques.

Xie et al. [37] developed an automatic temperature detection method based on Infrared Thermography (ITG) to overcome the challenges associated with traditional pig rectal temperature measurement. Automatic detection of six regions on the pig body surface (forehead, eyes, nose, ear root, back, and anus) was performed using an improved YOLOv5s model with BiFPN. After detection, the temperature values were automatically extracted. The proposed YOLOv5s-BiFPN model achieved optimal performance, with a mAP of 96.36%, a target detection speed of up to 100 frames per second, and a model size of 20 MB. Additionally, the variations in maximum temperature automatically extracted from the ear root and the forehead coincided with those obtained manually, and the temperature accuracy was ± 2 °C.

Wang et al. [38] proposed a method based on the detection model GG-YOLOv4 for the automatic detection of the ocular surface temperature of dairy cows from thermal images, with the aim of identifying health disorders. The model achieved a mAP of 96.88%, a detection speed of 40.33 frames per second, and a model size of 44.7 M. The comparison between the temperature values obtained with the model and the manually extracted values showed that the average absolute temperature extraction errors in the left and right eyes were 0.051 °C and 0.042 °C, respectively, and the average relative temperature extraction errors in the left and right eyes were 0.14% and 0.11%, respectively. The temperature accuracy was ± 2 °C.

The proposed model in this paper achieved a mean average precision (mAP) of 98.60%, outperforming the models developed by Xie et al. [37] and Wang et al. [38]. All methods mentioned above have the same temperature accuracy value (± 2 °C).

The algorithm proposed in this paper introduces innovative features to improve the detection of lizards and the extraction of their body temperature in a controlled laboratory environment. Firstly, this study significantly contributes to filling the notable gap in algorithm development and research regarding automatic and non-invasive methods for lizard detection and body temperature extraction in controlled laboratory environments. Secondly, employing a non-invasive and automatic method for extracting the body temperature of lizards in a controlled laboratory environment minimises potential harm and stress to the animals, thereby promoting a more efficient and humane way of monitoring lizard body temperature. Thirdly, due to the scarcity of publicly available lizard datasets, a dataset was created from scratch, providing a valuable resource for training and potentially benefiting future lizard-related research. Lastly, the simultaneous use of two cameras (RGB and thermal camera) significantly enhances the accuracy of lizard detection and enables precise temperature extraction.

In the dual-camera system, the RGB camera allows the detection of the lizard and its body parts using YOLOv5, and the thermal camera allows reading of the respective temperature of those parts. After calibration, the images from both cameras are properly localised, making it possible to determine accurately the lizard's position in the thermal image through a coordinate conversion from the RGB image to the thermal image. Based on the colour temperature scale present in the thermal image, the temperature values are then extracted. Therefore, this approach enables the automated and non-invasive extraction of the lizard's body temperature.

4. Conclusions

The work presented in this paper concerns the development of a system capable of detecting the lizard and its body parts, subsequently acquiring their respective temperature values. This method provides biologists with a faster and non-intrusive way to measure lizard body temperature in a controlled laboratory setting, allowing researchers to monitor the temperature preferences of lizards and enabling detailed observations of their thermoregulation strategies. By automating the temperature acquisition process, this method reduces stress and potential harm to the animals, offering a more ethical approach to studying lizards' behaviour.

This work can be divided into two main parts: the dataset creation and the detection of the lizard and its body parts; and the acquisition of the respective temperature values.

Since there were no datasets available online or in the Biology Laboratory, it was necessary and challenging to create a dataset from scratch, including creating a scenario, filming videos, obtaining frames from these videos, and labelling the images with each class.

The YOLOv5s (small) model was chosen because it is lightweight, has a fast inference time, and offers the best balance between training duration and the quality of the results obtained. When using the model to detect the lizard and its body parts, challenges were encountered in more complex images (images with noise), leading to some classes being incorrectly detected. However, the model correctly identified the lizard and its body parts in all images from the "test set", with confidence scores above 78% in which, in general, the "Lizard" (average of 96%) and "Tail" (average of 94%) classes presented the highest confidence scores, and the "Snout" class was the one with the lowest confidence score (average of 78%). The model achieved a precision of 90.00% and a recall of 98.80%. It can be concluded that the application of YOLOv5s for the detection of lizards and their body parts has demonstrated overall success.

The model was used to make detections only in RGB images and not in thermal images since it was trained only with RGB images. If the model was used in thermal images, it would generate erroneous detections because sometimes the lizard is not visible in the thermal image due to its temperature being equal to its background floor. The coordinate transformation from the RGB image to the thermal image proved to be effective, allowing the acquisition of the final temperature values of the lizard's body parts based on the colour temperature scale and the colour of the pixels present in the thermal image. The accuracy in acquiring the temperature values directly relied on the precise mapping of coordinates between the RGB and thermal images.

Overall, the system successfully achieves the intended end goal. However, it is important to highlight that there is still room for improvement.

Given the challenges encountered during the development of this work and the respective results obtained, a few proposals are presented to be implemented in future updates.

- Adaptation of the developed system to detect the body temperature of another species of animal kept in captivity.
- RGB and thermal cameras with better resolution.
- Obtain the values of additional parameters, such as emissivity and reflective temperature, that allow acquiring new information regarding the temperature measurement process, enabling a deeper analysis.

Author Contributions: Conceptualisation, A.L.A., A.F.R. and G.L.; methodology, A.L.A.; software, A.L.A. and A.F.R.; validation, A.L.A. and A.F.R.; formal analysis, A.F.R. and G.L.; investigation, A.L.A.; resources, A.F.R. and G.L.; data curation, A.L.A.; writing—original draft preparation, A.L.A.; writing—review and editing, A.F.R. and G.L.; visualisation, A.L.A.; supervision, A.F.R. and G.L.; project administration, A.F.R. and G.L. All authors have read and agreed to the published version of the manuscript.

Funding: This work has been supported by FCT—Fundação para a Ciência e a Tecnologia within the R&D Units Project Scope: UIDB/00319/2020.

Institutional Review Board Statement: Not applicable.

Informed Consent Statement: Not applicable.

Data Availability Statement: The original data presented in the study are openly available in OSF at DOI 10.17605/OSF.IO/UQYEG.

Acknowledgments: This work has been supported by the Laboratory of Automation and Robotics (LAR) of the University of Minho and the ALGORITMI research centre. The authors would also like to thank the Center for Biodiversity and Genetic Resources (CIBIO) of the University of Porto and Frederico Barroso for the collaboration in the acquisition of videos of the lizards used in this work.

Conflicts of Interest: The authors declare no conflicts of interest.

References

1. Jumper, J.; Evans, R.; Pritzel, A.; Green, T.; Figurnov, M.; Ronneberger, O.; Tunyasuvunakool, K.; Bates, R.; Žídek, A.; Potapenko, A.; et al. Highly Accurate Protein Structure Prediction with AlphaFold. *Nature* **2021**, *596*, 583–589. [CrossRef]
2. Jiménez-Luna, J.; Grisoni, F.; Schneider, G. Drug Discovery with Explainable Artificial Intelligence. *Nat. Mach. Intell.* **2020**, *2*, 573–584. [CrossRef]
3. Buchelt, A.; Adrowitzer, A.; Kieseberg, P.; Gollob, C.; Nothdurft, A.; Eresheim, S.; Tschiatschek, S.; Stampfer, K.; Holzinger, A. Exploring Artificial Intelligence for Applications of Drones in Forest Ecology and Management. *For. Ecol. Manag.* **2024**, *551*, 121530. [CrossRef]
4. Hurwitz, J.; Kirsch, D. Understanding Machine Learning. In *Machine Learning for Dummies*; IBM Limited Edition; John Wiley & Sons: Indianapolis, IN, USA, 2018; pp. 3–17.
5. Mueller, J.P.; Massaron, L. Descending the Gradient. In *Machine Learning for Dummies*, 2nd ed.; John Wiley & Sons: Indianapolis, IN, USA, 2021; pp. 139–151.
6. Burkov, A. Basic Practice. In *The Hundred-Page Machine Learning Book*; Andriy Burkov: Québec, QC, Canada, 2019; pp. 54–71.
7. Cunningham, P.; Cord, M.; Delany, S.J. Supervised Learning. In *Machine Learning Techniques for Multimedia*; Cord, M., Cunningham, P., Eds.; Cognitive Technologies; Springer: Berlin/Heidelberg, Germany, 2008; pp. 21–49. [CrossRef]
8. Lecun, Y.; Bottou, L.; Bengio, Y.; Haffner, P. Gradient-Based Learning Applied to Document Recognition. *Proc. IEEE* **1998**, *86*, 2278–2324. [CrossRef]
9. Krizhevsky, A.; Sutskever, I.; Hinton, G.E. ImageNet Classification with Deep Convolutional Neural Networks. *Commun. ACM* **2017**, *60*, 84–90. [CrossRef]
10. Zeiler, M.D.; Fergus, R. Visualizing and Understanding Convolutional Networks. In *Computer Vision–ECCV 2014*; Fleet, D., Pajdla, T., Schiele, B., Tuytelaars, T., Eds.; Lecture Notes in Computer Science; Springer International Publishing: Cham, Switzerland, 2014; Volume 8689, pp. 818–833. [CrossRef]

11. Szegedy, C.; Liu, W.; Jia, Y.; Sermanet, P.; Reed, S.; Anguelov, D.; Erhan, D.; Vanhoucke, V.; Rabinovich, A. Going Deeper with Convolutions. In Proceedings of the 2015 IEEE Conference on Computer Vision and Pattern Recognition (CVPR), Boston, MA, USA, 7–12 June 2015; pp. 1–9. [CrossRef]
12. He, K.; Zhang, X.; Ren, S.; Sun, J. Deep Residual Learning for Image Recognition. In Proceedings of the 2016 IEEE Conference on Computer Vision and Pattern Recognition (CVPR), Las Vegas, NV, USA, 27–30 June 2016; pp. 770–778. [CrossRef]
13. Simonyan, K.; Zisserman, A. Very Deep Convolutional Networks for Large-Scale Image Recognition. In *3rd International Conference on Learning Representations (ICLR 2015)*; Computational and Biological Learning Society: San Diego, CA, USA, 2015; pp. 1–14.
14. Tan, M.; Le, Q. EfficientNet: Rethinking Model Scaling for Convolutional Neural Networks. In Proceedings of the 36th International Conference on Machine Learning; PMLR: Long Beach, CA, USA, 2019; pp. 6105–6114.
15. Liu, W.; Anguelov, D.; Erhan, D.; Szegedy, C.; Reed, S.; Fu, C.-Y.; Berg, A.C. SSD: Single Shot MultiBox Detector. In *Computer Vision—ECCV 2016*; Leibe, B., Matas, J., Sebe, N., Welling, M., Eds.; Lecture Notes in Computer Science; Springer International Publishing: Cham, Switzerland, 2016; Volume 9905, pp. 21–37. [CrossRef]
16. Lin, T.-Y.; Goyal, P.; Girshick, R.; He, K.; Dollar, P. Focal Loss for Dense Object Detection. In Proceedings of the 2017 IEEE International Conference on Computer Vision (ICCV), Venice, Italy, 22–29 October 2017; pp. 2999–3007. [CrossRef]
17. Tan, M.; Pang, R.; Le, Q.V. EfficientDet: Scalable and Efficient Object Detection. In Proceedings of the 2020 IEEE/CVF Conference on Computer Vision and Pattern Recognition (CVPR), Seattle, WA, USA, 13–19 June 2020; pp. 10778–10787. [CrossRef]
18. Hao, M.; Sun, Q.; Xuan, C.; Zhang, X.; Zhao, M.; Song, S. Lightweight Small-Tailed Han Sheep Facial Recognition Based on Improved SSD Algorithm. *Agriculture* 2024, 14, 468. [CrossRef]
19. Jia, J.; Fu, M.; Liu, X.; Zheng, B. Underwater Object Detection Based on Improved EfficientDet. *Remote Sens.* 2022, 14, 4487. [CrossRef]
20. Roy, A.M.; Bhaduri, J.; Kumar, T.; Raj, K. WilDect-YOLO: An Efficient and Robust Computer Vision-Based Accurate Object Localization Model for Automated Endangered Wildlife Detection. *Ecol. Inform.* 2023, 75, 101919. [CrossRef]
21. Hu, J.; Jagtap, R.; Ravichandran, R.; Sathya Moorthy, C.P.; Sobol, N.; Wu, J.; Gao, J. Data-Driven Air Quality and Environmental Evaluation for Cattle Farms. *Atmosphere* 2023, 14, 771. [CrossRef]
22. Jubayer, F.; Soeb, J.A.; Mojumder, A.N.; Paul, M.K.; Barua, P.; Kayshar, S.; Akter, S.S.; Rahman, M.; Islam, A. Detection of Mold on the Food Surface Using YOLOv5. *Curr. Res. Food Sci.* 2021, 4, 724–728. [CrossRef]
23. Long, W.; Wang, Y.; Hu, L.; Zhang, J.; Zhang, C.; Jiang, L.; Xu, L. Triple Attention Mechanism with YOLOv5s for Fish Detection. *Fishes* 2024, 9, 151. [CrossRef]
24. Ahmad, I.; Yang, Y.; Yue, Y.; Ye, C.; Hassan, M.; Cheng, X.; Wu, Y.; Zhang, Y. Deep Learning Based Detector YOLOv5 for Identifying Insect Pests. *Appl. Sci.* 2022, 12, 10167. [CrossRef]
25. Su, X.; Zhang, J.; Ma, Z.; Dong, Y.; Zi, J.; Xu, N.; Zhang, H.; Xu, F.; Chen, F. Identification of Rare Wildlife in the Field Environment Based on the Improved YOLOv5 Model. *Remote Sens.* 2024, 16, 1535. [CrossRef]
26. Qiao, Y.; Guo, Y.; He, D. Cattle Body Detection Based on YOLOv5-ASFF for Precision Livestock Farming. *Comput. Electron. Agric.* 2023, 204, 107579. [CrossRef]
27. Jiang, B.; Wu, Q.; Yin, X.; Wu, D.; Song, H.; He, D. FLYOLOv3 Deep Learning for Key Parts of Dairy Cow Body Detection. *Comput. Electron. Agric.* 2019, 166, 104982. [CrossRef]
28. Tannous, M.; Stefanini, C.; Romano, D. A Deep-Learning-Based Detection Approach for the Identification of Insect Species of Economic Importance. *Insects* 2023, 14, 148. [CrossRef]
29. Hamzaoui, M.; Ould-Elhassen Aoueileyine, M.; Romdhani, L.; Bouallegue, R. An Improved Deep Learning Model for Underwater Species Recognition in Aquaculture. *Fishes* 2023, 8, 514. [CrossRef]
30. Aota, T.; Ashizawa, K.; Mori, H.; Toda, M.; Chiba, S. Detection of Anolis Carolinensis Using Drone Images and a Deep Neural Network: An Effective Tool for Controlling Invasive Species. *Biol. Invasions* 2021, 23, 1321–1327. [CrossRef]
31. Guo, S.-S.; Lee, K.-H.; Chang, L.; Tseng, C.-D.; Sie, S.-J.; Lin, G.-Z.; Chen, J.-Y.; Yeh, Y.-H.; Huang, Y.-J.; Lee, T.-F. Development of an Automated Body Temperature Detection Platform for Face Recognition in Cattle with YOLO V3-Tiny Deep Learning and Infrared Thermal Imaging. *Appl. Sci.* 2022, 12, 4036. [CrossRef]
32. Zhang, B.; Xiao, D.; Liu, J.; Huang, S.; Huang, Y.; Lin, T. Pig Eye Area Temperature Extraction Algorithm Based on Registered Images. *Comput. Electron. Agric.* 2024, 217, 108549. [CrossRef]
33. Karameta, E.; Gavriilidi, I.; Sfenthourakis, S.; Pafilis, P. Seasonal Variation in the Thermoregulation Pattern of an Insular Agamid Lizard. *Animals* 2023, 13, 3195. [CrossRef] [PubMed]
34. Liu, S.; Qi, L.; Qin, H.; Shi, J.; Jia, J. Path Aggregation Network for Instance Segmentation. In Proceedings of the 2018 IEEE/CVF Conference on Computer Vision and Pattern Recognition, Salt Lake City, UT, USA, 18–23 June 2018; pp. 8759–8768. [CrossRef]
35. Wang, C.-Y.; Mark Liao, H.-Y.; Wu, Y.-H.; Chen, P.-Y.; Hsieh, J.-W.; Yeh, I.-H. CSPNet: A New Backbone That Can Enhance Learning Capability of CNN. In Proceedings of the 2020 IEEE/CVF Conference on Computer Vision and Pattern Recognition Workshops (CVPRW), Seattle, WA, USA, 14–19 June 2020; pp. 1571–1580. [CrossRef]
36. Terven, J.; Córdova-Esparza, D.-M.; Romero-González, J.-A. A Comprehensive Review of YOLO Architectures in Computer Vision: From YOLOv1 to YOLOv8 and YOLO-NAS. *Mach. Learn. Knowl. Extr.* 2023, 5, 1680–1716. [CrossRef]

37. Xie, Q.; Wu, M.; Bao, J.; Zheng, P.; Liu, W.; Liu, X.; Yu, H. A Deep Learning-Based Detection Method for Pig Body Temperature Using Infrared Thermography. *Comput. Electron. Agric.* **2023**, *213*, 108200. [CrossRef]
38. Wang, Y.; Kang, X.; Chu, M.; Liu, G. Deep Learning-Based Automatic Dairy Cow Ocular Surface Temperature Detection from Thermal Images. *Comput. Electron. Agric.* **2022**, *202*, 107429. [CrossRef]

Disclaimer/Publisher's Note: The statements, opinions and data contained in all publications are solely those of the individual author(s) and contributor(s) and not of MDPI and/or the editor(s). MDPI and/or the editor(s) disclaim responsibility for any injury to people or property resulting from any ideas, methods, instructions or products referred to in the content.

Article

Elevating Detection Performance in Optical Remote Sensing Image Object Detection: A Dual Strategy with Spatially Adaptive Angle-Aware Networks and Edge-Aware Skewed Bounding Box Loss Function

Zexin Yan, Jie Fan, Zhongbo Li * and Yongqiang Xie *

Institute of System Engineering, Academy of Military Sciences, Beijing 100141, China; yanzexin@outlook.com (Z.Y.); hanyovladscarlet@mail.nwpu.edu.cn (J.F.)
* Correspondence: lzb05296@163.com (Z.L.); yqxie2021@outlook.com (Y.X.)

Abstract: In optical remote sensing image object detection, discontinuous boundaries often limit detection accuracy, particularly at high Intersection over Union (IoU) thresholds. This paper addresses this issue by proposing the Spatial Adaptive Angle-Aware (SA^3) Network. The SA^3 Network employs a hierarchical refinement approach, consisting of coarse regression, fine regression, and precise tuning, to optimize the angle parameters of rotated bounding boxes. It adapts to specific task scenarios using either class-aware or class-agnostic strategies. Experimental results demonstrate its effectiveness in significantly improving detection accuracy at high IoU thresholds. Additionally, we introduce a Gaussian transform-based IoU factor during angle regression loss calculation, leading to the development of Edge-aware Skewed Bounding Box Loss (EAS Loss). The EAS loss enhances the loss gradient at the final stage of angle regression for bounding boxes, addressing the challenge of further learning when the predicted box angle closely aligns with the real target box angle. This results in increased training efficiency and better alignment between training and evaluation metrics. Experimental results show that the proposed method substantially enhances the detection accuracy of ReDet and ReBiDet models. The SA^3 Network and EAS loss not only elevate the mAP of the ReBiDet model on DOTA-v1.5 to 78.85% but also effectively improve the model's mAP under high IoU threshold conditions.

Keywords: optical remote sensing; object detection; discontinuous boundary; rotation equivariant

Citation: Yan, Z.; Fan, J.; Li, Z.; Xie, Y. Elevating Detection Performance in Optical Remote Sensing Image Object Detection: A Dual Strategy with Spatially Adaptive Angle-Aware Networks and Edge-Aware Skewed Bounding Box Loss Function. *Sensors* **2024**, *24*, 5342. https://doi.org/10.3390/s24165342

Academic Editor: Man Qi, Matteo Dunnhofer

Received: 11 July 2024
Revised: 4 August 2024
Accepted: 13 August 2024
Published: 18 August 2024

Copyright: © 2024 by the authors. Licensee MDPI, Basel, Switzerland. This article is an open access article distributed under the terms and conditions of the Creative Commons Attribution (CC BY) license (https://creativecommons.org/licenses/by/4.0/).

1. Introduction

In the field of optical remote sensing image target detection, the random orientation of almost all objects in the image often leads to horizontal bounding boxes encompassing a significant amount of irrelevant background when annotating detected objects. In comparison, rotated bounding boxes can greatly improve the pixel area ratio between objects and background within the selected region, better accommodating the annotation of targets in arbitrary directions. They provide accurate information about the direction, position, and size of the targets, enabling the model to perform better in target detection tasks. Existing methods for rotated object detection are mainly built upon generic object detection approaches. They redefine the representation of detection boxes, introduce additional angular dimensions for rotated detection boxes, and optimize them through distance loss. Rotated object detection methods have played a crucial role in various fields, including text detection [1–6], face recognition [7–9], and remote sensing image target detection [10–12].

However, introducing angle parameters also introduces uncertainty into the detection task, with the most prominent issue being the discontinuous boundary problem. Additionally, because the evaluation metric for the final model is based on IoU values rather than angular differences, optimizing the predicted box angles through distance loss introduces a mismatch between loss calculation and evaluation metrics. These two issues have led to suboptimal performance of existing models under high IoU threshold conditions (IoU > 0.5). Achieving higher accuracy in rotated object detection has become a significant research focus.

This section first introduces the discontinuous boundary problem, describing the origins of the discontinuity in parameterized regression. It then discusses the mismatch between the calculation of angles in the loss function and evaluation metrics. Following that, it surveys the progress of domestic and international research and concludes by introducing the research objectives of this paper.

1.1. Discontinuous Boundary Issue

Rotated bounding boxes are commonly represented in two ways: the five-parameter representation [1,5,11–15] and the eight-parameter representation [2,16–18]. Both representations can describe the position, size, and rotation angle of a rotated bounding box. The five-parameter representation is often preferred due to its simplicity, intuitive nature, and the reduced number of parameters for computation and storage. In this representation, scholars have devised various methods for representing rotated boxes, establishing a range for the angle in radians, binding it closely with the height and width of the detection box to avoid the confusion of multiple angle values caused by the angular periodicity.

Due to different definitions of the θ range, three major categories of rotated box definitions have gradually emerged in rotated object detection. The OpenCV definition [5,12–15], abbreviated as OC, sets the angle range as angle $\in (0, 90°]$, $\theta \in (0, \frac{1}{2}\pi]$, where the acute angle between the positive x-axis and the edge defined as the width is positive. The long edge definition [5,11] defines the long edge as the width and the short edge as the height. The angle θ is the angle between the long edge width and the x-axis. Based on the angle range, it is further divided into le135 and le90 definitions. The angle ranges are set as angle $\in [-45°, 135°)$ or $\theta \in [-\frac{\pi}{4}, \frac{3\pi}{4})$ for le135, and angle $\in [-90°, 90°)$ or $\theta \in [-\frac{\pi}{2}, \frac{\pi}{2})$ for le90.

In parameterized regression-based rotational detection methods, as illustrated in Figure 1, significant differences exist between the ideal regression path and the actual regression path under the le90 definition and the counter-clockwise (CCW) representation. When regressing candidate boxes, if the ideal regression path (indicated by the dark red dashed arrow in Figure 1a) is followed, the candidate box rotates clockwise by $\frac{1}{8}\pi$ to approximate the true target box. Although the predicted box has an $IoU \approx 1$ with the true target box, the loss value is significantly greater than 0. This is because the angle of the predicted box changes from $-\frac{1}{2}\pi$ to $-\frac{5}{8}$, increasing the discrepancy with the true target box's angle of $\frac{3}{8}\pi$. Due to the periodic nature of angles, although the boxes are almost coincident visually, the angle difference leads to an increase in the loss value. In this situation, the model needs to regress along a more distant path, namely, counter-clockwise rotation by $\frac{7}{8}\pi$ (indicated by the dashed red arrow in Figure 1c), achieving an $IoU \approx 1$ and a loss value $Loss \approx 0$ to generate the final predicted box.

In summary, the discontinuous boundary problem is primarily caused by the angular periodicity and the exchangeability of the long and short sides in the five-parameter representation. The ideal regression path exceeds the predefined range, encountering boundary issues, leading to an increase in regression loss.

Faced with this challenge, the transformation of predicted boxes under the OpenCV definition involves both the exchange of width and height, as well as angle transformation, resulting in an overall complexity that may lead to less accurate predictions in such conditions [19]. In contrast, the long edge definition only necessitates considering a broad range of angle transformations when addressing the issue of discontinuous boundaries. To align with the module design in the subsequent sections of this article, the le90 definition is adopted as the angle range definition for rotated bounding boxes. Even in the case of the four-point representation method, discontinuous boundary problems may arise due to the order of corner points, a matter which will not be further elaborated upon here.

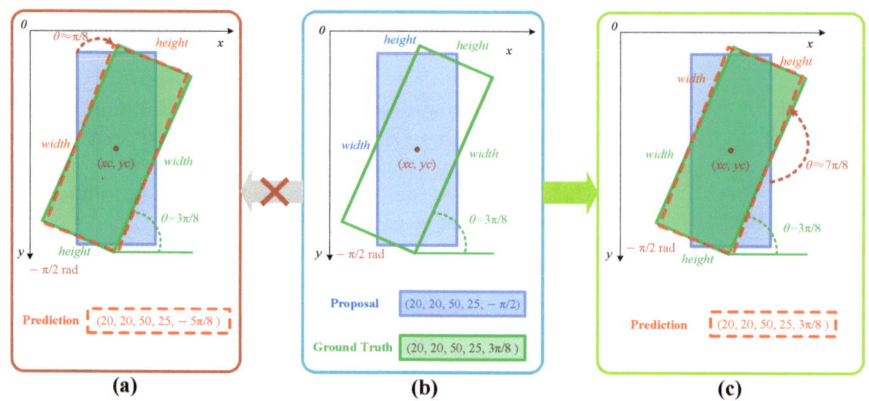

Figure 1. Prediction box generation method under le90 definition and CCW representation conditions. (**a**) Ideal regression path for generating prediction boxes. (**b**) Initial position of the proposal. (**c**) Actual regression path for generating prediction boxes.

1.2. Mismatch between Loss and Evaluation Metrics

Over the years, the primary approach to improving the detection accuracy of rotated object detection models has been to propose better network architectures and more effective strategies for extracting reliable local features. However, there is a largely overlooked optimization path that has received limited research attention in the field: designing a novel loss function to replace the commonly used regression loss function, aiming to enhance the model's detection performance. In object detection, the Intersection over Union (IoU) between the detection boxes generated by the model and the ground truth boxes is a crucial evaluation metric used to calculate the mAP. Therefore, in rotated object detection, special consideration needs to be given to minimizing angle errors and maximizing IoU values.

Most existing state-of-the-art rotated object detection models largely adopt the widely used Smooth L1 loss [20,21] as the loss function for the regression branch [22,23], following common practice in general object detection. This loss calculates the value based on the numerical differences between two bounding boxes to minimize coordinate, aspect ratio, and angle deviations. As depicted in Figure 2, the curve illustrates that the angle difference between two rectangles with an aspect ratio of 1:6 and overlapping centroids does not show a linear relationship with IoU values. When the angle difference is less than 20°, a smaller angle difference results in a larger change in IoU. Specifically, when the angle difference is within 12.96°, further reduction in the angle difference leads to a sharp increase in IoU values. However, at this point, the loss value of the model's angle difference diminishes linearly. This scenario causes the model to lack sufficient loss gradient when the predicted box's angle is close to the angle of the real target box. As a result, the model fails to learn how to generate more accurately angled detection boxes, preventing the detection boxes from further aligning accurately, which is clearly not an ideal outcome.

In summary, the Smooth L1 loss is not sufficiently sensitive to small angle errors, and a good local optimum based on the Smooth L1 loss may not necessarily be a local optimum for IoU. This makes it challenging to effectively guide the model to maximize the IoU values between the detection boxes and the real target boxes.

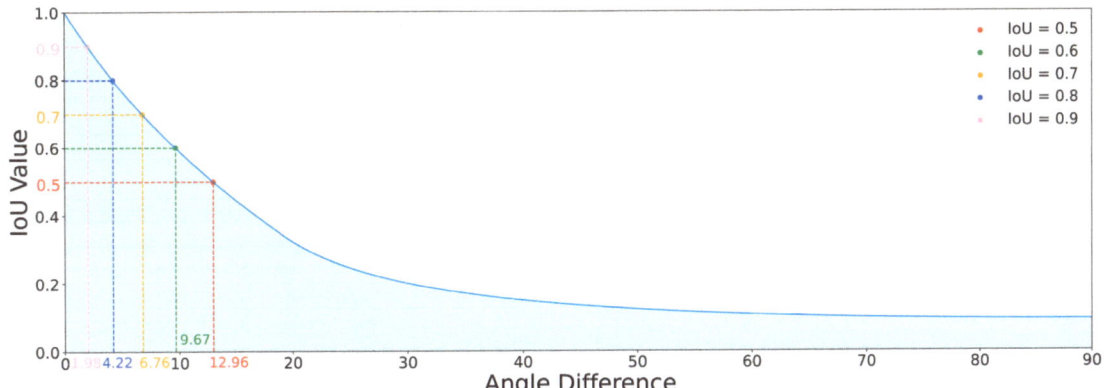

Figure 2. Relationship curve between angular difference and IoU for rectangles with a 1:6 aspect ratio and overlapping centroids.

1.3. Related Work

Due to the complexity of scenes in remote sensing images, including small sizes, large aspect ratios, randomly and densely distributed objects, and arbitrarily oriented objects, rotation object detection models based on bounding box regression continue to dominate due to their higher accuracy and robustness. Parameterized regression methods are mainly divided into two branches.

Single-stage: DAL [24] proposes dynamic adaptive learning to dynamically select high-quality anchor boxes for accurate object detection, significantly reducing the number of predefined anchor boxes. R3Det [12] introduces a feature refinement module based on RetinaNet, using a progressive regression approach for fast and accurate object detection. It also suggests a skewed IoU (SkewIoU) loss to mitigate sensitivity to angle transformations in targets with large aspect ratios. RSDet [17] regresses rotated bounding boxes using four points and introduces a modulation rotation loss to address the discontinuity issue in existing losses when facing discontinuous boundaries. S2A-Net [25] introduces a Feature Alignment Module (FAM) for adaptive alignment with high-quality anchor boxes and an Orientation-aware Detection Module (ODM) for encoding orientation information, alleviating inconsistencies between the classification and regression branches. CSL [26] transforms angle prediction from a regression problem into a classification task to address the issue of discontinuous boundaries.

Two-stage: ICN [10] designs a joint image cascade network to extract multiscale semantic features and optimize regression losses. RoI Transformer [11] predicts a coarse rotated RoI based on RPN in its first stage and refines the prediction using RoI Align in the second stage, resulting in more accurate rotated RoIs and improving both efficiency and accuracy compared to rotated RPN. SCRDet [13] uses a sampling fusion network to enhance sensitivity to small objects and proposes supervised pixel attention networks and channel attention networks for joint detection of small and cluttered objects. It combines Smooth L1 loss with IoU factors to address discontinuous boundary issues in rotated bounding boxes. Gliding Vertex [18] accurately describes target orientation by sliding along each corresponding edge of a horizontal bounding box, introducing an area ratio factor between the target and its horizontal bounding box to guide the model in predicting quadrilateral detection boxes accurately. Oriented R-CNN [22] introduces a midpoint offset representation on Faster R-CNN, modifying the output parameters of the RPN regression branch from 4 to 6 to achieve rotated candidate boxes, significantly improving detection accuracy and computational efficiency. These are mainstream state-of-the-art two-stage detection models proposed in the last three years, inheriting the architecture of Faster R-CNN [21]. ReDet [23] introduces the concept of E(2)-Equivariant Steerable CNNs [27] (E(2)-CNNs) into object detection based on RoI Transformer, utilizing E2-CNNs to rewrite ResNet50 as ReResNet50. It redesigns the Rotation-invariant Region of Interest Align (RiRoIAlign) module, aligning in both channel and spatial dimensions to obtain rotation-invariant features. ReBiDet [28], based on ReDet, enhances feature fusion through the ReBiFPN module, balances the difficulty and proportion of positive samples during training using the DPRL module, and optimizes anchor box sizes and aspect ratios to further improve the model's detection performance.

Existing object detection methods more or less involve the study of discontinuous boundary issues and mismatches between loss calculation and evaluation metrics. However, researchers have primarily concentrated on improving detection accuracy at IoU = 0.5, with limited attention to enhancing accuracy at IoU = 0.55 and above. These two issues result in suboptimal performance of current models at high IoU thresholds (greater than 0.5). As detection accuracy at IoU = 0.5 approaches a bottleneck, achieving higher precision in rotational object detection is poised to become a key focus of future research.

1.4. Goal of the Research

This paper addresses two primary issues: the discontinuous boundary problem and the mismatch between loss calculation and evaluation metrics. Solutions are proposed based on the ReBiDet model, and our work is organized into the following two aspects:

1.4.1. Discontinuous Boundary Issue

We conduct a thorough analysis of the discontinuous boundary problem, focusing on its origins in parameterized regression. Our analysis reveals that the core issue is the inaccurate localization of detection boxes due to large-angle variations during the regression of rotated bounding boxes. To address this problem, we propose the Spatially Adaptive Angle-Aware (SA^3) Network, a cascaded structure designed to handle the large-angle variations caused by discontinuous boundaries. This network enhances the model's ability to adapt to complex scenarios, thereby improving the accuracy of object detection.

1.4.2. Mismatch between Loss Calculation and Evaluation Metrics

To tackle the issue of traditional loss functions being insensitive to Intersection over Union (IoU) values, we propose the Edge-aware Skewed Bounding Box Loss (EAS Loss). This novel loss function addresses the nonlinear decay of the loss value when the predicted box's angle closely matches the target box's angle. By incorporating this loss function, we aim to improve the alignment between the loss calculation and IoU values, enhancing overall model performance.

Additionally, we perform ablation studies and extensive comparative experiments with state-of-the-art models on two datasets to validate the effectiveness of the proposed modules and approaches.

2. Methods

2.1. Spatially Adaptive Angle-Aware Network

2.1.1. Spatially Adaptive Angle-Aware Network Structure

The discontinuous boundary issue fundamentally results in imprecise localization of the final detection box due to the wide range of angle variations during rotated bounding box regression. Initial rotated object detection models, such as those based on Faster R-CNN [21], incorporated an angle parameter during the bounding box regression stage. However, the precision of the prediction boxes generated by a single regression was unsatisfactory. Even in general object detection, studies have shown that when dealing with candidate boxes of varying quality generated by the RPN module, a single regression function struggles to align all predicted bounding boxes accurately with the ground truth. Subsequently, the RoI Transformer [11] proposed a two-stage regression approach to produce more accurate prediction boxes. This strategy performed well under the condition of IoU = 0.5 and has been widely accepted by researchers, continuing to be adopted by state-of-the-art rotated object detection models. However, in rotated object detection, the introduction of the parameter θ exacerbates the discontinuous boundary issue. As a result, even with two regression functions, it remains challenging to achieve precise regression tasks in rotated object detection when IoU > 0.5.

Inspired by the Cascade R-CNN [29] approach, this subsubsection decomposes the regression task for rotated bounding boxes into three stages. Specifically, three dedicated regression functions $\{f_1, f_2, f_3\}$ are designed for each stage. These regression functions optimize the bounding boxes generated at their respective stages. The overall regression equation is as follows:

$$F(x, b) = f_3 \circ f_2 \circ f_1(x, b), \tag{1}$$

Here, $F(x, b)$ represents the regression equation, x denotes the image features corresponding to the candidate bounding boxes, and f_1, f_2, f_3 represent the regression functions for the three stages. It is important to note that each regression function f_t ($t = 1, 2, 3$) in the cascade is specifically trained for the optimization of the corresponding bounding box b^t ($t = 1, 2, 3$) at its respective stage. This approach gradually improves the generated predicted bounding boxes through coarse regression, fine-tuning, and precision refinement. The entire model learns automatically based on the training sample set $\{G_i, B_i\}$, requiring no further manual intervention, achieving the goal of adaptive angle regression. The regression process is illustrated in Figure 3. In an ideal scenario, the true target box has an angle of $\theta = \frac{7}{16}\pi$. After being filtered by the RPN module, a proposal box is generated that overlaps the centroid of the ground truth box, matching its width and height, but with an angle of $\theta = -\frac{1}{2}\pi$. Initially, Stage 1 performs a large-angle transformation, coarsely regressing the proposal box to the light blue Coarse Regression box. Subsequently, Stages 2 and 3 refine the regression further, resulting in the Refined Predicted box that closely aligns with the ground truth box.

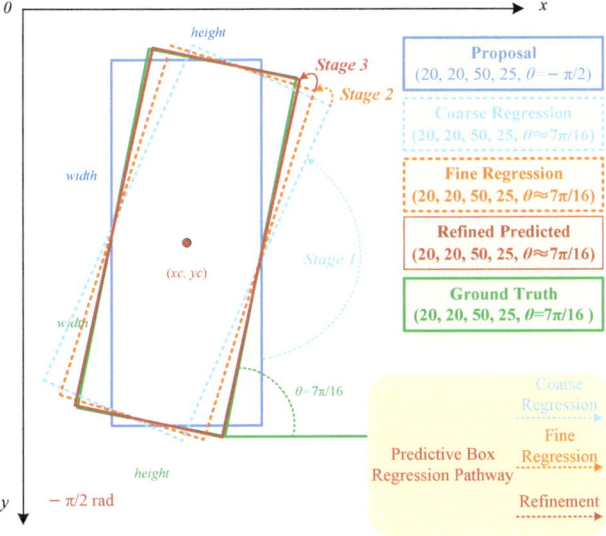

Figure 3. Process of generating predicted boxes by the Spatially Adaptive Angle-aware Network.

Based on the concept of adaptive angle-regression bounding boxes mentioned above, we have designed the SA^3 Network, as depicted in Figure 4. The input to the SA^3 Network is the candidate bounding boxes generated by RPN, and the output consists of the predicted bounding boxes and the classification of those boxes. The processing flow of the SA^3 Network is as follows: **Stage One: Rotation Bounding Box Regression.** The primary function is to regress horizontal candidate bounding boxes generated by RPN into rotated bounding boxes. The process involves filtering horizontal candidate bounding boxes through the DPRL sampler. Subsequently, RoI Align extracts the feature maps of RoIs.

Following the practices of many scholars in the field [11,20,21,23,30], these feature maps are resized to 7×7 (a compromise between computational complexity, feature expressiveness, and simplicity of network design) to facilitate subsequent classification and bounding box regression operations. These fixed-size feature map blocks are fed into the Rotated BBox Head for convolution and fully connected operations, resulting in the regression-generated rotated predicted bounding boxes and their respective classifications; **Stages Two and Three: Rotation Bounding Box Angle Fine-tuning.** The main function is to fine-tune various parameters of the rotated bounding boxes generated in the first stage. The process is similar to the first stage, where rotated bounding boxes are filtered through the Rotated DPRL sampler. Next, the RiRoI Align extracts feature maps corresponding to the rotated candidate bounding boxes. These feature maps are processed through the Rotated BBox Head for convolution and fully connected operations, producing refined rotated predicted bounding boxes and their respective classifications.

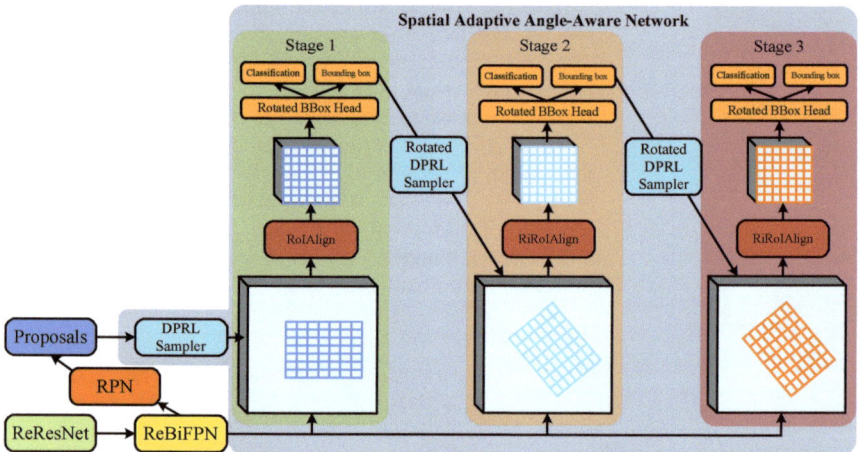

Figure 4. Spatially Adaptive Angle-aware Network structure.

2.1.2. Boundary Box Regression Class Strategy Selection

Boundary box regression class strategies include class-agnostic and class-aware strategies. Both have their advantages and disadvantages, requiring consideration of specific application scenarios.

Class-Agnostic Boundary Box Regression Strategy. From a global perspective, in remote sensing images, the orientation of objects of each category is randomly distributed. Given the widely varying aspect ratios of the bounding boxes corresponding to all objects, it is not meaningful to consider the category of objects in the feature map during the coarse regression stage. Therefore, for the regression function $f_1(x, b)$ in the first stage, a class-agnostic strategy is adopted for training the bounding box regression function. The network structure is depicted in Figure 5. The class-agnostic [31] strategy regresses bounding boxes without distinguishing specific categories during the process. This approach enhances recall rates, particularly in cases where there is incomplete annotation in the dataset, improving the model's robustness. Importantly, it significantly reduces the computational parameter volume of the bounding box regression branch. However, the drawback is that the class-agnostic strategy regresses all objects that the network deems possibly foreground, without determining their specific category. This can lead to multiple detection boxes corresponding to a single complex-patterned object and inaccurate classification. To mitigate the impact on the final detection accuracy, class-aware strategies are employed in the subsequent two stages of regression functions.

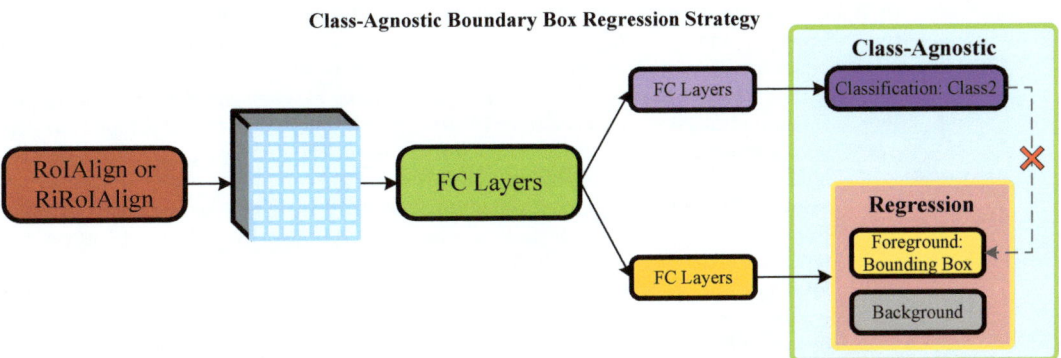

Figure 5. Structure diagram of the class-agnostic strategy regression function in the coarse regression stage.

Some studies in the general object detection field have demonstrated the effective enhancement of model detection performance using class-agnostic strategies [31]. Experimental findings indicate that, in cases of severe dataset category imbalance, the class-agnostic strategy effectively mitigates the impact of dataset incompleteness on the performance of the object detection model.

Class-Aware Boundary Box Regression Strategy. Unlike images captured from a horizontal perspective, objects of different categories in remote sensing images exhibit significant differences in the aspect ratios of their true target boxes. For instance, the aspect ratio of bridges and docks may exceed 1:10, while objects like planes and helicopters may have a ratio of 1:1. Directly applying the research experience of general object detection without considering the actual situation of optical remote sensing images may not be wise. Therefore, in the design of the regression functions $f_2, f_3(x, b)$ for the fine-tuning and precision refinement stages, a class-aware strategy is employed. The network structure is shown in Figure 6.

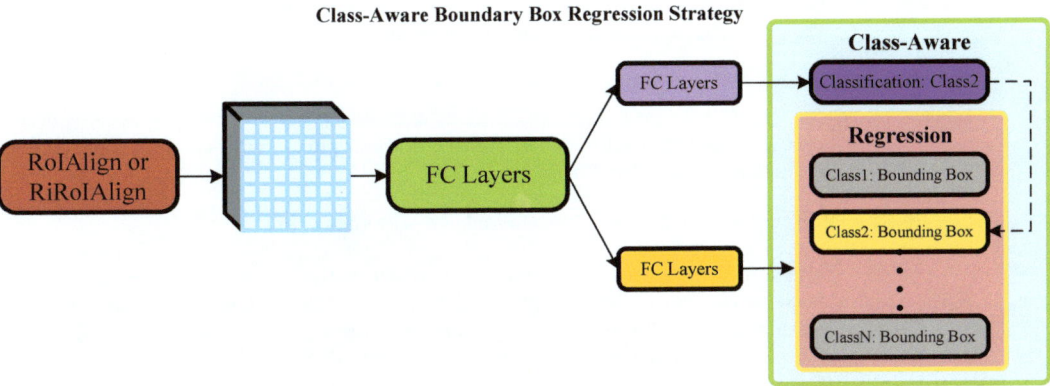

Figure 6. Structure diagram of class-aware strategy regression function in the fine-tuning and precision refinement stages.

When using a class-aware strategy to regress bounding boxes, the regression network individually infers each category by traversing all categories, computing bounding boxes for each category in the feature map block, and outputting the coordinates of bounding boxes for all categories. Then, based on the output results of the classification network branch, the bounding boxes corresponding to the category of the classification network branch are indexed, labeled as detection boxes, and other irrelevant category bounding boxes are discarded, obtaining the final detection results. The advantage of this approach is that it allows the model to undergo fine training for each category, serving as the basis for fine regression and refinement learning, ultimately enabling the model to generate accurate detection boxes for each category's object. The drawback is that for a dataset with annotated N categories of objects, the computational complexity of the regression network branch is N times that of the class-agnostic strategy, making it less efficient. However, for this project, improving detection accuracy through algorithm design takes precedence, while also considering a reduction in computational complexity. As per the empirical knowledge of generative artificial intelligence, the size of the computed parameter quantity is positively correlated with the actual performance of the model. In experiments, it was observed that although the class-aware strategy incurs a higher computational cost, it does not necessarily achieve better detection performance when the number of categories in the dataset is imbalanced or the total number of targets is small.

In summary, when using the SA^3 Network, one should consider whether the number of targets in each category in the dataset is balanced, the total number of targets, and other practical issues. A comprehensive judgment should be made regarding whether to adopt a class-aware or class-agnostic strategy during the fine regression and refinement stages of the SA^3 Network. Inappropriate strategy selection may result in a significant reduction in the model's detection performance.

2.2. Edge-Aware Skewed Bounding Box Loss

A persistent issue in the design of regression losses for rotated object detection is the inconsistency between model training metrics and final evaluation metrics. This inconsistency acts as a bottleneck, necessitating the design of a loss function specifically tailored for rotated object detection. Such a loss function must consider the unique characteristics of this task, especially in terms of the Intersection over Union (IoU), to enhance the model's sensitivity to changes in angles.

To address this challenge, we introduce a novel regression loss, named the EAS loss. This loss function takes into account the IoU value between the predicted bounding box and the ground truth. The EAS loss effectively mitigates the inconsistency between training and evaluation metrics, enabling the model to accurately regress angles, particularly when the predicted box is close to the ground truth.

2.2.1. EAS Loss Design

To enhance the sensitivity to small angle deviations and improve metric consistency, a natural idea is to use IoU when regressing the angle θ. In this case, the loss calculation is defined as in Equation (2):

$$\text{Loss}_{IoU} = -\log(\text{eps} + \text{IoU}), \tag{2}$$

However, this introduces a new problem: when the centroid of the predicted box is far from the centroid of the ground truth box, the IoU value will be very small, leading to slow angle regression by the model. Assuming the aspect ratio of the box is 1:6, as shown in Figure 7, even when the centroid of the predicted box and the ground truth box coincide, the slope of the Loss_{IoU} value becomes flat when the angle difference θ exceeds 60°. This reduces the regression efficiency of the model for larger angle differences.

To address this, we introduce an IoU factor into the angle regression loss L_θ^* calculation, as shown in Equation (3):

$$L_\theta = L_\theta^* - \log(\text{eps} + \text{IoU}), \tag{3}$$

Here, L_θ^* adopts the calculation method of the Smooth L1 loss, leading to the ideal EAS loss calculation Equation (4):

$$\text{Loss}_{\text{EAS}}(d_\theta, t_\theta, \beta, \gamma) = \begin{cases} 0.5 \cdot \left(\frac{|d_\theta - t_\theta|^2}{\beta} - \gamma \cdot \log(\text{eps} + \text{IoU}) \right), & \text{if } |d_\theta - t_\theta| < \beta \\ |d_\theta - t_\theta| - 0.5 \cdot \beta - \gamma \cdot \log(\text{eps} + \text{IoU}), & \text{otherwise} \end{cases}, \quad (4)$$

Here, eps $= 10^{-6}$, and β and γ are adjustable variables. When $\beta = 1$ and $\gamma = 1/9$, the relationship between Loss$_{\text{EAS}}$ and the angle difference θ is depicted in Figure 8.

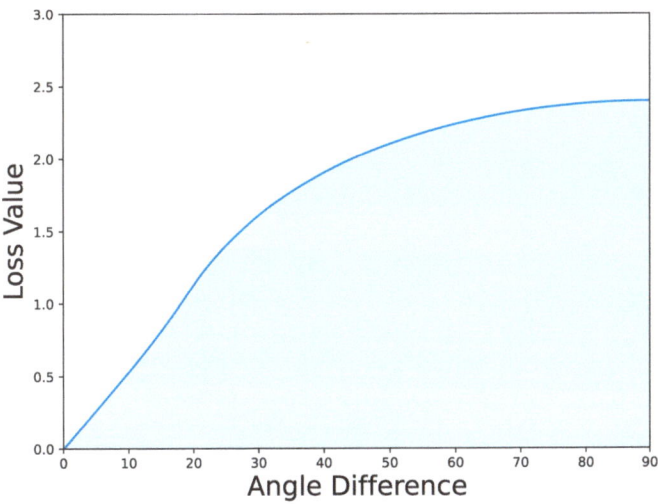

Figure 7. Curve depicting the relationship between IoU loss and angle difference.

When the centroids of the predicted box and the ground truth box do not coincide, the EAS loss, incorporating the IoU factor in the angle regression loss component, prevents ineffective rotations of the predicted box. When the centroids of the predicted box and the ground truth box do coincide, the IoU factor allows the model to quickly and accurately learn to generate more precise predicted boxes as the predicted and ground truth angles approach each other. This makes it an ideal angle regression loss function.

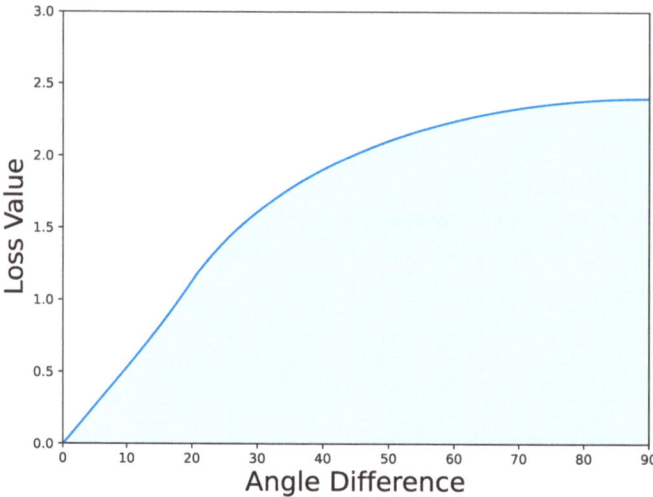

Figure 8. Curve depicting the relationship between EAS loss and angle.

The next issue is how to calculate the IoU value. Here, the IoU between the rotated bounding boxes is termed SkewIoU to distinguish it from horizontal IoU. The definition of SkewIoU is the same as IoU, as shown in Equation (5):

$$\text{SkewIoU}(P, G) = \frac{\text{Area}(P \cap G)}{\text{Area}(P \cup G)}, \tag{5}$$

where P and G are two rotated bounding boxes. To compute their intersection and union areas accurately, skewed bounding boxes are treated as polygons, and the convex polygon intersection calculation method [5] is employed to precisely calculate SkewIoU. This process involves sorting and combining the coordinates of the vertices and intersection points of the rotated boxes, followed by the application of the triangulation method to calculate the area of the intersection region. The overall computation is relatively complex, with sorting and triangulation being the primary time-consuming factors. In practical applications, libraries such as OpenCV and Shapely are often employed to calculate SkewIoU. The specific implementation and optimizations in these libraries can impact performance, but the fundamental principles of computational complexity remain unchanged.

In general object detection, the IoU loss has long been a focus for effectively mitigating the inconsistency between evaluation metrics (dominated by IoU) and regression loss calculation methods. However, in rotated object detection, due to the computationally expensive nature of SkewIoU calculation, it is primarily used for validation and evaluation and has not been widely adopted in loss functions.

2.2.2. Gaussian Transformation for Approximate IoU Calculation

Recent studies, such as PIoU [32], projection IoU [33], GWD [34], and KLD [35], have explored methods to simulate the approximate SkewIoU loss. Among these, GWD and KLD introduced a Gaussian modeling approach, simplifying the complex SkewIoU calculation into a more efficient process. However, their methods involve nonlinear transformations and hyperparameters in the final loss function design using Gaussian distribution distance metrics, making them not fundamentally SkewIoU. KFIoU [36] is a loss function based on the approximation of SkewIoU and center-point distance, but its performance in the proposed model in this paper is not ideal. As shown in Table 1, after incorporating the KFIoU loss, the model's AP50, AP75, and mAP all experienced varying degrees of decline. In the original paper, KFIoU achieved satisfactory results with backbone networks like ResNet-152 with large parameter sizes. In contrast, the model in this paper utilizes the

ReResNet-50 backbone network with relatively fewer computational parameters, which might be one of the reasons for the performance degradation.

Table 1. Experimental results of the ReBiDet model and KFIoU loss on the DOTA-v1.5 dataset.

Method	mAP	AP50	AP75
ReBiDet [28]	41.15	69.48	42.54
ReBiDet + KFIoU	36.70	68.40	33.62
ReBiDet * [28]	49.26	77.96	52.53
ReBiDet * + KFIoU	46.09	77.06	48.14

* Indicates training and testing on DOTA-v1.5 with random rotation augmentation and offline multi-scale augmentation.

Despite KFIoU's poor performance in ReBiDet, a simpler and more efficient method for approximating SkewIoU is worth considering. This method converts two rotated bounding boxes into Gaussian distributions, representing the overlapping region with a simpler mathematical formula and calculating the area of the minimum enclosing rectangle of this region. This approach significantly simplifies the original geometric computation while maintaining a high approximation to the exact IoU value [36]. This method adheres to the SkewIoU calculation process, is mathematically rigorous, and does not introduce additional hyperparameters. Although the accuracy of the SkewIoU values obtained is relatively low, precise SkewIoU calculation is not strictly necessary in loss functions; minor errors are tolerable as long as the trend of the approximated SkewIoU matches the true value. Below are the basic steps and derivations for approximating SkewIoU using the Gaussian transformation method [36]:

Transforming Rotated Rectangular Boxes into Gaussian Distributions. To begin, the rotated rectangular boxes are transformed into Gaussian distributions $G(\mu, \Sigma)$. Each rectangular box is represented by two parameters: the covariance matrix Σ and the center coordinates μ. For a rotated rectangular box $B(x, y, w, h, \theta)$, $\mu = \begin{bmatrix} x \\ y \end{bmatrix}$, and the covariance matrix is calculated as shown in Equation (6):

$$\Sigma = R \Lambda R^T, \quad (6)$$

where $R = \begin{bmatrix} \cos\theta & -\sin\theta \\ \sin\theta & \cos\theta \end{bmatrix}$, $\Lambda = \frac{1}{4}\begin{bmatrix} w^2 & 0 \\ 0 & h^2 \end{bmatrix}^2$. The final form of the covariance matrix Σ is given in Equation (7):

$$\Sigma = \begin{bmatrix} \frac{w^2}{4}\cos^2\theta + \frac{h^2}{4}\sin^2\theta & \frac{(w^2-h^2)}{4}\cos\theta\sin\theta \\ \frac{(w^2-h^2)}{4}\cos\theta\sin\theta & \frac{w^2}{4}\sin^2\theta + \frac{h^2}{4}\cos^2\theta \end{bmatrix}, \quad (7)$$

The transformed Gaussian distribution is illustrated in Figure 9.

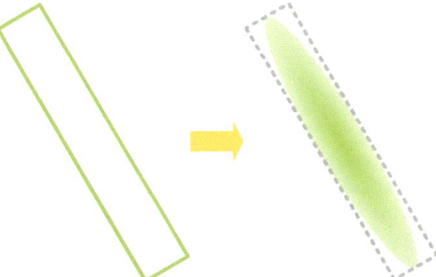

Figure 9. Transformed Gaussian distribution of rotated rectangular boxes.

Calculating Overlapping Area of Two Gaussian Distributions. The overlapping area of two Gaussian distributions is computed by multiplying the Gaussian distributions of the predicted box P and the true target box G using the Kalman filter's multiplication rule. Specifically, the predicted box's Gaussian distribution $G_P(\mu_P, \Sigma_P)$ is treated as the predicted value, and the true target box's Gaussian distribution $G_G(\mu_G, \Sigma_G)$ is treated as the observed value. This yields the approximate Gaussian distribution $G_I(\mu_I, \Sigma_I)$ for the overlapping region I, as shown in Equation (8):

$$\alpha G_I(\mu_I, \Sigma_I) = G_P(\mu_P, \Sigma_P) \cdot G_G(\mu_G, \Sigma_G), \tag{8}$$

Here, $\alpha G_I(\mu_I, \Sigma_I)$ does not have a probability sum of 1 and is not a standard Gaussian distribution. The coefficient α can be expressed by Equation (9):

$$\alpha = G_\alpha(\mu_G, \Sigma_P + \Sigma_G) = \frac{1}{\sqrt{\det(2\pi(\Sigma_P + \Sigma_G))}} e^{-\frac{1}{2}(\mu_P - \mu_G)^T (\Sigma_P + \Sigma_G)^{-1}(\mu_P - \mu_G)}, \tag{9}$$

When $\mu_P - \mu_G \approx 0$, α can be approximated as a constant. Since the EAS loss is computed separately for the centroids of the bounding boxes, we can consider the scenario where the centroids of the predicted box P and the ground truth box G coincide, i.e., $\mu_P = \mu_G$. In this case, the parameters μ_I and Σ_I of the Gaussian distribution $G_I(\mu_I, \Sigma_I)$ for the overlap region I are calculated using Equation (10):

$$\mu_I = \mu_P + K(\mu_G - \mu_P), \quad \Sigma_I = \Sigma_P - K\Sigma_P, \tag{10}$$

where $K = \Sigma_P(\Sigma_P + \Sigma_G)^{-1}$. When the central points of the predicted box P and the true target box G are close to overlapping ($\mu_P = \mu_G$), the center point μ_I of the Gaussian distribution for the overlap region I coincides with the central points of the predicted box P and the true target box G, as shown in Figure 10. The covariance matrix Σ_I of the overlapping area I can be approximated using Equation (11):

$$\Sigma_I = \Sigma_P \left(1 - \frac{\Sigma_P}{\Sigma_P + \Sigma_G}\right) = \frac{\Sigma_P \Sigma_G}{\Sigma_P + \Sigma_G}, \tag{11}$$

It can be observed that the covariance Σ_I of the overlapping area I is influenced by the covariance Σ_P and Σ_G of the Gaussian distributions of the predicted box P and the true target box G.

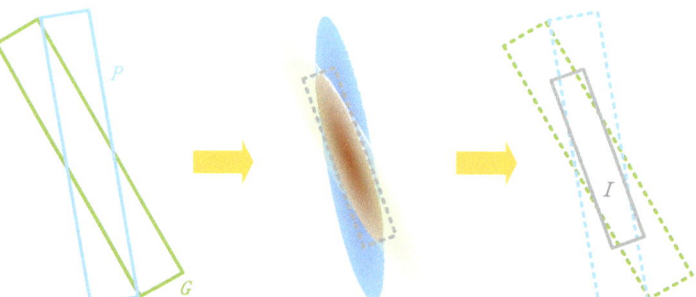

Figure 10. The overlapping area I of two Gaussian distributions.

Calculating Areas of Externally Circumscribed Rectangles. The areas of the minimum enclosing rectangles for the predicted box P, the ground truth box G, and the overlap region I are calculated using their respective Gaussian distributions. Since the dimensions of the predicted box P and the ground truth box G are known, $S_P(\Sigma_P)$ and $S_G(\Sigma_G)$ can be directly calculated. The area $S_I(\Sigma_I)$ can be conveniently derived from the covariance matrix Σ_I, as shown in Equation (12):

$$S_I(\Sigma_I) = 2^2 \sqrt{\prod_i \text{eig}(\Sigma_I)} = 4 \cdot \left|\Sigma_I^{1/2}\right| = 4 \cdot \sqrt{|\Sigma_I|}, \qquad (12)$$

Calculating the Approximate SkewIoU. The approximate SkewIoU (SkewIoU$_{\text{approx}}$) is calculated as follows:

$$\text{SkewIoU}_{\text{approx}} = \frac{S_I(\Sigma_I)}{S_P(\Sigma_P) + S_G(\Sigma_G) - S_I(\Sigma_I)}, \qquad (13)$$

However, $S_I(\Sigma_I)$ is not the exact area of the overlap of the predicted box P and the true target box G. Since $\alpha G_I(\mu_I, \Sigma_I)$ is not a standard Gaussian distribution and α is treated as a constant in the calculation, Xue Yang et al. derived that the upper bound of $\frac{S_I(\Sigma_I)}{S_P(\Sigma_P) + S_G(\Sigma_G) - S_I(\Sigma_I)}$ is $\frac{1}{3}$ [36]. Using this upper bound, a linear transformation is applied to expand its value range to [0, 1], resulting in the approximate SkewIoU:

$$\text{SkewIoU}_{\text{approx}} = 3 \cdot \frac{S_I(\Sigma_I)}{S_P(\Sigma_P) + S_G(\Sigma_G) - S_I(\Sigma_I)}, \qquad (14)$$

This approximation method for SkewIoU$_{\text{approx}}$ demonstrates high consistency with the true SkewIoU$_{\text{plain}}$ while being an ideal method for practical applications [36]. Despite the complexity of the derivation, the implementation is straightforward in the EAS loss function, where SkewIoU$_{\text{approx}}$ is used as a factor for angle regression without introducing additional hyperparameters and not participating in the backpropagation process.

The EAS loss function is given by:

$$\text{Loss}_{\text{EAS}}(d_\theta, t_\theta, \beta, \gamma) = \begin{cases} 0.5 \cdot \left(\frac{|d_\theta - t_\theta|^2}{\beta} - \gamma \cdot \log(\varepsilon + \text{SkewIoU}_{\text{approx}})\right) & \text{if } |d_\theta - t_\theta| < \beta \\ |d_\theta - t_\theta| - 0.5 \cdot \beta - \gamma \cdot \log(\varepsilon + \text{SkewIoU}_{\text{approx}}) & \text{otherwise} \end{cases}, \qquad (15)$$

where $\varepsilon = 10^{-6}$, and β and γ are adjustable parameters.

3. Experimental Results

3.1. Datasets

3.1.1. DOTA-v1.5 Dataset

DOTA [37,38], released by Wuhan University in January 2018, currently comprises three versions: DOTA-v1.0, DOTA-v1.5, and DOTA-v2.0.

DOTA-v1.0 [37] includes 2806 images sourced from various platforms, such as Google Earth, JL-1, and GF-2. This dataset contains 188,282 annotated objects across 15 different categories: Plane (PL), Baseball Diamond (BD), Bridge (BR), Ground Track Field (GTF), Small Vehicle (SV), Large Vehicle (LV), Ship (SH), Tennis Court (TC), Basketball Court (BC), Storage Tank (ST), Soccer-Ball Field (SBF), Roundabout (RA), Harbor (HA), Swimming Pool (SP), and Helicopter (HC). The annotations are quadrilaterals defined by four points $\{x_1, y_1; x_2, y_2; x_3, y_3; x_4, y_4\}$. However, DOTA-v1.0 misses many small-sized objects (approximately 10 pixels or smaller), leading to incomplete annotations. This limitation may result in inaccurate evaluations of object detection model performance [28].

DOTA-v1.5 [38], built upon DOTA-v1.0, expands the number of annotations to 402,089. It also includes additional annotations for the small objects omitted in DOTA-v1.0 and introduces a new category, Container Crane (CC), making it a more challenging dataset. The dataset presents the following difficulties: (1) the pixel size of images varies significantly, ranging from 800 × 800 to 4000 × 4000, complicating training on certain GPUs; (2) a high proportion of objects are small, with 98% of objects being smaller than 300 pixels and 57% smaller than 50 pixels [37], resulting in a substantial scale variation between tiny and large objects, complicating detection.

Due to the limited research published using DOTA-v2.0, cross-performance comparisons are challenging; thus, this paper does not utilize DOTA-v2.0. Instead, DOTA-v1.5 is employed for testing and analyzing the proposed methods, as it addresses the shortcomings of DOTA-v1.0 and remains the most commonly used public object detection dataset in the remote sensing field. It is noteworthy that the creators of the DOTA dataset have not publicly released the annotations for the test set, requiring researchers to upload their detection results for evaluation, thereby restricting comprehensive analysis of experimental results.

3.1.2. DFShip Dataset

The DFShip dataset is a fine-grained optical remote sensing ship dataset released by the Big Data and Decision-Making (National) Laboratory for the 2023 National Big Data and Computational Intelligence Challenge [39]. It comprises 41,495 images, all annotated with ship targets. The training set includes 30,285 images with corresponding target annotation files, totaling 120,605 annotated targets across 133 fine-grained categories. This makes DFShip the most detailed and extensive fine-grained ship dataset currently available, posing a significant challenge to the performance of object detection models. Both the preliminary and final test sets consist of 11,210 images, each presenting varying levels of detection difficulty. The organizers did not provide annotation files for the test sets; participants must test locally and upload the packaged test results to a designated server for validation. Since the organizers did not provide a specific name for this dataset, we refer to it as DFShip in this paper.

3.2. Setup

The experimental setup closely follows the approach described in the ReBiDet paper, with several key upgrades. The GPU configuration has been enhanced from 2 NVIDIA GTX 3090 Ti to 2 NVIDIA GTX 4090 GPUs. Additionally, the software stack has been updated: CUDA has been upgraded from version 11.8 to 12.0, PyTorch from version 1.11.0 to 1.13.1, torchvision from version 0.12.0 to 0.14.1, and the MMRotate framework from version 0.3.2 to 0.3.4. These updates ensure improved compatibility and facilitate the reproducibility of experiments across different environments.

During training, data augmentation techniques, such as horizontal and vertical flips, were applied. The batch size per GPU was set to 2, resulting in a total batch size of 4. The network was optimized using the Stochastic Gradient Descent (SGD) algorithm with a momentum of 0.9 and a weight decay of 0.0001. For the Region Proposal Network (RPN), the IoU threshold for positive samples was set to 0.7. The horizontal box Non-Maximum Suppression (NMS) threshold was set to 0.7, while the rotation box NMS threshold was set to 0.1.

The DOTA-v1.5 dataset consists of a total of 2806 images, with 1411 images in the training set, 458 images in the validation set, and 937 images in the test set. To ensure fairness in comparison experiments, we followed the practices of many scholars in the field and processed the data accordingly. Since the images in the dataset have varying sizes, the original images from the DOTA-v1.5 dataset were cropped to a size of 1024 × 1024 pixels with a stride of 824 pixels. After cropping, the DOTA-v1.5 dataset contains 15,749 images in the training set, 5297 images in the validation set, and 10,833 images in the test set. Both the training and validation sets were used for model training. Additionally, we performed multiple scale augmentation on the dataset, resizing the original 2806 images to three scales: 0.5, 1.0, and 1.5. These resized images were then cropped to 1024 × 1024 pixels with a stride of 524 pixels, resulting in a final dataset of 416,651 images for training and 71,888 images for testing. The model was trained for 12 epochs, with an initial learning rate of 0.01, and it was divided by 10 at epochs 9 and 11.

The DOTA-v1.5 dataset consists of a total of 2806 images, with 1411 images in the training set, 458 images in the validation set, and 937 images in the test set. To ensure fairness in comparative experiments, we followed established practices and processed the data accordingly. Given the varying sizes of images in the dataset, the original images were cropped to a size of 1024 × 1024 pixels with a stride of 824 pixels. After cropping, the DOTA-v1.5 dataset was expanded to include 15,749 images in the training set, 5297 images in the validation set, and 10,833 images in the test set. Both the training and validation sets were utilized for model training. Additionally, multiple scale augmentations were performed, resizing the original 2806 images to three scales: 0.5, 1.0, and 1.5. These resized images were cropped to 1024 × 1024 pixels with a stride of 524 pixels, resulting in a final dataset of 416,651 images for training and 71,888 images for testing. The model was trained for 12 epochs, with an initial learning rate of 0.01, which was reduced by a factor of 10 at epochs 9 and 11.

For the DFShip dataset, the original images are typically 1024 × 1024 in size and do not require additional cropping. Since the competition organizer's validation server only provides mAP at an IoU threshold of 0.5, the original training set was randomly split in a 4:1 ratio for this study. This resulted in a training set with 24,228 images and 96,644 annotated targets, and a test set with 6057 images and 23,961 annotated targets. The model was trained for 36 epochs, with an initial learning rate of 0.01, which was reduced by a factor of 10 at epochs 24 and 33.

3.3. Results and Analysis

3.3.1. Ablation Experiments of SA^3 Network

To evaluate the effectiveness of the proposed SA^3 Network, ablation experiments were conducted using ReDet and ReBiDet as baseline models. These experiments aimed to assess the impact of incorporating class-aware and class-agnostic bounding box regression strategies.

In these ablation experiments, the SA^3 Network was integrated into both ReDet and ReBiDet models to objectively evaluate its contribution. Due to the absence of ground truth annotations for the test set of the DOTA dataset, the trained models' results were packaged and uploaded. The detection metrics were then verified by the DOTA-v1.5 dataset authors using their designated servers.

Table 2 presents a performance comparison of the ReDet and ReBiDet models with and without the SA^3 Network. The integration of the SA^3 Network results in improved detection performance for both models. The mean average precision (mAP) reported in the table is computed using the COCO evaluation method, which averages the average precision (AP) values across all possible IoU thresholds. AP50 and AP75 refer to the AP at IoU thresholds of 0.5 and 0.75, respectively.

Table 2. Experimental results of the SA^3 Network with two strategies on the DOTA-v1.5 dataset.

Method	mAP	AP50	AP75
ReDet [23]	-	66.86	-
ReDet *	41.00	68.02	42.10
ReDet * + SA^3 Class-aware	41.23	68.94	42.94
ReDet * + SA^3 Class-agnostic	41.91	69.30	44.48
ReBiDet [28]	41.15	69.48	42.54
ReBiDet + SA^3 Class-aware	41.32	70.16	42.94
ReBiDet + SA^3 Class-agnostic	42.48	71.21	44.13

* Denotes the detection results reproduced on the experimental platform in this paper. Note: The training and inference of models in this table were directly conducted using DOTA-v1.5 for both training and inference, without applying random rotation augmentation and offline multi-scale augmentation to the dataset.

The results indicate that integrating the SA^3 Network leads to improvements in mAP, AP50, and AP75 for both the ReDet and ReBiDet models. For ReDet, the class-aware strategy increases AP50 by 0.92% and AP75 by 0.84%. The class-agnostic strategy, however, shows even more significant improvements, with AP50 increasing by 1.28% and AP75 by 2.38%. Similarly, for ReBiDet, the class-aware strategy results in AP50 increasing by 0.68% and AP75 by 0.4%, while the class-agnostic strategy yields an increase of 1.33% in AP50 and 1.59% in AP75.

These results demonstrate that the SA^3 Network significantly enhances detection accuracy, particularly at higher IoU thresholds. This improvement underscores the network's ability to optimize model precision for predicting rotated bounding boxes.

Despite the expectation that the class-aware strategy would provide more accurate results due to its tailored approach for each object class, the experimental outcomes show the opposite. The class-agnostic strategy performs better, which can be attributed to the imbalanced class distribution in the DOTA-v1.5 dataset. Some classes, such as Container Crane (CC), have very few instances compared to others like Small Vehicle (SV), which has a large number of instances. Figure 11 illustrates this imbalance, with CC having only 283 instances and SV having 295,272 instances. In such cases, the class-agnostic strategy is more suitable.

To further investigate the effectiveness of the SA^3 Network and the rationale behind strategy selection, we performed random rotation and offline multi-scale augmentations on the DOTA-v1.5 dataset. The class distribution after augmentation, shown in Figure 12, remains uneven but with an increased minimum instance count for Container Crane (CC), which rises to 1719. This theoretically enhances training effectiveness.

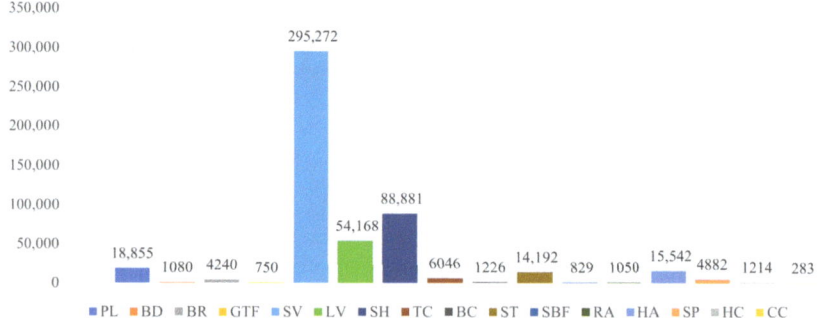

Figure 11. Distribution of various object classes in the DOTA-v1.5 training and validation sets.

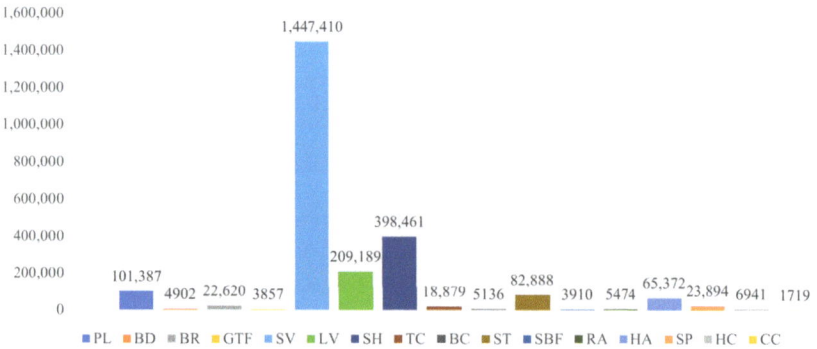

Figure 12. Distribution of various object classes in the augmented DOTA-v1.5 training and validation sets.

Table 3 displays the ablation experiment results using the augmented DOTA-v1.5 dataset. With the SA^3 Network, the class-aware strategy shows significant improvements over the ReDet baseline, with AP50 increasing by 1.00% and AP75 by 1.67%. The class-agnostic strategy results in AP50 increasing by 0.68% and AP75 by 1.64%. For ReBiDet, the class-aware SA^3 Network leads to an increase of 0.54% in AP50 and 0.85% in AP75, contrasting with the previous results where the class-agnostic strategy showed higher effectiveness.

Table 3. Experimental results of the SA^3 Network with two strategies on the augmented DOTA-v1.5 dataset.

Method	mAP	AP50	AP75
ReDet [23]	-	76.80	-
ReDet *	48.13	76.95	50.99
ReDet * + SA^3 Class-aware	49.38	77.95	52.66
ReDet * + SA^3 Class-agnostic	49.02	77.63	52.63
ReBiDet [28]	49.26	77.96	52.53
ReBiDet + SA^3 Class-aware	49.58	78.50	53.38
ReBiDet + SA^3 Class-agnostic	49.07	77.84	52.11

* Denotes the detection results reproduced on the experimental platform in this paper. Note: The training and inference of models in this table were conducted using DOTA-v1.5, with both random rotation augmentation and offline multi-scale augmentation applied during training and inference.

The augmented dataset, despite its imbalances, now provides sufficient instances for each class, making the class-aware strategy more effective. Therefore, while the class-agnostic strategy of the SA^3 Network is more suitable for datasets with significant class imbalances, the class-aware strategy proves advantageous for datasets with adequate instances for each class, even if imbalances exist.

3.3.2. Ablation Experiments of EAS Loss

Initially, we performed ablation experiments using ReDet as the baseline model to assess the effectiveness of the EAS loss function. Table 4 compares the performance of the ReDet and ReDet + SA^3 models, both with and without the EAS loss. This preliminary verification indicates that the proposed EAS loss function improves the model's detection performance. The evaluation metrics are defined as previously described. The results reveal that both baseline models, ReDet and ReDet + SA^3, exhibit varying degrees of improvement in mAP, AP50, and AP75 with the adoption of the EAS loss. Specifically, ReDet shows a 0.60% increase in AP50 and a 0.18% increase in AP75. For ReDet + SA^3, the class-aware strategy significantly enhances performance, with AP50 increasing by 1.10% and AP75 by 1.60%.

Table 4. EAS loss experiment results based on the ReDet model in the DOTA-v1.5 dataset.

Method	mAP	AP50	AP75
ReDet [23]	-	66.86	-
ReDet *	41.00	68.02	42.10
ReDet * + EAS	41.27	68.62	42.28
ReDet * + SA^3 Class-aware	41.23	68.94	42.94
ReDet * + SA^3 Class-aware + EAS	42.34	70.04	44.54
ReDet * + SA^3 Class-agnostic	41.91	69.30	44.48
ReDet * + SA^3 Class-agnostic + EAS	42.32	68.91	44.81

* Indicates the detection results reproduced on the experimental platform in this paper. Note: The training and inference of the models in this table were conducted directly using the DOTA-v1.5 dataset, without applying random rotation augmentation or offline multi-scale enhancement. The mAP in the table follows the COCO calculation method, representing the average AP across all possible IoU thresholds. AP50 refers to the AP when the IoU threshold is 0.5, while AP75 denotes the average precision at an IoU threshold of 0.75.

Further ablation experiments were conducted using ReBiDet as the baseline model to corroborate the effectiveness of the EAS loss function. Table 5 presents the performance comparison of the ReBiDet + SA^3 models with and without EAS loss. Similar to the ReDet results, the ReBiDet + SA^3 model exhibits notable improvements under the class-aware strategy when utilizing EAS loss: AP50 increases by 1.11%, AP75 by 1.38%, and mAP by 1.63%. The class-agnostic strategy shows smaller improvements, with AP75 and mAP increasing by 0.80% and 0.35%, respectively, while AP50 decreases by 0.62%, consistent with the results obtained with the ReDet model. This observed phenomenon is not coincidental. The EAS loss function facilitates more comprehensive training of the model's bounding box regression branch, whereas the class-agnostic strategy treats all categories as a single class, which can be less effective when training samples are insufficient. This fundamental conflict between the mechanisms explains the reduced generalization capability of the SA^3 Network under the class-agnostic strategy when influenced by EAS loss.

These experimental results indicate that the EAS loss function enhances the training effectiveness of the model's bounding box regression branch and optimizes the SA^3 Network, particularly improving the learning efficiency of the class-aware strategy. This ensures that the model can be effectively trained even with imbalanced categories and fewer samples.

We then evaluated the performance of the EAS loss function on a dataset with a larger number of training samples by applying random rotation and offline multi-scale augmentation to the DOTA-v1.5 dataset.

Table 5. Experimental results of EAS loss on the ReBiDet model in the DOTA-v1.5 dataset.

Method	mAP	AP50	AP75
ReBiDet + SA^3 Class-aware	41.32	70.16	42.94
ReBiDet + SA^3 Class-aware + EAS	42.95	71.28	44.33
ReBiDet + SA^3 Class-agnostic	42.48	71.21	44.13
ReBiDet + SA^3 Class-agnostic + EAS	42.83	70.59	44.93

Note: The training and inference of the models in this table were conducted directly using DOTA-v1.5, without applying random rotation augmentation or offline multi-scale enhancement.

Initially, ablation experiments were conducted using ReDet + SA^3 as the baseline model. Given that the class-agnostic strategy performs relatively poorly on the unaugmented DOTA-v1.5 dataset, we employed the class-aware strategy for SA^3 in this case. Table 6 compares the performance of the ReDet + SA^3 models with and without EAS loss. The results show slight improvements across all performance indicators: AP50 increased by 0.25%, AP75 by 0.16%, and mAP by 0.36%. These improvements are significantly less than those observed with the unaugmented DOTA-v1.5 dataset.

Table 6. Experimental results of EAS loss on the ReDet model in the augmented DOTA-v1.5 dataset.

Method	mAP	AP50	AP75
ReDet [23]	-	76.80	-
ReDet *	48.13	76.95	50.99
ReDet * + SA^3	49.38	77.95	52.66
ReDet * + SA^3 + EAS	49.74	78.20	52.82

* indicates the reproduction of detection results on the experimental platform in this paper. Note: the SA^3 Networks in this table adopt the class-aware strategy, and the model is trained and tested using random rotation augmentation and offline multi-scale enhancement on the DOTA-v1.5 dataset.

Similarly, ablation experiments using ReBiDet + SA^3 as the baseline model were conducted, maintaining the class-aware strategy for SA^3 as explained above. Table 7 presents a performance comparison of the ReBiDet + SA^3 models with and without EAS loss. As with the ReDet + SA^3 model, the EAS loss yields minor improvements: AP50 increased by 0.35%, AP75 decreased by 0.12%, and mAP increased by 0.55%. These improvements are again notably smaller compared to those seen with the unaugmented DOTA-v1.5 dataset.

Table 7. Experimental results of EAS loss on the ReBiDet model in the augmented DOTA-v1.5 dataset.

Method	mAP	AP50	AP75
ReBiDet [28]	49.26	77.96	52.53
ReBiDet + SA^3	49.58	78.50	53.38
ReBiDet + SA^3 + EAS	50.13	78.85	53.26

Note: the SA^3 Network in this table adopts the class-aware strategy, and the model is trained and tested using random rotation augmentation and offline multi-scale enhancement on the DOTA-v1.5 dataset.

The diminished advantage of the EAS loss function with an increased number of effective training samples is consistent with our previous observations. This effect is evident in both the ReDet + SA^3 and ReBiDet + SA^3 models, with the latter exhibiting stronger feature extraction capabilities. The results suggest that enhanced feature extraction and sample selection strategies cannot maintain the significant advantage of EAS loss. Nevertheless, the performance of both ReDet/ReBiDet + SA^3 models still benefits from the use of EAS loss.

In summary, the EAS loss function effectively improves the learning efficiency of the model's regression branch for bounding box angle regression. It ensures adequate training even with imbalanced categories and fewer samples. Specifically, the class-aware strategy of the SA^3 Network benefits significantly from the EAS loss, enhancing the performance metrics, such as AP50, AP75, and mAP. In scenarios with insufficient training samples for certain categories, the SA^3 Network does not need to choose between class-aware and class-agnostic strategies. As the number of effective training samples increases and each category has sufficient instances, the advantage of the EAS loss diminishes but remains relative to the Smooth L1 loss. These results validate the effectiveness of the EAS loss function in various scenarios.

3.3.3. Multi-IoU Threshold Comparison Experiment on the DFShip Dataset

The DOTA series datasets, depending on the version, include 15 or 16 categories, such as airplanes and cars, with ships being one of the categories. To further demonstrate the effectiveness of the proposed solutions, design concepts, and research methods in addressing ship detection issues, we conducted validation experiments on the DFShip dataset, which specializes in ship detection. This subsection provides a horizontal comparison of the ReBiDet + SA^3 + EAS model proposed in this paper with other state-of-the-art models on the DFShip dataset. We selected and reproduced several leading models from the DOTA series datasets for this comparison, including CSL [26], R3Det [12], S2A-Net [25], Oriented R-CNN [22], and our baseline model ReDet [23]. These models are among the most advanced in the field. Due to the unavailability of annotated test data from the official source, we split the original training set into training and test sets in a 4:1 ratio for the experiments, focusing on the precision of generated bounding boxes, specifically mAP at high IoU thresholds.

Table 8 presents the experimental results. At low threshold conditions, such as IoU = 0.5, the performance advantage of our proposed model over other advanced models is not significant. This is partly due to the dataset characteristics, which include clear images, accurate annotations, and a large number of training samples, all of which generally enhance the detection performance of the models. However, under high threshold conditions, such as IoU = 0.75, our model demonstrates a clear advantage with a detection accuracy of 98.28%, outperforming all other models in the comparison and surpassing the baseline model ReDet by 2.22%. The detection accuracy at IoU = 0.75 is only 0.56% lower than at IoU = 0.5.

Table 8. Horizontal comparison of model performance on the DFShip dataset at IoU thresholds from 0.5 to 0.95.

IoU	CSL	R3Det	S2A-Net	ReDet	Oriented R-CNN	ReBiDet + SA^3 + EAS
0.50	88.14	95.94	97.72	98.76	98.52	98.81
0.55	87.33	95.61	97.59	98.73	98.52	98.81
0.60	86.06	95.17	97.38	98.62	98.41	98.76
0.65	83.63	94.35	96.96	98.32	98.31	98.68
0.70	78.96	92.22	96.07	97.89	98.04	98.56
0.75	70.99	87.60	94.27	96.06	97.56	98.22
0.80	54.92	78.25	89.01	91.45	95.79	97.20
0.85	33.81	57.78	71.88	78.58	87.30	93.18
0.90	13.10	25.62	35.37	44.43	55.52	77.05
0.95	1.89	1.73	3.18	5.44	4.66	21.30
mAP	59.88	72.43	77.94	80.83	83.26	88.06

Note: The mAP is calculated using the all-point interpolation method. The red bold font represents the highest detection accuracy value at the same threshold in the horizontal comparison.

At an even higher threshold of IoU = 0.85, the detection accuracy of ReBiDet + SA^3 + EAS remains robust at 93.18%, while the performance of all other models significantly declines.

At IoU = 0.95, ReBiDet + SA^3 + EAS maintains a detection accuracy of 19.18%, whereas the other models' accuracies drop to single digits. Figure 13 visually presents the comparison results from Table 8. These results confirm that our model not only achieves high detection accuracy but also generates more precise bounding boxes. This validates our research approach and demonstrates the effectiveness of the design methods and strategies employed.

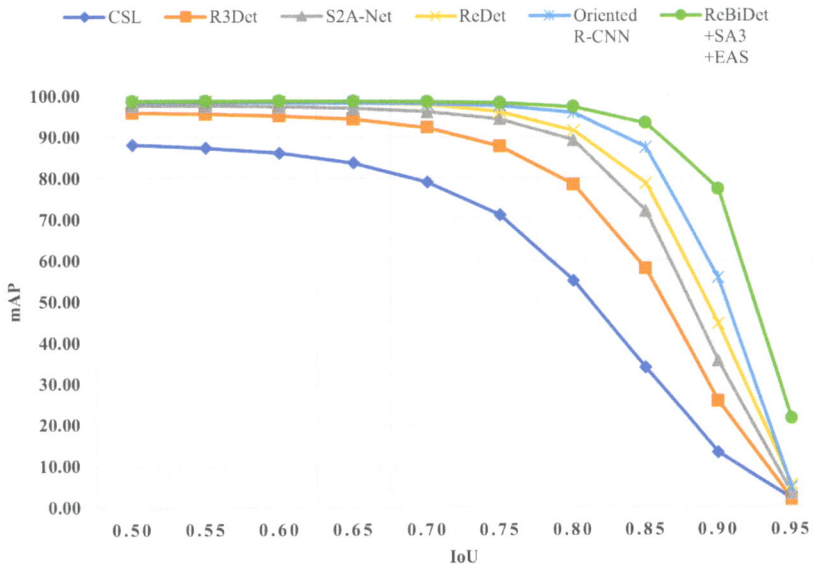

Figure 13. Horizontal comparison of model performance on the DFShip dataset at IoU thresholds from 0.5 to 0.95. The mAP is calculated using the all-point interpolation method.

3.3.4. Results

This subsubsection presents a comprehensive comparison of the ReDet and ReBiDet models, enhanced with the SA^3 Network and EAS loss, against several state-of-the-art models on the DOTA-v1.5 dataset. The objective is to evaluate the effectiveness of the proposed methods by performing a horizontal comparison with other advanced models published in recent years. All comparison models are sourced from reputable journals and conferences, and their performance metrics are cited directly.

Table 9 provides a detailed comparison of the ReDet and ReBiDet models integrated with the SA^3 Network and EAS loss against other state-of-the-art models on the DOTA-v1.5 dataset. On the unaugmented DOTA-v1.5 dataset, the ReBiDet + SA^3 + EAS model, using the class-aware strategy, surpasses all compared models with a mean average precision (mAP) of 71.28%. This is slightly higher than the ReBiDet + SA^3 model utilizing the class-agnostic strategy. Notably, ReBiDet + SA^3 + EAS ranks within the top three for average precision (AP) values across 15 subcategories and achieves first place in 10 subcategories. It excels in categories with a high aspect ratios, such as bridges (BR) and harbors (HA), as well as in categories with low aspect ratios, such as small vehicles (SV) and large vehicles (LV), which share some characteristics with ships.

When evaluated on the augmented DOTA-v1.5 dataset, which includes random rotation and offline multi-scale enhancements, the ReBiDet + SA^3 + EAS model demonstrates superior performance, achieving a mAP of 78.85%. It ranks in the top three for accuracy in 10 subcategories, including ships (SH). These results underscore the effectiveness of the proposed approach in significantly enhancing the target detection performance of existing models in optical remote sensing images.

Table 9. Comparisons with state-of-the-art methods on DOTA-v1.5 OBB task.

Method	PL	BD	BR	GTF	SV	LV	SH	TC	BC	ST	SBF	RA	HA	SP	HC	CC	mAP
single-scale:																	
RetinaNet-O [40]	71.43	77.64	42.12	64.65	44.53	56.79	73.31	90.84	76.02	59.96	46.95	69.24	59.65	64.52	48.06	0.83	59.16
FR-O [37]	71.89	74.47	44.45	59.87	51.28	68.98	79.37	90.78	77.38	67.50	47.75	69.72	61.22	65.28	60.47	1.54	62.00
Mask R-CNN [30]	76.84	73.51	49.90	57.80	51.31	71.34	79.75	90.46	74.21	66.07	46.21	70.61	63.07	64.46	57.81	9.42	62.67
HTC [29]	77.80	73.67	51.40	63.99	51.54	73.31	80.31	90.48	75.12	67.34	48.51	70.63	64.84	64.48	55.87	5.15	63.40
CMR [29]	77.77	74.62	51.09	63.44	51.64	72.90	79.99	90.35	74.90	67.58	49.54	72.85	64.19	64.88	55.87	3.02	63.41
DAFNe [41]	-	-	-	-	-	-	-	-	-	-	-	-	-	-	-	-	64.76
FR OBB [37] + RT [11]	71.92	76.07	51.87	69.24	52.05	75.18	80.72	90.53	78.58	68.26	49.18	71.74	67.51	65.53	62.16	9.99	65.03
ReDet [23]	79.20	82.81	51.92	71.41	52.38	75.73	80.92	90.83	75.81	68.64	49.29	72.03	73.36	70.55	63.33	11.53	66.86
ReDet + SA^3	80.12	83.54	54.08	72.56	52.76	77.18	87.63	90.87	84.05	74.25	62.07	73.77	75.71	65.61	67.31	7.26	69.30
ReBiDet [28]	80.54	82.90	53.62	74.55	52.55	79.65	87.53	90.84	84.57	72.93	65.02	73.05	75.87	65.56	65.18	7.32	69.48
ReDet + SA^3 + EAS	80.21	84.25	53.50	72.53	52.74	77.04	87.81	90.88	83.91	69.34	64.26	73.33	75.84	66.24	64.80	9.09	70.04
ReBiDet + SA^3	80.70	83.67	54.89	74.58	57.99	79.93	88.38	90.87	85.03	74.26	66.06	73.15	76.79	70.09	67.32	15.70	71.21
ReBiDet + SA^3+EAS	80.12	83.79	56.21	73.10	58.27	80.52	88.09	90.89	84.12	74.00	67.20	75.04	77.04	70.72	67.98	13.33	71.28
multi-scale:																	
DAFNe [41]	-	-	-	-	-	-	-	-	-	-	-	-	-	-	-	-	71.99
OWSR * [42]	-	-	-	-	-	-	-	-	-	-	-	-	-	-	-	-	74.90
RTMDet-R-tiny * [43]	88.14	83.09	51.80	77.54	65.99	82.22	89.81	90.88	80.54	81.34	64.64	71.51	77.13	76.32	72.11	46.67	74.98
RTMDet-R-s * [43]	88.14	85.82	52.90	82.09	65.58	81.83	89.78	90.82	83.31	82.47	68.51	70.93	78.00	75.77	73.09	47.32	76.02
RTMDet-R-m * [43]	89.07	86.71	52.57	82.47	66.13	82.55	89.77	90.88	84.39	83.34	72.93	73.03	77.82	75.98	80.21	42.00	76.65
ReDet * [23]	88.51	86.45	61.23	81.20	67.60	83.65	90.00	90.86	84.30	75.33	71.49	72.06	78.32	74.73	76.10	46.98	76.80
ReBiDet * [28]	86.23	85.89	61.99	82.41	67.86	83.94	89.78	90.88	86.37	83.70	72.12	77.58	78.38	73.24	75.01	52.05	77.96
RTMDet-R-l * [43]	89.31	86.38	55.09	83.17	66.11	82.44	89.85	90.84	86.95	83.76	68.35	74.36	77.60	77.39	77.87	60.37	78.12
ReBiDet * + SA^3	88.76	86.35	60.97	81.93	73.39	84.26	90.05	90.88	87.20	83.30	72.19	77.07	78.67	72.62	72.94	55.34	78.50
ReBiDet * + SA^3 + EAS	86.84	85.58	62.23	82.60	68.05	83.92	89.83	90.90	86.91	83.23	73.62	76.64	78.54	72.12	77.53	63.05	78.85

* indicates multi-scale training and testing. Note. The RetinaNet OBB (RetinaNet-O) [40], Faster R-CNN OBB (FR-O) [37], Mask R-CNN [30], and Hybrid Task Cascade (HTC) [29] results are based on a reproduced version of DOTA-v1.5 [38] and have been used by some scholars [23,43]. "Single-scale" indicates the model is directly trained and tested on the DOTA-v1.5 dataset, while "multi-scale" indicates the model is trained and tested using random rotation and offline multi-scale enhancement on DOTA-v1.5. For ease of reading and comparison, the first, second, and third highest values in each column are marked in red, yellow, and green, respectively, and are bolded.

4. Discussion

The experimental results confirm that the proposed SA^3 Network demonstrates excellent precision and localization performance in detecting rotated objects. The EAS loss function, designed to enhance edge perception of inclined bounding boxes, significantly improves the model's learning efficiency in rotated object detection tasks. This leads to more accurate localization of detection boxes, highlighting the effectiveness of the design approach and methodology presented in this paper.

The SA^3 Network incorporates a cascaded regression branch. Initially, a coarse regression branch converts horizontal bounding boxes generated by the Region Proposal Network (RPN) into rotated bounding boxes based on the features of the detected objects. Depending on the task scenario, the network employs either a class-aware or class-agnostic strategy. Fine regression and refinement branches further optimize the angle parameters of the rotated bounding boxes, thereby enhancing the fitting accuracy of the final detection boxes.

The inclusion of EAS loss in the angle regression branch introduces an Intersection over Union (IoU) factor, which mitigates the mismatch between traditional loss functions and evaluation metrics. This results in overall improvements across various detection accuracy metrics. The EAS loss adjusts the function gradient of the angle regression branch, addressing the issue of limited learning when the predicted box angle is close to the true target angle. Consequently, the model becomes more sensitive towards the end of the angle regression branch, producing more accurate predicted box angles and significantly improving the AP75 accuracy metric.

Figure 14 illustrates the detection results for the same port remote sensing image. In Figure 14a, the detection results of the ReBiDet model are shown. The ReBiDet model struggles with accurate bounding box localization for elongated objects, such as ships, due to issues with discontinuous boundaries. Figure 14b displays the results after integrating the SA^3 Network. It is evident that ReBiDet + SA^3 generates more precise bounding boxes for ships and docks, and improves detection confidence for the ships in the upper-right corner and the dock in the middle-lower part of the image. Additionally, the improved localization accuracy leads to a higher detection rate and a lower false detection rate. After incorporating the SA^3 Network, the harbor in the middle-upper part of the image is correctly detected, and previously misidentified objects on the harbor are now accurately classified as irrelevant.

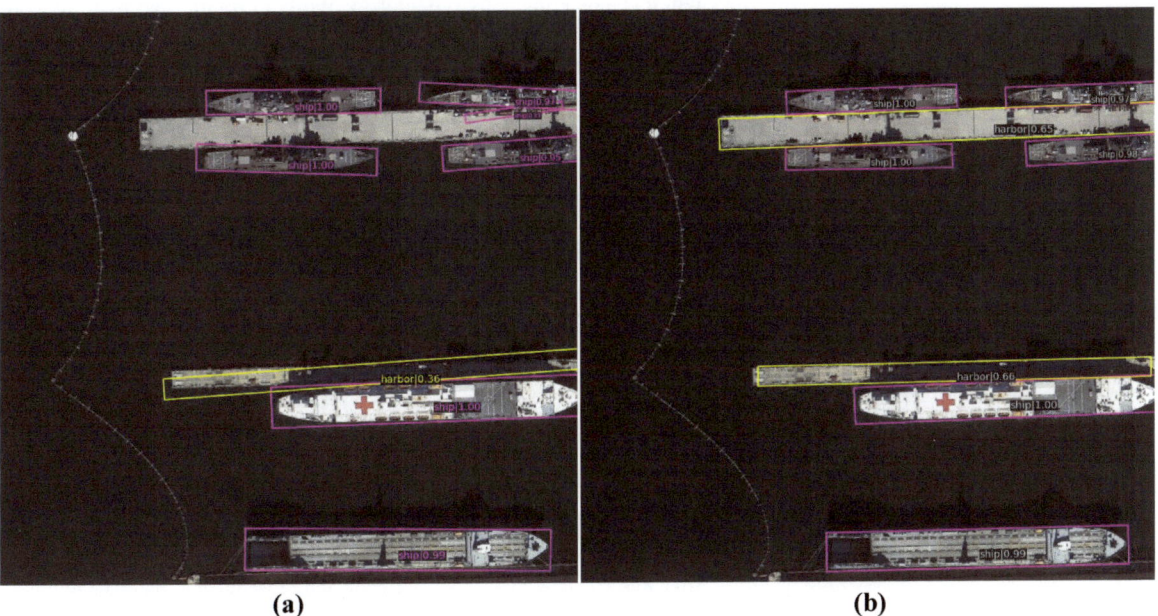

Figure 14. Detection results on an image from the DOTA-v1.5 dataset. (**a**) Detection result of ReBiDet. (**b**) Detection result of the proposed ReBiDet + SA^3.

However, the proposed method has some limitations concerning computational parameters and inference speed. The ReBiDet + SA^3 + EAS model increases the number of computational parameters by 13.99 M compared to the baseline ReBiDet model. While the EAS loss itself does not significantly contribute to this increase, the additional parameters primarily result from the SA^3 Network. This increase in computational parameters leads to an additional 3.5 ms per image on an RTX4090 platform, corresponding to a decrease of 1.8 FPS. Although these computational trade-offs are acceptable in many scenarios given the improvements in detection accuracy, reducing computational parameters and enhancing inference speed while maintaining high detection accuracy remains a critical area for future research.

In conclusion, the methods proposed in this paper are highly effective in improving the detection accuracy of randomly oriented objects, particularly those with elongated contour features, in remote sensing applications. These advancements provide valuable insights for enhancing ship detection capabilities in optical remote sensing images.

5. Conclusions

This paper addresses the challenges associated with significant deviations in final detection boxes due to a wide range of angles, as well as the imprecision inherent in traditional angle regression losses. To tackle these issues, we propose the SA^3 Network and EAS loss. The SA^3 Network employs a hierarchical regression structure that includes coarse, fine, and refinement stages to progressively optimize the angle parameters of rotated bounding boxes. The EAS loss introduces the SkewIoU factor, calculated using Gaussian transformation, to enhance the precision of angle regression losses. This approach improves both training efficiency and model performance, particularly under high IoU threshold conditions.

Experimental results validate the effectiveness of the SA^3 Network and EAS loss. The proposed methods significantly improve detection accuracy, especially for rotated objects in optical remote sensing images.

Future work will focus on enhancing the interpretability of rotation-equivariant convolutional neural networks. Understanding the operational mechanisms of these networks presents a significant research opportunity. Our goal is to gain a more intuitive understanding of the features extracted by these networks. We aim to explore techniques for visualizing rotation-equivariant features, which will facilitate a deeper analysis of their limitations and potential improvement strategies.

Author Contributions: Methodology, Z.Y. and J.F.; software, Z.Y.; supervision, Z.L. and Y.X.; writing—o draft, Z.Y.; writing—review and editing, Z.Y.; visualization, Z.Y. All authors have read and agreed to the published version of the manuscript.

Funding: This research received no external funding.

Institutional Review Board Statement: Not applicable.

Informed Consent Statement: Not applicable.

Data Availability Statement: The DOTA dataset was publicly released by Wuhan University in November 2017 and includes complete sets of training, validation, and testing images, as well as annotation files for the training and validation sets. It can be downloaded at this address: https://captain-whu.github.io/DOTA/dataset.html (accessed on 29 May 2024). Since the downloaded package does not include annotation files for the testing set, users need to submit their model's inference results to the Wuhan University official Evaluation Server to obtain the accuracy of the inference results, at this address: https://captain-whu.github.io/DOTA/evaluation.html (accessed on 29 May 2024). After this article is published, our code and other related materials will also be released.

Acknowledgments: This work was supported by the Institute of Systems Engineering, Academy of Military Sciences, and we would like to express our gratitude to Yongqiang Xie and Zhongbo Li, for their support and guidance throughout the project. Their contributions have been invaluable to the success of this study.

Conflicts of Interest: The authors declare no conflicts of interest.

Abbreviations

The following abbreviations are used in this manuscript:

IoU	Intersection over Union
SA^3	Spatial Adaptive Angle-Aware
EAS Loss	Edge-aware Skewed Bounding Box Loss
RPN	Region Proposal Network

References

1. Jiang, Y.; Zhu, X.; Wang, X.; Yang, S.; Li, W.; Wang, H.; Fu, P.; Luo, Z. R2CNN: Rotational Region CNN for Orientation Robust Scene Text Detection. *arXiv* **2017**, arXiv:abs/1706.09579.
2. Liao, M.; Shi, B.; Bai, X. TextBoxes++: A Single-Shot Oriented Scene Text Detector. *IEEE Trans. Image Process.* **2018**, *27*, 3676–3690. [CrossRef]
3. Liao, M.; Zhu, Z.; Shi, B.; Xia, G.s.; Bai, X. Rotation-Sensitive Regression for Oriented Scene Text Detection. In Proceedings of the 2018 IEEE/CVF Conference on Computer Vision and Pattern Recognition, Salt Lake City, UT, USA, 18–23 June 2018; pp. 5909–5918. [CrossRef]
4. Liu, X.; Liang, D.; Yan, S.; Chen, D.; Qiao, Y.; Yan, J. FOTS: Fast Oriented Text Spotting with a Unified Network. In Proceedings of the 2018 IEEE/CVF Conference on Computer Vision and Pattern Recognition, Salt Lake City, UT, USA, 18–23 June 2018; pp. 5676–5685. [CrossRef]

5. Ma, J.; Shao, W.; Ye, H.; Wang, L.; Wang, H.; Zheng, Y.; Xue, X. Arbitrary-Oriented Scene Text Detection via Rotation Proposals. *IEEE Trans. Multimed.* **2018**, *20*, 3111–3122. [CrossRef]
6. Zhou, X.; Yao, C.; Wen, H.; Wang, Y.; Zhou, S.; He, W.; Liang, J. EAST: An Efficient and Accurate Scene Text Detector. In Proceedings of the 2017 IEEE Conference on Computer Vision and Pattern Recognition (CVPR), Honolulu, HI, USA, 21–26 July 2017; pp. 2642–2651. [CrossRef]
7. Huang, C.; Ai, H.; Li, Y.; Lao, S. High-Performance Rotation Invariant Multiview Face Detection. *IEEE Trans. Pattern Anal. Mach. Intell.* **2007**, *29*, 671–686. [CrossRef] [PubMed]
8. Rowley, H.A.; Baluja, S.; Kanade, T. Rotation invariant neural network-based face detection. In Proceedings of the 1998 IEEE Computer Society Conference on Computer Vision and Pattern Recognition (Cat. No.98CB36231), Santa Barbara, CA, USA, 25 June 1998; pp. 38–44. [CrossRef]
9. Zhou, L.F.; Gu, Y.; Wang, P.S.P.; Liu, F.Y.; Liu, J.; Xu, T.Y. Rotation-Invariant Face Detection with Multi-task Progressive Calibration Networks. In *Pattern Recognition and Artificial Intelligence*; Springer International Publishing: Berlin/Heidelberg, Germany, 2020; pp. 513–524.
10. Azimi, S.M.; Vig, E.; Bahmanyar, R.; Körner, M.; Reinartz, P. Towards Multi-class Object Detection in Unconstrained Remote Sensing Imagery. In *Lecture Notes in Computer Science*; Springer: Berlin/Heidelberg, Germany, 2019.
11. Ding, J.; Xue, N.; Long, Y.; Xia, G.; Lu, Q. Learning RoI Transformer for Oriented Object Detection in Aerial Images. In Proceedings of the 2019 IEEE/CVF Conference on Computer Vision and Pattern Recognition (CVPR), Long Beach, CA, USA, 15–20 June 2019; pp. 2844–2853. [CrossRef]
12. Yang, X.; Liu, Q.; Yan, J.; Li, A. R3Det: Refined Single-Stage Detector with Feature Refinement for Rotating Object. *Proc. Aaai Conf. Artif. Intell.* **2021**, *35*, 3163–3171. [CrossRef]
13. Yang, X.; Yang, J.; Yan, J.; Zhang, Y.; Zhang, T.; Guo, Z.; Sun, X.; Fu, K. SCRDet: Towards More Robust Detection for Small, Cluttered and Rotated Objects. In Proceedings of the 2019 IEEE/CVF International Conference on Computer Vision (ICCV), Seoul, Republic of Korea, 27 October–2 November 2019; pp. 8231–8240.
14. Yang, X.; Sun, H.; Fu, K.; Yang, J.; Sun, X.; Yan, M.; Guo, Z. Automatic Ship Detection in Remote Sensing Images from Google Earth of Complex Scenes Based on Multiscale Rotation Dense Feature Pyramid Networks. *Remote. Sens.* **2018**, *10*, 132. [CrossRef]
15. Yang, X.; Yan, J.; Yang, X.; Tang, J.; Liao, W.; He, T. SCRDet++: Detecting Small, Cluttered and Rotated Objects via Instance-Level Feature Denoising and Rotation Loss Smoothing. *IEEE Trans. Pattern Anal. Mach. Intell.* **2022**, *45*, 2384–2399. [CrossRef] [PubMed]
16. Liu, Y.; Zhang, S.; Jin, L.; Xie, L.; Wu, Y.; Wang, Z. Omnidirectional scene text detection with sequential-free box discretization. In Proceedings of the 28th International Joint Conference on Artificial Intelligence, Macao, China, 10–16 August 2019; AAAI Press: Washington, DC, USA, 2019.
17. Qian, W.; Yang, X.; Peng, S.; Yan, J.; Guo, Y. Learning Modulated Loss for Rotated Object Detection. *Proc. Aaai Conf. Artif. Intell.* **2021**, *35*, 2458–2466. [CrossRef]
18. Xu, Y.; Fu, M.; Wang, Q.; Wang, Y.; Chen, K.; Xia, G.S.; Bai, X. Gliding Vertex on the Horizontal Bounding Box for Multi-Oriented Object Detection. *IEEE Trans. Pattern Anal. Mach. Intell.* **2021**, *43*, 1452–1459. [CrossRef]
19. Yang, X.; Zhang, G.; Yang, X.; Zhou, Y.; Wang, W.; Tang, J.; He, T.; Yan, J. Detecting Rotated Objects as Gaussian Distributions and its 3-D Generalization. *IEEE Trans. Pattern Anal. Mach. Intell.* **2023**, *45*, 4335–4354. [CrossRef]
20. Girshick, R. Fast R-CNN. In Proceedings of the 2015 IEEE International Conference on Computer Vision (ICCV), Santiago, Chile, 7–13 December 2015; pp. 1440–1448. [CrossRef]
21. Ren, S.; He, K.; Girshick, R.; Sun, J. Faster R-CNN: Towards Real-Time Object Detection with Region Proposal Networks. *IEEE Trans. Pattern Anal. Mach. Intell.* **2017**, *39*, 1137–1149. [CrossRef] [PubMed]
22. Xie, X.; Cheng, G.; Wang, J.; Yao, X.; Han, J. Oriented R-CNN for Object Detection. In Proceedings of the 2021 IEEE/CVF International Conference on Computer Vision (ICCV), Montreal, QC, Canada, 10–17 October 2021; pp. 3500–3509. [CrossRef]
23. Han, J.; Ding, J.; Xue, N.; Xia, G. ReDet: A Rotation-equivariant Detector for Aerial Object Detection. In Proceedings of the 2021 IEEE/CVF Conference on Computer Vision and Pattern Recognition (CVPR), Nashville, TN, USA, 20–25 June 2021; pp. 2785–2794.
24. Ming, Q.; Zhou, Z.; Miao, L.; Zhang, H.; Li, L. Dynamic Anchor Learning for Arbitrary-Oriented Object Detection. *Proc. Aaai Conf. Artif. Intell.* **2021**, *35*, 2355–2363. [CrossRef]
25. Han, J.; Ding, J.; Li, J.; Xia, G.S. Align Deep Features for Oriented Object Detection. *IEEE Trans. Geosci. Remote. Sens.* **2022**, *60*, 1–11. [CrossRef]
26. Yang, X.; Yan, J.; He, T. On the Arbitrary-Oriented Object Detection: Classification Based Approaches Revisited. *Int. J. Comput. Vis.* **2022**, *130*, 1873–1874. [CrossRef]
27. Weiler, M.; Cesa, G. General E(2)-Equivariant Steerable CNNs. *arXiv* **2019**, arXiv:1911.08251. [CrossRef]
28. Yan, Z.X.; Li, Z.B.; Xie, Y.Q.; Li, C.Y.; Li, S.A.; Sun, F.W. ReBiDet: An Enhanced Ship Detection Model Utilizing ReDet and Bi-Directional Feature Fusion. *Appl. Sci.* **2023**, *13*, 25. [CrossRef]
29. Chen, K.; Pang, J.; Wang, J.; Xiong, Y.; Li, X.; Sun, S.; Feng, W.; Liu, Z.; Shi, J.; Ouyang, W.; et al. Hybrid Task Cascade for Instance Segmentation. In Proceedings of the 2019 IEEE/CVF Conference on Computer Vision and Pattern Recognition (CVPR), Long Beach, CA, USA, 15–20 June 2019; pp. 4969–4978. [CrossRef]
30. He, K.; Gkioxari, G.; Dollár, P.; Girshick, R. Mask R-CNN. *IEEE Trans. Pattern Anal. Mach. Intell.* **2020**, *42*, 386–397. [CrossRef]

31. Jaiswal, A.; Wu, Y.; Natarajan, P.; Natarajan, P. Class-agnostic Object Detection. In Proceedings of the 2021 IEEE Winter Conference on Applications of Computer Vision (WACV), Waikoloa, HI, USA, 3–8 January 2021; pp. 918–927. [CrossRef]
32. Chen, Z.; Chen, K.; Lin, W.; See, J.; Yu, H.; Ke, Y.; Yang, C. PIoU Loss: Towards Accurate Oriented Object Detection in Complex Environments. In Proceedings of the Computer Vision—ECCV 2020, Glasgow, UK, 23–28 August 2020; Springer International Publishing: Berlin/Heidelberg, Germany; pp. 195–211.
33. Zheng, Y.; Zhang, D.; Xie, S.; Lu, J.; Zhou, J. Rotation-Robust Intersection over Union for 3D Object Detection. In Proceedings of the Computer Vision—ECCV 2020, Glasgow, UK, 23–28 August 2020; Springer International Publishing: Berlin/Heidelberg, Germany; pp. 464–480.
34. Yang, X.; Yan, J.; Ming, Q.; Wang, W.; Zhang, X.; Tian, Q. Rethinking Rotated Object Detection with Gaussian Wasserstein Distance Loss. In Proceedings of the 2021 International Conference on Machine Learning, Virtual, 18– 24 July 2021.
35. Yang, X.; Yang, X.; Yang, J.; Ming, Q.; Wang, W.; Tian, Q.; Yan, J. Learning high-precision bounding box for rotated object detection via kullback-leibler divergence. *Adv. Neural Inf. Process. Syst.* **2021**, *34*, 18381–18394.
36. Yang, X.; Zhou, Y.; Zhang, G.; Yang, J.; Wang, W.; Yan, J.; Zhang, X.; Tian, Q. The KFIoU Loss for Rotated Object Detection. *arXiv* **2022**, arXiv:abs/2201.12558.
37. Xia, G.S.; Bai, X.; Ding, J.; Zhu, Z.; Belongie, S.; Luo, J.; Datcu, M.; Pelillo, M.; Zhang, L. DOTA: A Large-Scale Dataset for Object Detection in Aerial Images. In Proceedings of the 2018 IEEE/CVF Conference on Computer Vision and Pattern Recognition, Salt Lake City, UT, USA, 18–23 June 2018; pp. 3974–3983. [CrossRef]
38. Ding, J.; Xue, N.; Xia, G.S.; Bai, X.; Yang, W.; Yang, M.Y.; Belongie, S.; Luo, J.; Datcu, M.; Pelillo, M.; et al. Object Detection in Aerial Images: A Large-Scale Benchmark and Challenges. *IEEE Trans. Pattern Anal. Mach. Intell.* **2022**, *44*, 7778–7796. [CrossRef]
39. Data, B.; Laboratory, D.M.N. Fine-Grained Dense Ship Detection Task Based on High-Resolution Remote Sensing Visible Light Data. 2023. Available online: https://www.datafountain.cn/competitions/635/datasets (accessed on 31 July 2024).
40. Lin, T.Y.; Goyal, P.; Girshick, R.; He, K.; Dollár, P. Focal Loss for Dense Object Detection. In Proceedings of the 2017 IEEE International Conference on Computer Vision (ICCV), Venice, Italy, 22–29 October 2017; pp. 2999–3007. [CrossRef]
41. Lang, S.; Ventola, F.G.; Kersting, K. DAFNe: A One-Stage Anchor-Free Deep Model for Oriented Object Detection. *arXiv* **2021**, arXiv:2109.06148.
42. Li, C.; Xu, C.; Cui, Z.; Wang, D.; Jie, Z.; Zhang, T.; Yang, J. Learning Object-Wise Semantic Representation for Detection in Remote Sensing Imagery. In Proceedings of the IEEE/CVF Conference on Computer Vision and Pattern Recognition (CVPR) Workshops, Long Beach, CA, USA, 15–20 June 2019.
43. Lyu, C.; Zhang, W.; Huang, H.; Zhou, Y.; Wang, Y.; Liu, Y.; Zhang, S.; Chen, K. RTMDet: An Empirical Study of Designing Real-Time Object Detectors. *arXiv* **2022**, arXiv:abs/2212.07784.

Disclaimer/Publisher's Note: The statements, opinions and data contained in all publications are solely those of the individual author(s) and contributor(s) and not of MDPI and/or the editor(s). MDPI and/or the editor(s) disclaim responsibility for any injury to people or property resulting from any ideas, methods, instructions or products referred to in the content.

Article

Efficient Model Updating of a Prefabricated Tall Building by a DNN Method

Chunqing Liu [1], Fengliang Zhang [1], Yanchun Ni [2,3,*], Botao Ai [1], Siyan Zhu [1], Zezhou Zhao [1] and Shengjie Fu [1]

[1] School of Civil and Environmental Engineering, Harbin Institute of Technology, Shenzhen 518055, China; chunqingliu98@outlook.com (C.L.); zhangfengliang@hit.edu.cn (F.Z.); 22s054013@stu.hit.edu.cn (Z.Z.); 22s054016@stu.hit.edu.cn (S.F.)

[2] College of Civil Engineering, Tongji University, Shanghai 200092, China

[3] Guangdong Provincial Key Laboratory of Intelligent and Resilient Structures for Civil Engineering, Harbin Institute of Technology, Shenzhen 518055, China

* Correspondence: yanchunni@tongji.edu.cn

Abstract: The significance of model updating methods is becoming increasingly evident as the demand for greater precision in numerical models rises. In recent years, with the advancement of deep learning technology, model updating methods based on various deep learning algorithms have begun to emerge. These methods tend to be complicated in terms of methodological architectures and mathematical processes. This paper introduces an innovative model updating approach using a deep learning model: the deep neural network (DNN). This approach diverges from conventional methods by streamlining the process, directly utilizing the results of modal analysis and numerical model simulations as deep learning input, bypassing any additional complex mathematical calculations. Moreover, with a minimalist neural network architecture, a model updating method has been developed that achieves both accuracy and efficiency. This distinctive application of DNN has seldom been applied previously to model updating. Furthermore, this research investigates the impact of prefabricated partition walls on the overall stiffness of buildings, a field that has received limited attention in the previous studies. The main finding was that the deep neural network method achieved a Modal Assurance Criterion (MAC) value exceeding 0.99 for model updating in the minimally disturbed 1st and 2nd order modes when compared to actual measurements. Additionally, it was discovered that prefabricated partitions exhibited a stiffness ratio of about 0.2–0.3 compared to shear walls of the same material and thickness, emphasizing their role in structural behavior.

Keywords: model updating; DNN; sensors; partition walls

1. Introduction

The prevalence of prefabricated construction in China has increased significantly over the past decade. This growth can be attributed to several factors, including the rapid construction time, eco-friendliness, industrialization, and standardization. China's slowing population growth and aging demographics are pushing up labor costs, particularly affecting the traditional labor-intensive construction. In areas with high labor costs, prefabricated construction is now cheaper than traditional methods. With cost-effectiveness and sustainability [1–3], prefabrication is becoming a dominant trend. Structurally prefabricated components offer rotational connections differing from traditional cast-in-place or welded joints. Seismic performance research highlights the need for robust connection design in prefab construction. Innovations like the new prefabricated self-centering steel frames and modular precast shear wall systems have demonstrated superior dynamic response and energy dissipation in tests and analyses [4–6].

A substantial body of research has emerged in recent years on the seismic performance of prefabricated partitions in buildings. Zhai et al. [7] found that prefabricated reinforced concrete (RC) shear walls with different infills, particularly integrated shear walls with RC

Citation: Liu, C.; Zhang, F.; Ni, Y.; Ai, B.; Zhu, S.; Zhao, Z.; Fu, S. Efficient Model Updating of a Prefabricated Tall Building by a DNN Method. *Sensors* **2024**, *24*, 5557. https://doi.org/10.3390/s24175557

Academic Editor: Hossam A. Gabbar

Received: 4 August 2024
Revised: 22 August 2024
Accepted: 25 August 2024
Published: 28 August 2024

Copyright: © 2024 by the authors. Licensee MDPI, Basel, Switzerland. This article is an open access article distributed under the terms and conditions of the Creative Commons Attribution (CC BY) license (https://creativecommons.org/licenses/by/4.0/).

infilling, exhibit superior shear bearing capacity, stiffness, energy dissipation, and seismic performance. The main focus of research on partition walls is not on mechanical properties. Many studies in this field have chosen to focus on the performance of partitions for daily use such as sound insulation [8–10], thermal performance [11–14], and environmental performance [15–18]. Moreover, Li et al. [19] noted that non-structural infill walls could increase a building's overall stiffness by 60%, suggesting a similar effect for partition walls. Despite advances in finite element analysis, significant discrepancies remain between model results and actual measurements, as highlighted by Yang et al. [20], who attributed these differences to the simplification of structural behavior, discretization of continuous systems, and physical parameter errors. Therefore, effective seismic resistance solutions, more accurate methods to reflect dynamic characteristics, and improved monitoring and maintenance of prefabricated structures have become key research focuses.

Since the concept of model updating for finite element models was proposed, the earlier adopted methods have been mainly classified as a deterministic model updating method. This type of method tries to get the calculated structural responses, modal shapes, frequencies, and other parameters as close as possible to the measured data by adjusting the structural parameters, as cited in the literature [21,22]. This method is usually an ill-posed inverse problem in model updating, and the incompleteness of data and the complexity of the structure frequently make the results of this updating method inconsistent and incomplete, as mentioned in paper [23].

Subsequently, a model updating method based on the Bayesian formula was proposed. Katafygiotis et al. [24] provide a comprehensive explanation of this structural model updating method, taking the measured data of the structure as inputs to calculate the posterior probability distribution function (PDF) in the Bayesian formula, and the maximum value of the parameterized posterior probability distribution is then taken as the updated parameter. In order to determine the posterior PDF in the Bayesian formula with the parameter to be updated θ, the Markov chain Monte Carlo (MCMC) method was used in the literature [25]. This method can be used regardless of whether the problem is identifiable or not, and it can obtain the most likely value of the updated parameters and quantitatively evaluate the uncertainty of this value. Boulkaibet et al. [26] employed the mixed Markov chain theory to update the structural parameters and provide a new evaluation formula for Markov chain convergence. In order to improve the efficiency of this calculation method, Zhang et al. [27] incorporated the Metropolis–Hastings (MH) algorithm. Yang et al. [20] considered and reduced the impact of white noise on the calculated posterior probability distribution function when using this method for model updating of the coupled plate system.

In recent years, there has been a rapid increase in the number of publications focusing on deep learning-based methods for model updating, showcasing the potential of deep learning in the domain of model updating. Lee et al. [28] have presented a novel methodology for structural damage detection, which leverages finite element model updating to establish a reference model that encapsulates the target structure's characteristics. This approach addresses the limitations of traditional simulation-based damage detection by incorporating measured responses and employing DNN to identify the extent and location of structural damage. Utilizing an inverse eigenvalue problem approach and DNN, Gong and Park [29] innovatively updated finite element models with high accuracy, as evidenced by the dynamic updating of a suspension bridge model.

Employing a multi-fidelity deep neural network (MF-DNN) for surrogate modeling, Torzoni et al. [30] introduced a methodology-enhancing real-time structural health monitoring by effectively locating and quantifying damage through an MCMC-informed probability update, leveraging high- and low-fidelity simulated datasets for comprehensive sensor data mimicry. In summary, deep learning—a swiftly evolving methodology—is increasingly applied for various purposes within the SHM field, such as encompassing seismic response modeling and finite element model updating [31–33].

This study introduces a novel method for modal updating, termed Deep Neural Network Model Updating (DNNMU), which is an algorithm that, at its core, utilizes the DNN algorithm, and intuitively and simply uses the input and output of finite element models as training data to achieve its model updating objective. The efficacy of the method is demonstrated through its application in a case study involving a prefabricated dormitory, which is located on the Shenzhen campus of the Harbin Institute of Technology.

A comprehensive ambient vibration test was conducted on this building using four three-axial accelerometers. The Bayesian operational modal analysis method [30–36] was utilized to identify the building's modal parameters, which mainly included natural frequency, damping ratio, and modal shape. A finite element model of the building was established for model updating. The analysis results were compared with the calculation results, empirical formulas, and measured results of the design institute. The stiffness of the partition wall was updated, and the approximate ratio of the stiffness provided by the partition wall to the partition wall's inherent stiffness was determined. The efficacy and accuracy of the DNNMU method in model updating were validated by comparing the updating accuracy with that of another similar research.

2. Methodology

Deep Neural Network Model Updating is an innovative model updating method proposed in this work. The fundamental principle of the DNNMU method involves the systematic generation of a series of numerical simulation outcomes. These outcomes result from the iterative computation of parameters that have been randomly modified. The generated data are then utilized to train a deep learning model, which is subsequently employed to predict the parameters in need of adjustment, thereby facilitating the refinement of the model. When contrasted with conventional methodologies, this strategy is characterized by its straightforwardness, intuitiveness, and superior efficiency. It demonstrates commendable adaptability across a diverse array of model updating scenarios. Moreover, the embedded deep learning model within the DNNMU framework is designed to be modular, allowing for substitution to meet specific user requirements without compromising the method's applicability.

2.1. The Compatibility of DNN with Model Updating

In numerical simulation, the process of solving and calculating correctly modeled finite element models that accurately reflect real-world problems can be regarded as a function. It can be treated as a function because it fits the definition: each set of inputs, such as finite element software models and parameters, yields a unique set of solutions, like node displacement, element internal force, and structural mode.

In the context of model updating, the function is applied to the parameters involved in the updating process. Denoting the numerical simulation process as a function $S(\cdot)$, it can be expressed as:

$$S(E_1, E_2, \ldots, E_n) = f, \Phi \qquad (1)$$

where f and Φ are the analytical natural frequency and modal shape of the finite element model, respectively. $E_1, E_2, \ldots, E_n$ stands for the elastic modulus of different structures, which are the updating target. Note that, for the structural mass, it is commonly taken as known since it is easier to estimate their values. When other conditions are the same, the ratio of elastic modulus is equal to the ratio of stiffness.

The reverse process of S in Equation (1) could be written as:

$$E_1, E_2, \ldots, E_n = R(f, \Phi) \qquad (2)$$

R is referred to as the inverse process of S rather than the inverse function because it cannot be rigorously demonstrated that every set of inputs to R has a unique set of outputs corresponding to it. Although considering the practical significance of the variables around this equation, R is likely to be a function.

This work adopts the Modal Assurance Criterion (MAC) as the evaluation standard for vibration modes:

$$mac_{ij} = \frac{\left(\hat{v}_i^T v_j\right)^2}{\left(\hat{v}_i^T \hat{v}_i\right)\left(v_j^T v_j\right)} \qquad (3)$$

where the mode shapes under the ith mode obtained from numerical simulation are denoted as v_i, while the mode shapes under the ith mode obtained from field tests are denoted as $\hat{v}_i^T$.

Assuming that the selected modal ranks are the top n ranks, then mac_{ij} refers to the element in the ith row and jth column of the square matrix $MAC_{n \times n}$. The mac value between two vectors is employed to gauge the approximation level between the vectors. A mac value closer to 1 suggests a higher degree of similarity between the vectors (equal to 1 when the vectors are identical, always less than 1 otherwise). In the following discussion, a mac mentioned without a subscript default to being the same rank, that is, mac_{ii}.

Recall that the core objective of model updating is to identify an optimal set of parameters for adjustment, thereby aligning the results from finite element model calculations as closely as possible with the real-world test results, f and Φ. As described in the MCMC model updating method in reference [27], a set of parameters is generated through a predefined PDF. These parameters are then input into the model, yielding a set of outputs and an associated output distribution function. This output distribution function is then compared to the actual values to derive the distribution function of another set of inputs. This iterative process continues until a superior set of inputs is identified, which makes the mac value approach 1 as closely as possible.

The DNNMU algorithm introduced in this study deviates from traditional methods, instead opting to leverage DNN to approximate the inverse process R. The approximation process involves the generation of sufficient training sets by the iterative invocation of the function computed by the finite element software. Each sample within these training sets includes features and labels. Here, the label corresponds to the input parameter $E_1, E_2, \ldots, E_n$ for each computation, and the feature equates to the output f, Φ derived from that computation. Evidently, as the feature corresponds to the output of the function S and the label to the function's input, the neural network trained using these samples will approximate the inverse process R. Consequently, for each input set (f, Φ), it can infer the parameter $E_1, E_2, \ldots, E_n$ that was used in the computation beforehand.

2.2. Algorithm Architecture

The updating algorithm, depicted in Figure 1, consists of two primary components: dataset generation and neural network training. These components can operate independently of each other.

The first part, dataset generation, involves creating a dataset for neural network training. This dataset is generated by continuously producing random parameters per predefined rules. These parameters are then utilized in numerical modal analysis with finite element software. After each calculation, the parameters, frequencies, and mode shapes involved are stored as a set of samples. In these samples, the parameters serve as labels, while the frequencies and mode shapes are the features. Subsequently, a subset of these samples is later randomly selected during the training phase to act as a validation set.

The second part encompasses the training of the neural network. This phase uses the dataset generated in the first component. While this dataset is designed with DNN characteristics in mind, it is not the only viable option. Other feasible alternatives include a range of deep learning and broader machine learning algorithms. The specifics of the neural network architecture employed in this study will be discussed in detail in the subsequent section.

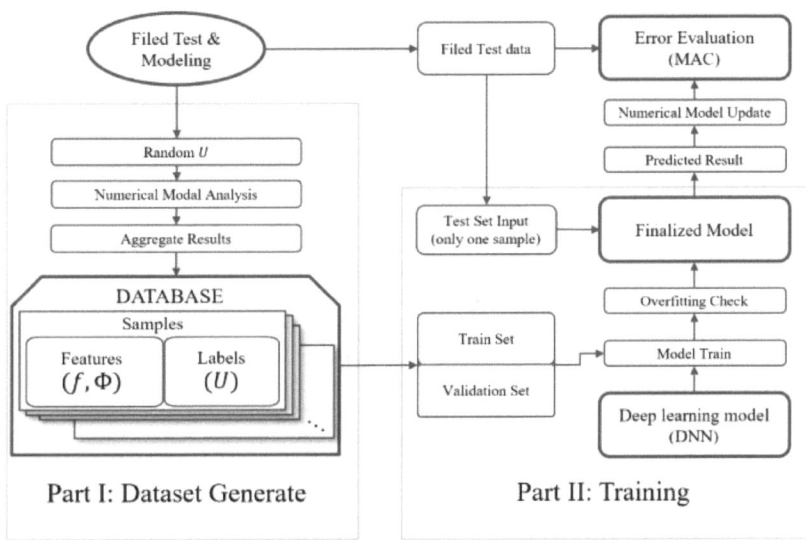

Figure 1. DNNMU algorithm architecture.

Upon conclusion of the second component, a trained DNN neural network is acquired. The outcomes from the modal identification of measured data are also compiled in the form of labels in the samples. These are entered into the neural network to yield the network's predicted parameters. These predicted parameters are then re-introduced into the finite element model for further modal analysis. The comparison of the output results with the measured data forms the basis for evaluating the updating effect of the neural network using the MAC.

In summary, the DNNMU algorithm exhibits high versatility. Within the scope of model updating, parameters that need adjustment invariably exist, serving as the labels in the algorithm's sample set. Moreover, parameters for assessing the updating outcomes also exist, acting as the features within the samples. This study presents a specific instance within this algorithmic framework, with the choice of DNN explained in Section 4.2. It is noted that this framework retains its applicability across diverse model updating scenarios, such as those involving bridges, geotechnical structures, or aircraft wings, and is compatible with various updating algorithms.

2.3. Generation and Characteristic Analysis of Training Sets

The construction of a DNN requires consideration of the network's purpose and the characteristics of the dataset. The function of the neural network is to fit a complex process from a one-dimensional input to a one-dimensional output. After generating parameters using the method described in Section 4.1, which will be detailed later, each set of parameters is inputted into a finite element model to conduct a modal analysis, producing the following computational results: frequencies and mode shapes. Assuming the number of modes analyzed is denoted as n_m and the number of nodes involved in the model updating is n_p, the calculated frequencies for each modal order are:

$$f_i, \ i = 1, 2, \ldots, n_m \tag{4}$$

The mode shape matrix for the jth node under the ith mode is:

$$\varphi_{ij} = \begin{bmatrix} dx \\ dy \\ dz \end{bmatrix}, \ i = 1, 2, \ldots, n_m, \ j = 1, 2, \ldots, n_p \tag{5}$$

The mode shape matrix for the jth node under the ith mode is written as a 3×1 column vector, and after concatenation end-to-end, two features for training are formed.

In the preprocessing of parameters, we first consider the mode shape Φ. This feature reflects the relative relationships among all nodes under this mode. While its absolute size holds no meaning, its relative ratio—which represents the shape of structural vibration—is significant and thus should remain unchanged. This is achieved by normalization.

In contrast, the frequency f, another feature, has meaning both in its absolute magnitude and in the proportional relationships between different frequencies. Consequently, it is neither subjected to relative scaling nor translation. For this project, the first-order frequency measures approximately 0.7 Hz and the overall frequency value is close to 1. Therefore, we did not process the frequency. However, in other projects, physical quantities with properties like frequency may require absolute scaling, especially if their magnitudes are significantly larger or smaller than 1.

Finally, we considered the output parameters, which represent the material's stiffness. Both the absolute size and relative ratio of this physical quantity hold practical significance. As such, we performed absolute scaling, setting the scaling factor as the reciprocal of the maximum initial value, 2×10^{10}.

The fitting objective should also be considered. The mode shapes calculated from the finite element model are guaranteed to be smooth curves, while the measured data often have obvious non-smooth points due to errors, as shown in Figures 1 and 2. Assume that when previously generated, the set of all possible parameters is R_I, and the set of all possible output mode shapes is R_O. The inverse process that needs to be fitted clearly satisfies:

$$R : R_O \to R_I \tag{6}$$

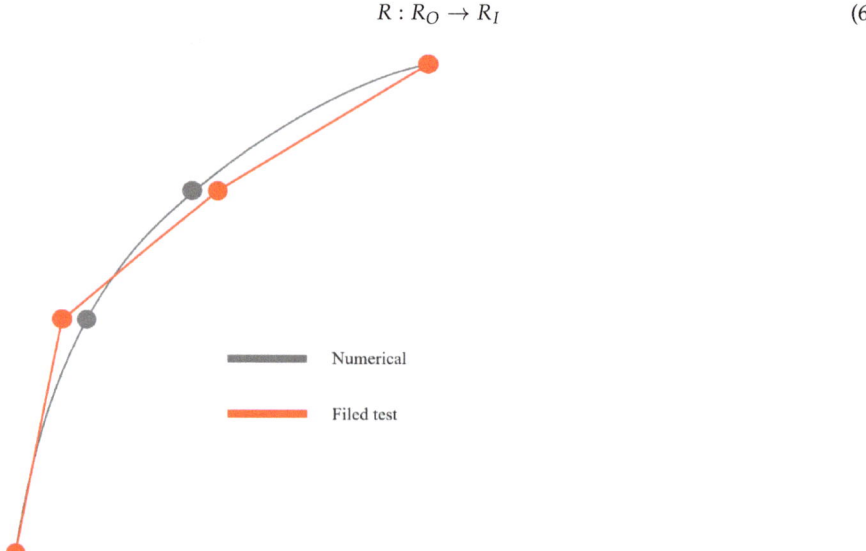

Figure 2. The difference between measured and numerical simulation modal shape.

Suppose the matrix composed of the true values of the parameters to be updated is $E'_{n \times 1}$. During the training of the neural network, the function fitting only occurs within the ranges of these two sets, R_O and R_I. However, because the measured data are not smooth, there is still an error between the finite element simulation and the actual situation. Even if the true value $E'_{n \times 1}$ of the parameters to be updated belongs to the set R_I, the measured result Φ of the mode shape is most likely not in the set R_O, but in the interval not covered by the training set.

Since the function f is obviously continuous, the fact that the measured result is not in the interval will not result in the failure of function prediction, but it may cause a large deviation in the updating result after the neural network overfits the function, or even be inconsistent with the reality. Moreover, the finite element model itself has various differences from the actual building. When overfitting is performed, it is equivalent to requiring the finite element to infinitely approximate the true value. Then, the other aspects of the error between the finite element model and the actual building will be increasingly reflected in the updated parameters, causing the parameters to vary greatly. Therefore, when constructing the neural network, it is crucial to avoid overfitting and control the degree of fitting.

3. Field Vibration Test for a Prefabricated Building

3.1. Target Building and Experimental Equipment

The Dormitory Building named Liyuan No. 6 at the Shenzhen campus of Harbin Institute of Technology (as shown in Figure 3) has a cross shape in the horizontal direction. The external length of the main structure is 32.1 m, and the external width is 29.8 m. The building consists of one underground floor and thirty above-ground floors, with an eave height of 97.80 m (measured from ±0.00). The seismic fortification intensity is 7 degrees, and the designed earthquake group is the first group, with a structural safety level of two. The lateral resistance components of the structure are evenly distributed, and the dimensions in the two main axis directions are similar, which can make the modal shape challenging to identify due to the closely spaced modes.

Figure 3. Target building. (The Chinese characters on the building are the building name "Liyuan").

In this test, four Fortimus seismometers were used, each unit containing an accelerometer, a data collector, and storage equipment, all from the seismic equipment company Güralp (Güralp Systems Ltd. 3 Midas House, Calleva Park Aldermaston, Reading RG7 8EA United Kingdom). The instrument photo is as shown in Figure 4. According to manufacturer data, all four accelerometers have a sensitivity of 0.112×10^{-6} m/s. The default sampling rate for the instruments is set at 200 Hz, significantly higher than the general natural frequency range of 0.1 Hz. Therefore, within the potential natural frequency range of the building, 200 Hz satisfies the Nyquist theorem. As such, the default sampling rate was used, and the instrument time was set to Greenwich Mean Time (GMT).

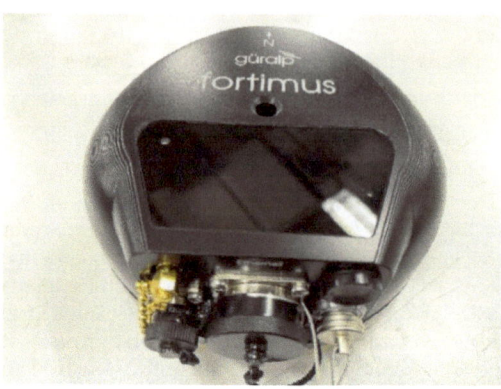

Figure 4. Photo of Fortimus seismometers.

3.2. Field Testing Arrangement

The testing methodology involved using four available instruments to perform a series of tests, aimed at obtaining the modal shapes at 30 different locations in the building. As depicted in Figure 5, the solid dots in (a) represent the measurement points, labeled Sn to denote the nth setup. The abbreviations TM, F, and S represent the terminal, floor, and plane measurement positions, respectively. These measurement points are arranged uniformly in the vertical direction; horizontally, due to the need for GPS synchronization of the instruments, we chose to conduct tests at positions 1S and 2S, and this was done on each floor.

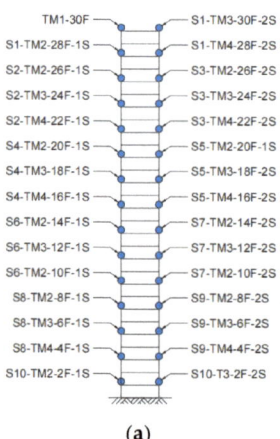

(a)

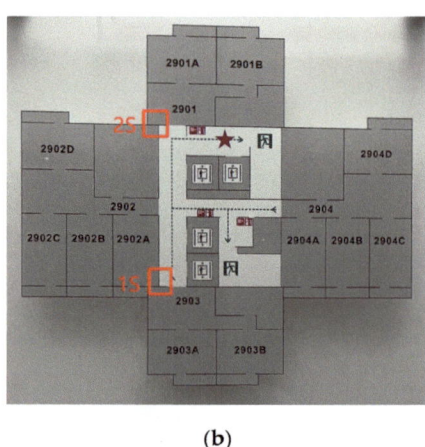

(b)

Figure 5. Test point locations. (**a**) Vertical measurement point position; (**b**) Vertical measurement point position.

Due to the limited number of instruments, the plan was proposed for 10 setups. One instrument (TM1) was always placed at a fixed point at the top of the building (30F) as a reference channel. The remaining three instruments were alternately arranged at other measurement points in the building to facilitate the use of old wires with GPS synchronization. The specific placement of each measurement point for every setup is summarized in Table 1.

Table 1. Test plan.

SETUP	Measurement Locations			
	TM1	TM2	TM3	TM4
1	30F-1S	28F-1S	30F-2S	28F-2S
2	30F-1S	26F-1S	24F-1S	22F-1S
3	30F-1S	26F-2S	24F-2S	22F-2S
4	30F-1S	20F-1S	18F-1S	16F-1S
5	30F-1S	20F-2S	18F-2S	16F-2S
6	30F-1S	14F-1S	12F-1S	10F-1S
7	30F-1S	14F-2S	12F-2S	10F-2S
8	30F-1S	8F-1S	6F-1S	4F-1S
9	30F-1S	8F-2S	6F-2S	4F-2S
10	30F-1S	2F-1S	2F-2S	

Prior to testing, the four instruments were placed together for a preliminary 20-min test to ensure their timing accuracy. Each subsequent array was then tested for 25 min. Throughout the testing process, the north direction on each instrument consistently pointed in the same direction, perpendicular to the building axis. For data processing, the north and east directions of each instrument were converted to x and y directions to align with the model. This test was completed on 1 September 2023.

3.3. Modal Analysis

Due to the large amount of data collected, only two representative examples of the acceleration time history of the building are presented in Figure 6. These data, obtained from field tests, were analyzed using the P-EM Bayesian operational modal analysis method [30] to identify the modal parameters of Liyuan No. 6. Compared with the conventional Bayesian modal analysis method, the P-EM approach offers superior performance in separating closely spaced modes and offers enhanced efficiency. It retains all the benefits of the standard Bayesian method, such as the ability for quantitative analysis of result uncertainty. The results of the modal analysis are presented in Figure 7 and Table 2.

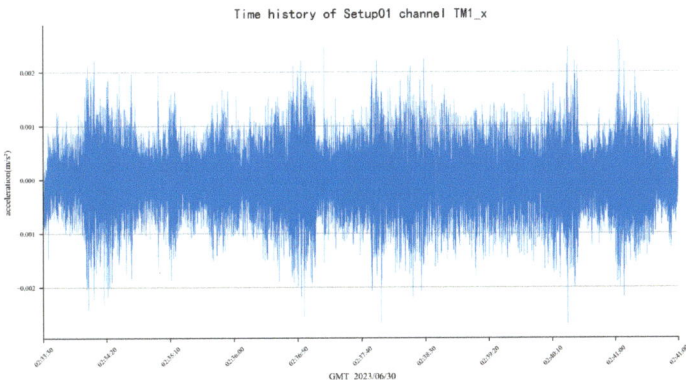

Figure 6. Examples of acceleration time history data.

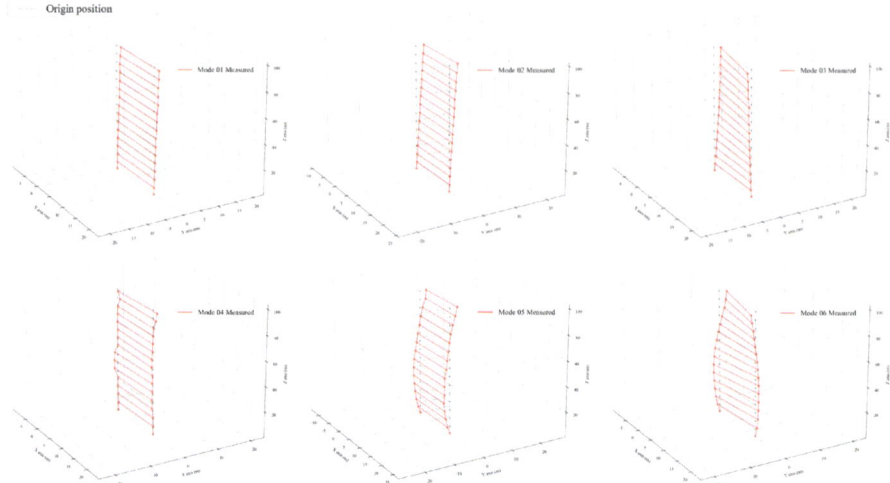

Figure 7. Modal shape results, Liyuan No. 6 dormitory.

Table 2. Modal identification results, Liyuan No. 6 dormitory.

Mode	Frequency		Damping Ratio		Prediction Error PSD		Modal Force PSD	
	MPV [Hz]	c.o.v. [%]	MPV [‰]	c.o.v. [%]	MPV [(µg)^2/Hz]	c.o.v. [%]	MPV [(µg)^2/Hz]	c.o.v. [%]
1	0.702	0.17	5.622	0.21	59.956	76.20	81.364	28.52
2	0.769	0.15	6.320	0.30	59.956	76.20	83.162	23.59
3	1.430	0.35	6.468	0.29	32.186	67.60	39.935	20.68
4	2.909	0.24	15.003	0.09	92.700	47.04	34.162	32.21
5	3.002	0.28	12.471	0.06	121.079	38.32	40.833	25.36
6	4.282	0.45	14.777	0.33	52.568	38.25	20.771	42.13

3.4. Finite Element Model and Numerical Analysis Results

The numerical simulation portion of this study was performed using ANSYS 2021 due to its superior simulation performance and its flexibility in secondary development. A finite element model of the building was established based on design drawings provided by the design institute. Two models were constructed for this study. Model 1 includes only structural components such as shear walls, beams, and floors. Model 2, on the other hand, includes partition walls as shear wall units in their original positions. Model 1 was used to compare the results of the modal analysis. Subsequent model updating was conducted using Model 2, which was built based on Model 1 by adding the partition wall.

Given that the basement does not participate in the modal analysis of the upper part of the building and could potentially generate local modal interference during the analysis process, it was excluded from the modeling. Similarly, to avoid potential local modal interference, the balcony was simplified as a beam load for modeling. This was achieved by considering the balcony board's thickness (120 mm) and length (1000 mm), while ignoring the weight of the upper railing. The concrete weight was taken as 24 KN/m^3, leading to a uniformly distributed load that is applied to the beam as follows:

$$p = 24 \times 0.12 \times 1 = 2.88 \, \text{KN/m} \tag{7}$$

Upon completion of the modeling (as shown in Figure 8), the modal analysis was performed using the subspace iteration method in ANSYS. The frequency range was set from 0.01 Hz to 20 Hz, and the damping ratio was 0.05.

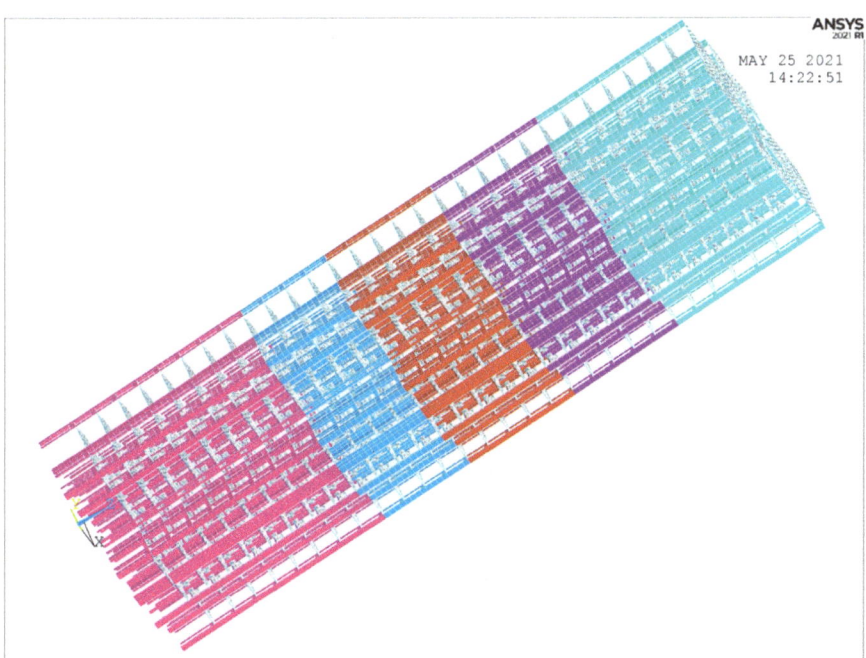

Figure 8. Finite element model without partition walls, Liyuan No. 6 dormitory.

The results of modal identification of Model 1 can be found in Table 3, which includes data from field tests and the design institute. Two field test results are presented in the table: one from Liyuan No. 6 and the other from Liyuan No. 7. Both buildings are dormitories and were constructed using nearly identical design drawings.

Table 3. Modal identification results of Liyuan No. 6.

Mode	Frequency [Hz]				Characteristics
	Filed Test	Numerical	Numerical by Design Institute	Filed Test of Liyuan No. 7	
1	0.702	0.399	0.375	0.752	BX1
2	0.769	0.435	0.393	0.822	BY1
3	1.430	0.477	0.478	1.483	T1
4	2.909	1.812	1.727	2.994	BX2
5	3.002	1.883	1.764	3.147	BY2
6	4.282	2.006	1.985	4.392	T2

As Table 3 illustrates, the two field test results are in close agreement, as are the numerical results. However, a significant discrepancy becomes apparent when comparing the field test results to the numerical results. The frequency of bending modes in the field test is nearly twice that of the numerical results, while the frequency of torsional modes in the field test is approximately one quarter that of the numerical results. In the numerical analysis, neither model considered the stiffness provided by the partition walls. Therefore, this study focuses on these walls for model updating. The goal was to investigate the ratio of the stiffness they provide in the building, compared to that provided by shear walls of the same thickness and location.

4. Application

4.1. Random Generation of Calculation Parameters

As discussed in the comparison and analysis of results in Section 3.4, the noticeable discrepancy between the field test modal identification results of the study's test subject—Liyuan No. 6—and the numerical results can be attributed to the overlooked stiffness of the partition walls. While there has been considerable research on the inherent stiffness and mechanical properties of precast concrete partition walls, studies on their influence on the stiffness of a finished building have been relatively scarce. This research seeks to identify an appropriate set of stiffness values for these partition walls through model updating, with the goal of adapting the finite element model of Building 6. Simultaneously, we hope to propose a reference value for the actual stiffness of precast concrete partition walls in high-rise buildings, extrapolated from neural network predictions of material stiffness.

Assume that the parameter to be updated is E_i, $(i = 1, 2, \ldots, n)$, and the empirical possible value for each parameter is $\overline{E}_i$, $(i = 1, 2, \ldots, n)$. Assume a class of probability density functions p that have similar property. Choose n of these p function, denoted as $p_i(i = 1, 2, \ldots, n)$. Each time a sample is selected, pick values $r_i(i = 1, 2, \ldots, n)$ that individually adhere to these n probability density functions. The parameters selected then are:

$$M_{n \times 1} = \begin{bmatrix} \overline{m}_1 \times r_1 \\ \overline{m}_2 \times r_2 \\ \vdots \\ \overline{m}_n \times r_n \end{bmatrix} = \begin{bmatrix} m_1 \\ m_2 \\ \vdots \\ m_n \end{bmatrix} \qquad (8)$$

In this study of the actual stiffness of partition walls in high-rise building, there was little prior experience to draw on. Therefore, the empirical value was selected as 2/3 of the concrete stiffness of the studied partition wall, and the probability density function was chosen as:

$$p = \begin{cases} 1.25, & x \in [0.2, 1] \\ 0, & x \notin [0.2, 1] \end{cases} \qquad (9)$$

Obviously, when choosing parameters, the selection of the probability density function p corresponding to each parameter is crucial. In general, the more certain the property that the parameter represents, the more accurate the empirical estimate, and consequently, the larger the value of p around $x = 1$ will be, and the opposite is true for less certain properties. This selection is due to the uncertainty of the partition wall stiffness, and the stiffness of the partition wall connected by splicing cannot approach its inherent stiffness. Once the parameters to be adjusted are generated, they can be incorporated into the program to modify the finite element model parameters and perform iterative calculations using the finite element software. The calculated results are then saved as a training dataset.

In the modeling process of partition walls, they were treated similarly to shear walls, modeled using shell units, with the elastic modulus of each partition wall serving as the updating parameter. Under consistent circumstances of connections and geometric dimensions, the stiffness contributed by the SHELL unit directly correlates with the elastic modulus. Therefore, the ratio of the stiffness provided by the partition wall unit in the model to that of a shear wall of identical thickness can be determined by dividing the parameter value, derived post-updating, by the original elastic modulus of the material. The subject of this study utilized precast concrete partition walls, of grade C30, with an elastic modulus of 30 GPa. The targets for updating are all partition walls situated beneath beams or between shear walls, serving as dividers. Considering the different positions and opening situations of the partition walls, they are categorized into eight types, along with the glass curtain wall on the first floor, resulting in a total of nine parameters to be updated, as illustrated in Table 4.

Table 4. Properties of partition walls to be updated.

Wall Type	Initial Elastic Modulus [Pa]	Poisson's Ratio	Density (Kg/m^3)
Curtain Wall	5.00×10^9	0.250	1000
Fire Wall without Openings	2.00×10^{10}	0.167	2420
Elevator Partition Wall	2.00×10^{10}	0.167	2420
Fire Wall with Openings	2.00×10^{10}	0.167	2420
External Partition Wall	2.00×10^{10}	0.167	2420
Windowed External Partition Wall	2.00×10^{10}	0.167	2420
Balcony External Partition Wall	2.00×10^{10}	0.167	2420
Internal Partition Wall	2.00×10^{10}	0.167	2420
Internal Partition Wall with Door	2.00×10^{10}	0.167	2420

In Table 4, it is not difficult to notice that the load-bearing structures of the building were not included in the scope of model updating. The reason for this is that, in general, the deviation of the elastic modulus of the load-bearing structures is about 5%, and in the subject of this paper, this deviation has a significantly smaller impact on the natural frequency of the structure compared to the partition walls. Since the error ranges of the two are clearly different, including the main load-bearing structure in the scope of consideration could likely lead to extremely unreasonable deviations due to the influence of the partition walls.

In the modeling, partition walls were treated as shear walls of equivalent thickness (without openings), with the material properties attributed to the respective wall positions. It is important to note that as this paper categorizes variations such as openings and construction methods, the updated elastic modulus does not truly represent a specific property of material, but rather serves as an equivalency value for the stiffness provided by the materials in the wall area. The updated elastic modulus will be referred to as the equivalent elastic modulus.

Upon incorporation of the partition walls, the Partition wall model in ANSYS of the first and second floors are shown in Figure 9. In the figure, different colors correspond to different materials, with examples including green for curtain walls, dark blue for external partition walls, light blue for windowed external partition walls, and yellow for unopened firewalls.

The selection of the initial elastic modulus for the concrete partition wall, as mentioned above, was set to two-thirds of the elastic modulus of C30 concrete, i.e., 2×10^{10} Pa. The elastic modulus of the curtain wall was tentatively set at 5×10^9 Pa due to lack of reference. It is worth noted that an overestimation of the elastic modulus is of little concern during this process. This is because if such a situation arises, the updated values for each elastic modulus will shift away from the pre-determined estimated range set by this research significantly. At that point, the necessary adjustments are made.

4.2. DNN Structure Design

Drawing upon previous discussions, the primary aim of the neural network construction in this research is to model the inverse function of calculations performed during modal analysis using finite element software. It is crucial that the network demonstrates a certain level of generalization ability to avoid overfitting. The final structure of the neural network used is shown in Figure 10.

The neural network is designed to take two distinct one-dimensional matrices as input: the frequencies and the mode shapes. Given the unique physical implications of these parameters, they should not be simply concatenated into a larger one-dimensional matrix. For instance, the frequency under a specific mode embodies comprehensive information about the integrated stiffness and shape of the entire building, while an element in the mode shape merely represents the relative displacement of a specific node in each direction under the corresponding mode. Therefore, a conventional deep neural network architecture is used to process the mode shapes, while the frequencies are processed in a shallower neural network that merges with the former in the final layer. Notably, in contrast to the network that processes mode shapes, the network designated for frequency processing does not employ dropout on its neurons.

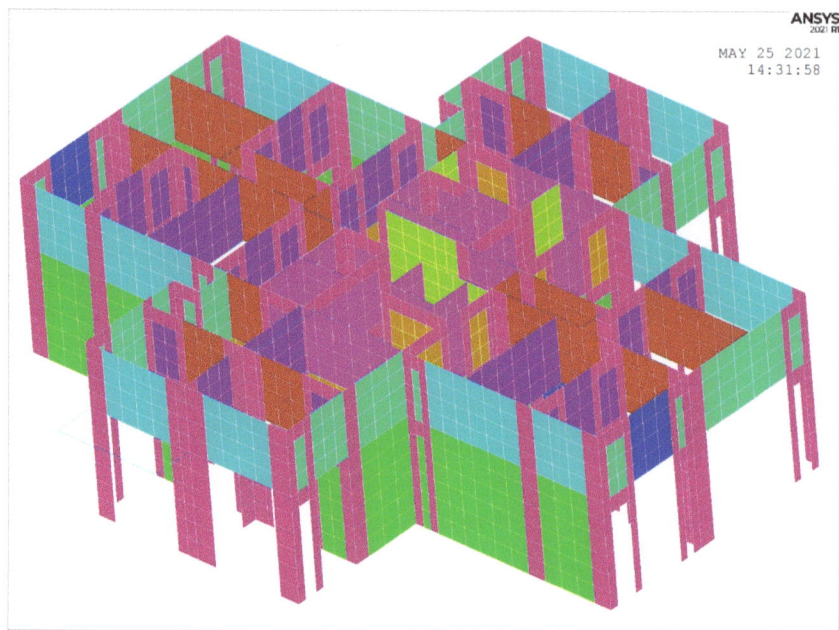

Figure 9. Partition wall model in ANSYS of the first and second floors.

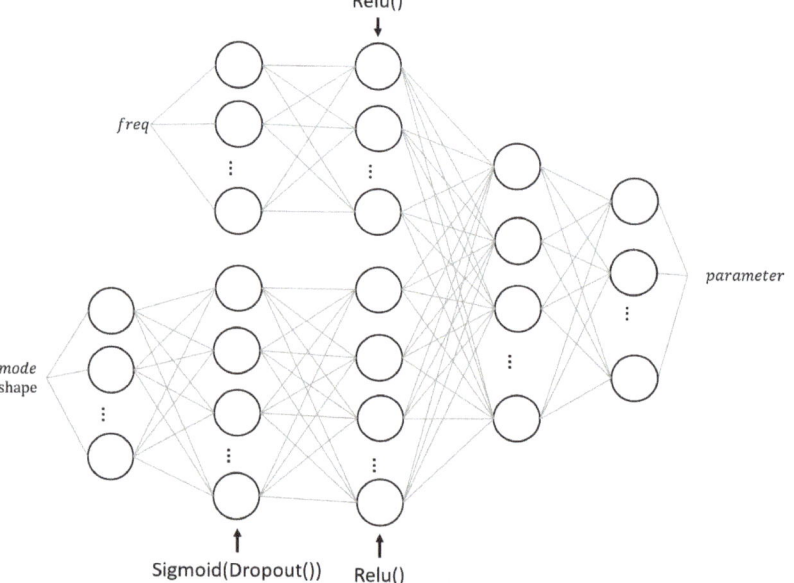

Figure 10. DNN structure.

Since all nine DNN output parameters represent the modulus of elasticity, the Mean Square Error (MSE) function serves as the loss function. Considering the real-world implications of the partition's modulus of elasticity, a coefficient of variation is incorporated as a penalty term. Notably, the first of the nine modified partitions is merely a curtain wall positioned on a single floor, and hence is excluded from the calculation of the coefficient of variation. The loss function is:

$$loss = MSEloss(target, predict) * \left(1 + c.o.v\left(\widehat{predict}\right)\right) \quad (10)$$

where *target* and *predict* respectively denote the label and the DNN output, and $\widehat{predict}$ signifies the parameters from the second to the ninth of the DNN output.

4.3. DNN Generalization Ability Test

As previously mentioned, the generalization capability of the DNN is crucial for the research. The probability density function used to generate parameters is defined in Equation (9).

With this probability density function, 5000 samples were generated. The selection of this number was based on papers [20,25] using the MCMC method, which also requires the comprehensiveness of samples. Sample sets of 2500, 5000, and 7500 were generated, and the results showed a slight improvement in accuracy when the quantity increased to 5000, with almost no change beyond that.

A broader probability density function was adopted to generate another 5000 samples:

$$\dot{p} = \begin{cases} 1/1.4, & x \in [0.1, 1.5] \\ 0, & x \notin [0.1, 1.5] \end{cases} \quad (11)$$

These samples were used as a validation set in the tests. At this point, if the computed loss value significantly increases after a certain number of epochs, it indicates that the neural network is overfitting in the region covered by the original probability density. The changes in loss after 600 epochs are displayed in Figure 11a,b. These figures use samples from the sample set as the validation set and incorporate a training set characterized by a wide parameter range.

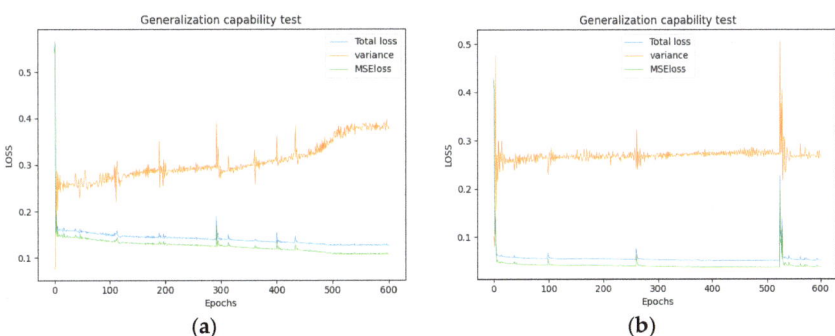

Figure 11. The loss curve of the DNN under the validation set of the same range (**a**) and a broader range (**b**).

The figure illustrates that within 600 epochs, the neural network progressively improves its generalization ability for samples beyond the training parameter range. This improvement is evidenced by a steady decrease in MSE loss throughout the training, suggesting no overfitting within the finite element solution's scope. However, predictably, the equivalent elastic modulus of several partition walls exhibits increasing variability as training progresses. As previously analyzed, this variation stems from the neural network overly fitting the inverse process of the finite element calculation, leading to a shortfall in the generalization capacity for the inputted measured modes. Based on these observations, an epoch iteration count of 200 was deemed appropriate for this case.

4.4. Model Updating of a Prefabricated Building

Upon finalizing the model using the DNNMU method, the predicted parameters were generated, specifically the elastic modulus. These parameters were then incorporated into the finite element model. Subsequently, a numerical modal analysis was conducted on this updated model to yield the modal parameters. The resulting values were compared with measured results. Table 5 illustrates the comparison between the updated calculated frequencies and the measured frequencies, along with the MAC matrix. Figure 12 presents the comparison of mode shapes.

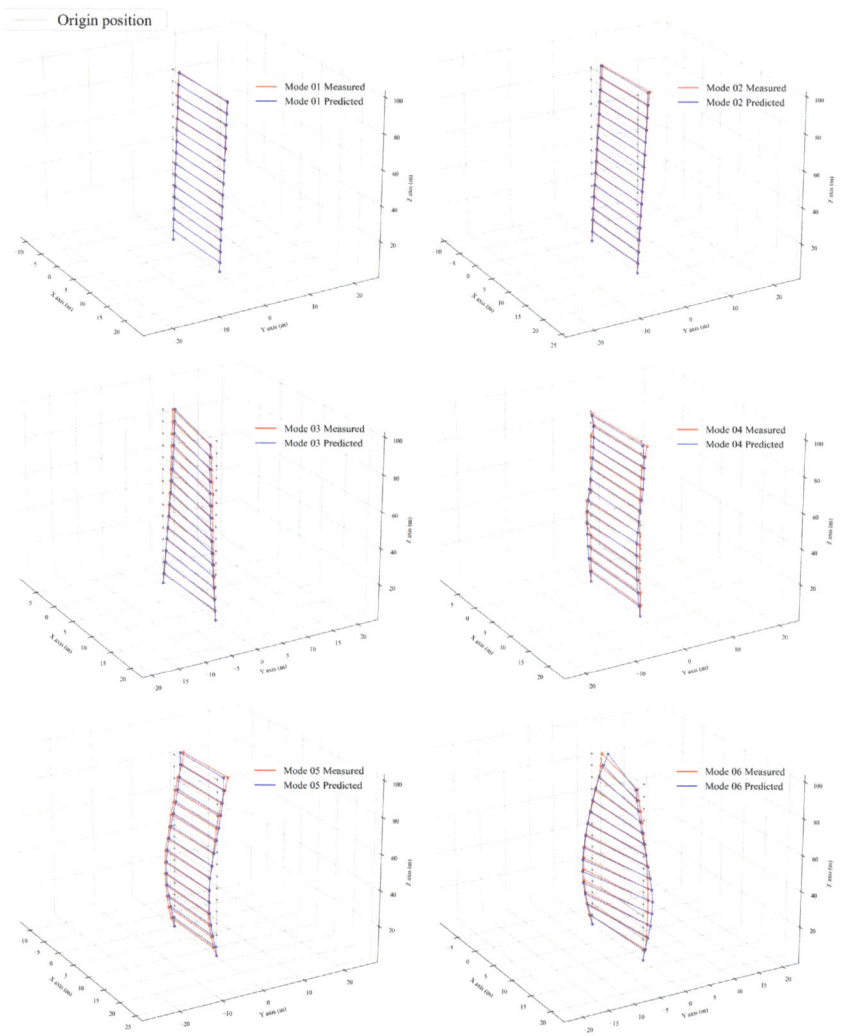

Figure 12. Modal shape comparison.

Table 5. Results comparison.

	Mode 01	Mode 02	Mode 03	Mode 04	Mode 05	Mode 06
Measured f	0.7029	0.7672	1.4180	2.9080	3.0071	4.2914
Updated f	0.7097	0.7619	1.3842	2.9616	3.0840	4.1090
c.o.v	0.97%	0.68%	2.38%	1.84%	2.56%	4.25%
Characteristic	BX1	BY1	T1	BX2	BY2	T2
mac	0.9932	0.0046	0.6019	0.0439	0.0220	0.1062
	0.0094	0.9973	0.0766	0.0284	0.0003	0.0201
	0.3986	0.0052	0.9619	0.0018	0.0216	0.0738
	0.0770	0.0058	0.0396	0.9587	0.3529	0.5800
	0.0095	0.0768	0.0020	0.3658	0.9709	0.1527
	0.1605	0.0454	0.1274	0.2655	0.0011	0.8921

The first observation from the results is that the discrepancy between the sixth order modal frequency prediction of this model updating method and the actual measured data are within a 5% margin. Subsequently, the updated modal shapes results are examined. As depicted in Figure 12, the MAC values of the updated first-order bending mode for the building along the x and y axes are 0.993 and 0.997, respectively, while the first-order torsional mode has a MAC value of 0.96. Clearly, the second order updating of the bending and torsional modes are not as effective as the first order, with the bending mode at approximately 0.95 and the torsional mode near 0.9. The rationale for this is evident: the mode shapes derived from the finite element model are invariably smooth curves, whereas the modal analysis results of the study's subject, Liyuan No. 6, obviously show significant noise interference. This issue is especially pronounced in higher-order modes, resulting in significant non-smooth points in the measured mode shapes. Consequently, regardless of the outstanding performance of the updating algorithm, the MAC value will inevitably be limited under these conditions and cannot closely approach 1. Compared to the article [37] published in 2019 using the MCMC method, the model correction results of that article also achieve a relative frequency error of less than 5%. The MAC values for the first-order modes (corresponding to Mode 01–Mode 03 in this paper) are generally greater than 0.93, with the highest being 0.989. The MAC values for the second-order modes (Mode 04–Mode 06) are around 0.9. In comparison, our method shows similar performance on the torsional modes, specifically Mode 03 and Mode 06, but demonstrates significant advantages in the remaining modes.

When examining the MAC values over six modal shapes, it is evident that less smooth actual test modal shapes correspond to lower MAC values. The updated modal shapes strike a balance, accommodating the noise of the measured modal shapes, with their overall form reflecting the measured modal shapes after noise removal. Consequently, the author claims that the DNNMU model updating method's efficacy is most prominent in the two smoothest measured modal shapes in this dataset—BX1 and BY1—where the two modal shapes of the updated finite element model align almost perfectly with the measured results. In comparison, articles [25–27], which also conduct model updating on buildings, generally report a MAC value reaching 0.95 in the model updating results, with the torsional mode having a lower MAC value. However, this method demonstrates superior accuracy in the updating results.

To verify the robustness, 20 groups of 2500 samples were randomly selected from a total of 7500 samples and used to perform model updating. The average coefficient of variation for the nine parameters to be updated (elastic modulus) was 9.24%, and the average coefficient of variation for the modal frequencies of the corrected model was 2.86%. Considering the inevitable differences in the elastic modulus for each category, which arise because this paper classifies the partitions within a 30-story building into only nine categories, the coefficient of variation result is not considered high. Additionally, since the objects of adjustment are partition walls, the variance in the degree to which they are fixed to the main structure also affects the updated elastic modulus result.

In terms of efficiency, compared to traditional MCMC methods or any iteration-based methods, the greatest advantage of the method proposed in this article lies in its efficiency. The most time-consuming step in model updating is the calculation of the numerical model, as it often requires tens of thousands of calculations on the model. As shown in Figure 1, part I of the DNNMU method requires thousands of calculations on each parameter probability density function to generate and store a dataset. This process only needs to be carried out once, rather than having to perform thousands of numerical simulations in each iteration as traditional MCMC methods do. When part II conducts the training of the neural network, thanks to the simple structure of the neural network itself, the training can be completed within minutes even just using a CPU. For example, if a finite element software simulation takes 20 s per run, an MCMC method using 2000 sample points and iterating ten times would need approximately five and a half hours. Although our method requires 5000 finite element simulations, it only needs half the time of the former. Once a sufficient number of sample points are obtained, each model updating with the DNNMU method only requires training, which takes less than 3 min due to the simplicity of the DNN architecture used, thus making it highly efficient. Compared to article [38], which uses an ensemble learning decision tree, the training and model adjustment time of that research on a simple laboratory structure is 70 s, which can be said to be nearly equivalent to the efficiency of proposed method.

The predicted elastic modulus of each partition walls is shown in Table 6. Based on the model updating outcomes, the stiffness offered by the prefab partition within the building approximates 25–45% of that provided by the shear wall of equivalent material and thickness. Incorporating these partitions results in nearly doubling the first two modal frequencies of the complete building model. This underscores the significant influence of the stiffness provided by the partitions on the dynamic properties of the structure.

Table 6. Updated elastic modulus result.

Wall Type	Initial Elastic Modulus [Pa]	Initial Elastic Modulus [Pa]	Ratio (%)
Curtain Wall	5.00×10^9	2.72×10^9	54.40
Fire Wall without Openings	2.00×10^{10}	9.06×10^9	45.30
Elevator Partition Wall	2.00×10^{10}	6.94×10^9	34.70
Fire Wall with Openings	2.00×10^{10}	8.25×10^9	41.25
External Partition Wall	2.00×10^{10}	8.71×10^9	43.55
Windowed External Partition Wall	2.00×10^{10}	5.46×10^9	27.30
Balcony External Partition Wall	2.00×10^{10}	5.78×10^9	28.90
Internal Partition Wall	2.00×10^{10}	7.33×10^9	36.65
Internal Partition Wall with Door	2.00×10^{10}	6.07×10^9	30.35

5. Conclusions

This paper presents the DNNMU method, which uses environmental vibration test data and modal analysis of a prefabricated building to improve model updating. By analyzing the first six modal parameters and comparing them with finite element models and design calculations, the study confirms the critical influence of partition walls on building dynamics.

The main finding was that the DMU achieves modal frequency deviations within 5% and mac values predominantly over 0.9; in particular, for the first two modes with minimal noise interference, the mac values achieved were 0.993 and 0.997.

The originality of this paper lies in the distinctive application of deep neural networks (DNN) to the field of structural model updating, resulting in a streamlined and efficient methodology that requires very few hyperparameters to be adjusted, making it highly adaptable. This approach requires only a single large-scale numerical simulation to generate the dataset, followed by a relatively simple DNN architecture to implement the updates.

In addition, the variance of 0.399 Hz and 0.702 Hz in natural frequency between the building modeled and the actual structure in the study was higher that is commonly found. This study found that prefabricated partitions have a stiffness ratio of about 0.2–0.3 compared to shear walls of the same material and thickness, emphasizing their role in structural behavior.

A drawback of this method is its relative lack of interpretability compared to MCMC or with machine learning methods more broadly. While this research focuses on a single structure, the results lay the groundwork for future studies on the DNNMU method's application across diverse buildings and partition types, and to explore the utility of different deep learning models for model updating. This opens up prospects for advancing building design and performance. Moreover, the DNN network architecture utilized in this study invites further exploration to identify more suitable, effective, and accurate deep learning models for model updating.

The DNNMU has the potential to be applied to other structures. If the evaluation metrics for the model updating results are still the modal frequencies and mode shapes (MAC matrix) as in this paper, then the DNN architecture in the method does not need any major modifications; simply replacing the dataset will be sufficient. If the evaluation metrics change, it may be necessary to adjust the DNN structure to accommodate indicators with different characteristics. Furthermore, replacing the DNN in the method with another deep learning model is also feasible. The new model should have the capability to distinguish between different indicators and the ability to prevent overfitting.

Author Contributions: Conceptualization, C.L., F.Z. and Y.N.; Methodology, C.L.; Project Administration, F.Z. and Y.N.; Software, C.L.; Supervision, F.Z. and Y.N.; Validation, C.L., F.Z., B.A., S.Z., Z.Z. and S.F.; Visualization, C.L.; Writing—Original Draft, C.L.; Writing—Review and Editing, C.L., Y.N. and F.Z. All authors have read and agreed to the published version of the manuscript.

Funding: This research was partly funded by the Guizhou Provincial Key Technology R&D Program (No.: [2021] General 357; [2023] General 424), Guangdong Provincial Key Laboratory of Intelligent and Resilient Structures for Civil Engineering [No.2023B1212010004], and the National Natural Science Foundation of China [No.52378312].

Data Availability Statement: Data will be made available on request.

Conflicts of Interest: The authors declare no conflict of interest.

References

1. Liu, S.; Li, Z.; Teng, Y.; Dai, L. A dynamic simulation study on the sustainability of prefabricated buildings. *Sustain. Cities Soc.* **2022**, *77*, 103551. [CrossRef]
2. Xie, L.; Chen, Y.; Xia, B.; Hua, C. Importance-performance analysis of prefabricated building sustainability: A case study of Guangzhou. *Adv. Civ. Eng.* **2020**, *2020*, 8839118. [CrossRef]
3. Navaratnam, S.; Satheeskumar, A.; Zhang, G.; Nguyen, K.; Venkatesan, S.; Poologanathan, K. The challenges confronting the growth of sustainable prefabricated building construction in Australia: Construction industry views. *J. Build. Eng.* **2022**, *48*, 103935. [CrossRef]
4. Magliulo, G.; Ercolino, M.; Petrone, C.; Coppola, O.; Manfredi, G. The Emilia earthquake: Seismic performance of precast reinforced concrete buildings. *Earthq. Spectra* **2014**, *30*, 891–912. [CrossRef]
5. Yun, C.; Chao, C. Study on seismic performance of prefabricated self-centering steel frame. *J. Constr. Steel Res.* **2021**, *182*, 106684. [CrossRef]
6. Zhao, B.; Wu, D.; Zhu, H. New modular precast composite shear wall structural system and experimental study on its seismic performance. *Eng. Struct.* **2022**, *264*, 114381. [CrossRef]
7. Zhai, X.; Zhang, X.; Cao, C.; Hu, W. Study on seismic performance of precast fabricated RC shear wall with opening filling. *Constr. Build. Mater.* **2019**, *214*, 539–556. [CrossRef]
8. Papaioannou, M.; Karagiannis, J.; Sotiropoulou, A.; Badogiannis, E.; Poulakos, G. Sound Insulation Performance of Prefabricated Concrete Wall Partitions in School Buildings in Athens. In Proceedings of the Euronoise 2018, Crete, Greece, 27–31 May 2018.

9. Sotiropoulou, A.; Karagiannis, I.; Papaioannou, M.; Badogiannis, E. Sound Insulation Performance of Prefabricated Concrete Partitions in Hellenic School Buildings. *J. Civ. Eng. Archit.* **2019**, *13*, 353–372.
10. Tsirigoti, D.; Giarma, C.; Tsikaloudaki, K. Indoor Acoustic Comfort Provided by an Innovative Preconstructed Wall Module: Sound Insulation Performance Analysis. *Sustainability* **2020**, *12*, 8666. [CrossRef]
11. Liu, P.; Luo, X.; Chen, Y.; Yu, Z.; Yu, D. Thermodynamic and acoustic behaviors of prefabricated composite wall panel. In *Structures*; Elsevier: Amsterdam, The Netherlands, 2020; Volume 28, pp. 1301–1313.
12. Zou, D.; Sun, C. Analysis for Thermal Performance and Energy-Efficient Technology of Prefabricated Building Walls. *Int. J. Heat Technol.* **2020**, *38*. [CrossRef]
13. Tawil, H.; Tan, C.G.; Sulong, N.H.R.; Nazri, F.M.; Sherif, M.M.; El-Shafie, A. Mechanical and thermal properties of composite precast concrete sandwich panels: A Review. *Buildings* **2022**, *12*, 1429. [CrossRef]
14. Perera, D.; Poologanathan, K.; Gatheeshgar, P.; Upasiri, I.R.; Sherlock, P.; Rajanayagam, H.; Nagaratnam, B. Fire performance of modular wall panels: Numerical analysis. In *Structures*; Elsevier: Amsterdam, The Netherlands, 2021; Volume 34, pp. 1048–1067.
15. Yu, S.; Liu, Y.; Wang, D.; Bahaj, A.S.; Wu, Y.; Liu, J. Review of thermal and environmental performance of prefabricated buildings: Implications to emission reductions in China. *Renew. Sustain. Energy Rev.* **2021**, *137*, 110472. [CrossRef]
16. Valencia-Barba, Y.E.; Gómez-Soberón, J.M.; Gómez-Soberón, M.C.; Rojas-Valencia, M.N. Life cycle assessment of interior partition walls: Comparison between functionality requirements and best environmental performance. *J. Build. Eng.* **2021**, *44*, 102978. [CrossRef]
17. Zhao, J.; Wang, Y.; Ma, Z. Factors influencing the environmental performance of prefabricated buildings: A case study of community A in Henan Province of China. *Nat. Environ. Pollut. Technol.* **2020**, *19*, 221–227.
18. Cortês, A.; Almeida, J.; Santos, M.I.; Tadeu, A.; de Brito, J.; Silva, C.M. Environmental performance of a cork-based modular living wall from a life-cycle perspective. *Build. Environ.* **2021**, *191*, 107614. [CrossRef]
19. Li, B.; Hutchinson, G.L.; Duffield, C.F. The influence of non-structural components on tall building stiffness. *Struct. Des. Tall Spec. Build.* **2011**, *20*, 853–870. [CrossRef]
20. Lam, H.F.; Yang, J.; Au, S.K. Bayesian model updating of a coupled-slab system using field test data utilizing an enhanced Markov chain Monte Carlo simulation algorithm. *Eng. Struct.* **2015**, *102*, 144–155. [CrossRef]
21. Ahmadian, H.; Mottershead, J.E.; Friswell, M.I. Regularisation methods for finite element model updating. *Mech. Syst. Signal Process.* **1998**, *12*, 47–64. [CrossRef]
22. Mottershead, J.E.; Friswell, M.I. Model updating in structural dynamics: A survey. *J. Sound Vib.* **1993**, *167*, 347–375. [CrossRef]
23. Katafygiotis, L.S.; Beck, J.L. Updating models and their uncertainties: Model identifiability. *J. Eng. Mech. ASCE* **1998**, *124*, 463–467. [CrossRef]
24. Katafygiotis, L.S.; Papadimitriou, C.; Lam, H.F. A probabilistic approach to structural model updating. *Soil Dyn. Earthq. Eng.* **1998**, *17*, 495–507. [CrossRef]
25. Beck, J.L.; Au, S.K. Bayesian updating of structural models and reliability using Markov chain Monte Carlo simulation. *J. Eng. Mech.* **2002**, *128*, 380–391. [CrossRef]
26. Boulkaibet, I.; Mthembu, L.; Marwala, T.; Friswell, M.I.; Adhikari, S. Finite element model updating using the shadow hybrid Monte Carlo technique. *Mech. Syst. Signal Process.* **2015**, *52*, 115–132. [CrossRef]
27. Zhang, J.; Wan, C.; Sato, T. Advanced Markov chain Monte Carlo approach for finite element calibration under uncertainty. *Comput.-Aided Civ. Infrastruct. Eng.* **2013**, *28*, 522–530. [CrossRef]
28. Lee, Y.; Kim, H.; Min, S.; Yoon, H. Structural damage detection using deep learning and FE model updating techniques. *Sci. Rep.* **2023**, *13*, 18694. [CrossRef] [PubMed]
29. Gong, M.; Park, W. Finite element model updating of structures using deep neural network. *KSCE J. Civ. Environ. Eng. Res.* **2019**, *39*, 147–154.
30. Torzoni, M.; Manzoni, A.; Mariani, S. A deep neural network, multi-fidelity surrogate model approach for Bayesian model updating in SHM. In *European Workshop on Structural Health Monitoring*; Springer International Publishing: Cham, Switzerland, 2022; pp. 1076–1086.
31. Sung, H.; Chang, S.; Cho, M. Component model synthesis using model updating with neural networks. *Mech. Adv. Mater. Struct.* **2023**, *30*, 400–411. [CrossRef]
32. Zhang, R.; Liu, Y.; Sun, H. Physics-guided convolutional neural network (PhyCNN) for data-driven seismic response modeling. *Eng. Struct.* **2020**, *215*, 110704. [CrossRef]
33. Seventekidis, P.; Giagopoulos, D.; Arailopoulos, A.; Markogiannaki, O. Structural Health Monitoring using deep learning with optimal finite element model generated data. *Mech. Syst. Signal Process.* **2020**, *145*, 106972. [CrossRef]
34. Zhu, Z.; Au, S.K.; Li, B.; Xie, Y.L. Bayesian operational modal analysis with multiple setups and multiple (possibly close) modes. *Mech. Syst. Signal Process.* **2021**, *150*, 107261. [CrossRef]
35. Brownjohn, J.M.W.; Au, S.K.; Zhu, Y.; Sun, Z.; Li, B.; Bassitt, J.; Hudson, E.; Sun, H. Bayesian operational modal analysis of Jiangyin Yangtze River bridge. *Mech. Syst. Signal Process.* **2018**, *110*, 210–230. [CrossRef]
36. Brownjohn, J.M.W.; Raby, A.; Au, S.K.; Zhu, Z.; Wang, X.; Antonini, A.; Pappas, A.; D'Ayala, D. Bayesian operational modal analysis of offshore rock lighthouses: Close modes, alignment, symmetry, and uncertainty. *Mech. Syst. Signal Process.* **2019**, *133*, 106306. [CrossRef]

37. Lam, H.F.; Hu, J.; Zhang, F.L.; Ni, Y.C. Markov chain Monte Carlo-based Bayesian model updating of a sailboat-shaped building using a parallel technique. *Eng. Struct.* **2019**, *193*, 12–27. [CrossRef]
38. Kamali, S.; Mariani, S.; Hadianfard, M.A.; Marzani, A. Inverse surrogate model for deterministic structural model updating based on random forest regression. *Mech. Syst. Signal Process.* **2024**, *215*, 111416. [CrossRef]

Disclaimer/Publisher's Note: The statements, opinions and data contained in all publications are solely those of the individual author(s) and contributor(s) and not of MDPI and/or the editor(s). MDPI and/or the editor(s) disclaim responsibility for any injury to people or property resulting from any ideas, methods, instructions or products referred to in the content.

Article

TrajectoryNAS: A Neural Architecture Search for Trajectory Prediction

Ali Asghar Sharifi [1], Ali Zoljodi [1] and Masoud Daneshtalab [1,2,*]

[1] School of Innovation, Design and Technology (IDT), Mälardalen University, 72123 Västerås, Sweden; ali.zoljodi@mdu.se (A.Z.)
[2] Department of Computer Systems, Tallinn University of Technology, 19086 Tallinn, Estonia
* Correspondence: masoud.daneshtalab@mdu.se

Abstract: Autonomous driving systems are a rapidly evolving technology. Trajectory prediction is a critical component of autonomous driving systems that enables safe navigation by anticipating the movement of surrounding objects. Lidar point-cloud data provide a 3D view of solid objects surrounding the ego-vehicle. Hence, trajectory prediction using Lidar point-cloud data performs better than 2D RGB cameras due to providing the distance between the target object and the ego-vehicle. However, processing point-cloud data is a costly and complicated process, and state-of-the-art 3D trajectory predictions using point-cloud data suffer from slow and erroneous predictions. State-of-the-art trajectory prediction approaches suffer from handcrafted and inefficient architectures, which can lead to low accuracy and suboptimal inference times. Neural architecture search (NAS) is a method proposed to optimize neural network models by using search algorithms to redesign architectures based on their performance and runtime. This paper introduces TrajectoryNAS, a novel neural architecture search (NAS) method designed to develop an efficient and more accurate LiDAR-based trajectory prediction model for predicting the trajectories of objects surrounding the ego vehicle. TrajectoryNAS systematically optimizes the architecture of an end-to-end trajectory prediction algorithm, incorporating all stacked components that are prerequisites for trajectory prediction, including object detection and object tracking, using metaheuristic algorithms. This approach addresses the neural architecture designs in each component of trajectory prediction, considering accuracy loss and the associated overhead latency. Our method introduces a novel multi-objective energy function that integrates accuracy and efficiency metrics, enabling the creation of a model that significantly outperforms existing approaches. Through empirical studies, TrajectoryNAS demonstrates its effectiveness in enhancing the performance of autonomous driving systems, marking a significant advancement in the field. Experimental results reveal that TrajcetoryNAS yields a minimum of 4.8 higger accuracy and 1.1* lower latency over competing methods on the NuScenes dataset.

Keywords: autonomous driving; neural architecture search; trajectory prediction; 3D point cloud

Citation: Sharifi, A.A.; Zoljodi, A.; Daneshtalab, M. TrajectoryNAS: A Neural Architecture Search for Trajectory Prediction. *Sensors* 2024, 24, 5696. https://doi.org/10.3390/s24175696

Academic Editors: Enrico Meli and Bogdan Smolka

Received: 15 July 2024
Revised: 27 August 2024
Accepted: 29 August 2024
Published: 1 September 2024

Copyright: © 2024 by the authors. Licensee MDPI, Basel, Switzerland. This article is an open access article distributed under the terms and conditions of the Creative Commons Attribution (CC BY) license (https://creativecommons.org/licenses/by/4.0/).

1. Introduction

Predicting future actions or states of objects around an intelligent system, such as an autonomous driving (AD) vehicle, is crucial in preventing disasters or crashes. Driving in the real world is a stochastic process due to the presence of other vehicles and pedestrians that can take their next step, resulting in accidents or congestion. AD systems require the crucial ability to predict the trajectory of surrounding objects [1–3]. Predicting accurately the trajectory of surrounding objects is important in simultaneous localization and mapping (SLAM) because it provides crucial information about static and dynamic objects and allows for the refinement of object locations based on these predicted trajectories [4]. To perform the task of predicting in self-driving vehicles, 2D and 3D data can be utilized. 3D data can usually be represented in different formats, including depth images, point clouds, meshes, and volumetric grids. The optical camera is usually good for classification tasks such as distinguishing the type of surrounding objects or detecting lane markers or traffic signs. While the performance of measuring distances and velocities is rather weak, this information can be retrieved well from radars. LIDARs are complementary to the other two sensors, showing competitive results. Distances and velocities can be estimated with very high accuracy. Therefore, it is the preferred representation for many scene-understanding-related applications such as autonomous driving and robotics.

Our paper presents TrajectoryNAS, an application-specific Neural Architecture Search (NAS) that aims to create a trajectory model with high accuracy and minimum displacement errors, both final and average (FDE and ADE (Section 4.2)). Our empirical studies reveal that accurate object detection is crucial to achieving precise trajectory predictions. Therefore, TrajectoryNAS is designed to localize objects with a minimum error and improve the accuracy of final trajectory predictions. Additionally, to minimize the time required for inference, the final objective of TrajectoryNAS is to reduce the model latency.

In conclusion, our contributions to this challenge can be summarized as follows:

- **Trajectory Prediction NAS:** TrajectoryNAS is a novel trajectory prediction for autonomous driving, being the first to implement neural architecture search (NAS) in an end-to-end manner. It integrates object detection, tracking, and predicting, addressing the complex interdependencies among these tasks and the challenges of point-cloud processing.
- **Hybrid Exploration and Exploitation:** We introduce a two-step process to efficiently handle the computational demands of NAS on large datasets. This approach first explores architectures using a mini dataset, which is 10× faster than the complete dataset, and then trains the selected architecture on the full dataset (exploitation), ensuring both scalability and accuracy.
- **Multi-Objective Architecture Search:** We introduce a multi-objective energy function to assess the proposed architecture in both an accuracy and latency manner.

2. Related Works

2.1. Trajectory Prediction

In this section, we provide a brief overview of the literature focused on predicting trajectories using point-cloud data. We begin by exploring cascade approaches (traditional approaches). In these approaches, the output of a detector serves as input to a tracker. The tracker's output is then used by a trajectory-predicting algorithm to estimate the anticipated movements of traffic participants in the upcoming seconds as in Figure 1 (top row). Following that, the state-of-the-art approaches that do detection, tracking, and predicting in an end-to-end manner are reviewed, depicted in Figure 1 (bottom row).

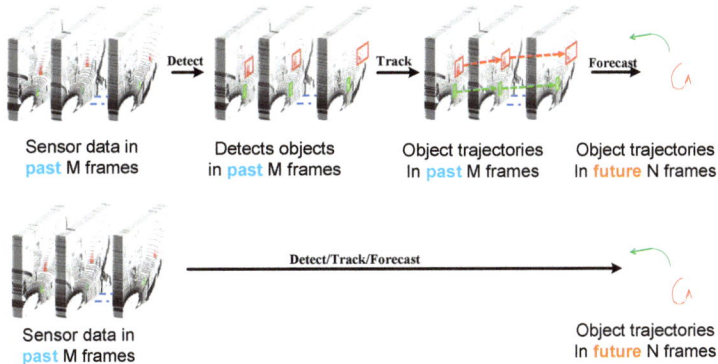

Figure 1. (Top Row) Cascade methods that independently address detection, tracking, and predicting, they inherently carry the risk of compounding errors throughout the pipeline. This originates from—

each sub-module's assumption of receiving perfect input, which rarely holds true in real-world applications. Consequently, errors introduced in earlier stages propagate and magnify downstream, potentially leading to inaccurate final outcomes. (Bottom Row) End-to-end methods that predict future movement directly from raw data, enabling end-to-end training and benefiting from the joint optimization of object detection, tracking, and prediction tasks.

2.1.1. Cascade Approaches

Traditional self-driving autonomy decomposes the problem into three subtasks (object detection, object tracking, and motion prediction) and relies on independent components that perform these subtasks sequentially. These modules are usually learned independently, and uncertainty is usually propagated [1]. In these methods, it is assumed that the exact paths taken by the agents are known. By examining the trajectory data over a short period of time, predictions can be made for future moments. For instance, the NuScenes [5] and Argoverse [6] datasets provide trajectories and their corresponding labels for this purpose.

Many of the approaches presented in the literature are based on neural networks that use recurrent neural networks (RNNs), which explicitly take into account a history composed of the past states of the agents [7]. In RNNs and their variants, memory is a single hidden state vector that encodes all the temporal information. Thus, memory is addressable as a whole, and it lacks the ability to address individual elements of knowledge [3]. Ref. [3] presents the memory-augmented neural trajectory predictor (MANTRA). In this model, an external, associative memory is trained to store useful and non-redundant trajectories. Instead of a single hidden representation addressable as a whole, the memory is element-wise addressable, permitting selective access to only relevant pieces of information at runtime.

Spatial and temporal learning will be two key components in prediction learning. Ignoring either information will lead to information loss and reduce the model's capability of context learning. Consequently, researchers are focusing on jointly learning RNN spatial and temporal information. Ref. [8] utilize rasterization to encode both the agents and high-definition map details, transforming corresponding elements such as lanes and crosswalks into lines and polygons of diverse colors. However, the rasterized image is an overly complex representation of environment and agent history and requires significantly more computation and data to train and deploy. In an effort to address this, VectorNet [9] proposes a vector representation to exploit the spatial locality of individual road components with graph neural networks. LaneConv [10] constructs a lane graph from vectorized map data and proposes LaneGCN to capture the topology and long dependency of the agents and map information. Both VectorNet [9] and LaneConv [10] can be viewed as extensions of graph neural networks in prediction with a strong capability to extract spatial locality. Nevertheless, both works fail to fully utilize the temporal information of agents with less focus on temporal feature extraction. In order to combine spatial and temporal learning in a flexible and unified framework, Ref. [11] proposes temporal point-cloud networks (TPCN). TPCN models the prediction learning task as joint learning between a spatial module and a temporal module.

Across a range of visual benchmarks, transformer-based models exhibit comparable or superior performance when compared to other network types like convolutional and recurrent neural networks [12]. This trend extends to trajectory prediction as well. Ref. [13] proposes a new transformer that simultaneously models the time and social dimensions. Their method allows an agent's state at one time to directly affect another agent's state in the future. In parallel, Ref. [14] develops an RNN-based approach for context-aware multi-modal behavior forecasting. The model input includes both a road network attention module and a dynamic interaction graph to capture interpretable geometric and social relationships.

As mentioned, cascade approaches in order to trajectory prediction are developed separately from their upstream perception. As a result, their performance degrades significantly when using real-world noisy tracking results as inputs. Ref. [15] presents a novel prediction framework that uses affinity matrices rather than tracklets as inputs, thereby completely removing the chances of errors occurring in data association and passing more information to prediction. To consider this propagation of errors, Ref. [15] applies three types of data augmentation to increase the robustness of prediction with respect to tracking errors. They inject identity switches (IDS), fragments (FRAG), and noise.

2.1.2. End-to-End Approaches

To prevent the propagation of errors and reduce inference time in traditional methods, as they learn independently, researchers [16–19] attempted to perform detection and tracking in an end-to-end manner. With the same purpose, Ref. [20] proposed a network that parallelized tracking and prediction using a graph neural network (GNN).

To our best knowledge, Fast and Furious (FaF) [21] proposes the first deep neural network capable of jointly performing 3D detection, tracking, and motion prediction using data captured by a 3D sensor. However, Ref. [21] limited its predictions to a mere 1 s duration. In contrast, IntentNet [22] enlarges the prediction horizon and estimates future high-level driver behavior. Ref. [23] moved a step further and performed detection, predicting, and motion planning jointly. Furthermore, Ref. [23] introduces an additional perception loss that encourages the intermediate representations to generate accurate 3D detections and motion prediction. This ensures the interoperability of these intermediate representations and enables significantly accelerated learning. The statistical interconnections among actors are overlooked by all the previously mentioned methods, and instead, they individually predict each trajectory using the provided features. Ref. [2] designed a novel network that explicitly takes into account the interactions among actors. To capture their spatial-temporal dependencies, Ref. [2] proposes a recurrent neural network with a transformer architecture.

Ref. [24] suggests a reversing of the detect-then-forecast pipeline rather than following the conventional sequence of detecting, tracking, and subsequently forecasting objects. Afterward, object detection and tracking are performed on the projected point-cloud sequences to obtain future poses. A notable advantage of this methodology lies in the comprehensive representation of predictions, incorporating details about RNNs and the background and foreground objects existing within the scene. Similarly, in a comparable fashion, FutureDet [25] directly predicts the future locations of objects observed at a specific time instead of predicting point-cloud sequences over time and then backcasting them to determine their origin in the current frame. This allows the model to reason about multiple possible futures by linking future and current locations in a many-to-one manner. This approach leverages existing LiDAR detectors to predict object positions in unseen future scans. Building upon the recently proposed CenterPoint LiDAR detector [17], FutureDet predicts not only future locations but also velocity vectors for each object in every frame between the current and final predicted future frame. This enables the model to estimate consistent object trajectories throughout the entire forecasting horizon. In the process of forecasting, it is essential to link all trajectories to the collection of object detections in the current (observed) LiDAR scan. For each future detection i, FutureDet computes the distance to every detection j from the previous timestep. Subsequently, for each i, FutureDet selects the most suitable j (permitting multiple-to-one matching).

Additionally, it is argued that current evaluation metrics for predicting directly from raw LiDAR data are inadequate as they can be manipulated by simplistic predictors, leading to inflated performance. These metrics, originally designed for trajectory-based prediction, do not effectively address the interconnected tasks of detection and forecasting. To overcome these limitations, a novel evaluation procedure is proposed by FutureDet. The new metric integrates both detection and forecasting tasks. Notably, this approach surpasses state-of-the-art methods without the necessity of object tracks or high definition (HD) maps as model inputs.

2.2. Neural Architecture Search

Optimizing model hyperparameters is an effective way to improve intelligent systems using automated machine learning (AutoML) [26]. Neural architecture search (NAS) is a subset of AutoML that aims to create efficient neural networks for complex learning tasks [27]. Early NAS methods used reinforcement learning (RL) [28,29] or evolutionary algorithms [30,31]. However, evaluating 20,000 neural architectures over four days requires remarkable computing capacity, such as 500 NVIDIA® GPUs used in this study were sourced from NVIDIA Corporation, which is headquartered in Santa Clara, CA, USA [28]. Recently, methods for differentiable neural architecture search (NAS) have been proven to achieve state-of-the-art results across various learning tasks [32–34]. DARTS [33] is a differentiable NAS method that uses the gradient descent algorithm to search and train neural architecture cells jointly. Despite the success of differentiable NAS methods in various domains [34], they suffer from inefficient training due to interfering with the training of different sub-networks each other [35]. Moreover, it has been proven that with equal search spaces and training setups, differentiable NAS methods converge to similar results [36].

Meta-heuristic-based NAS methods [37–39] benefit from fast and flexible algorithms to search a discrete search space. FastStereoNet [39] is a state-of-the-art meta-heuristic method that designs an accurate depth estimation pipeline. TrajectoryNAS is a fast multi-objective meta-heuristic NAS designed to optimize trajectory prediction approaches by searching a wider design space compared to differentiable methods or evolutionary NAS approaches.

3. TrajectoryNAS

Current trajectory prediction techniques rely on handcrafted neural network architectures. These models, while effective for tasks like 3D object detection, are suboptimal for trajectory prediction. Building on the success of neural architecture search (NAS), TrajectoryNAS offers an interactive approach to designing neural networks specifically for 3D trajectory prediction. However, it is important to note that training a trajectory prediction model is both costly and time-consuming, requiring approximately 12 GPU hours for a single model.

As a result, the NAS procedure becomes significantly slow, requiring approximately 1200 GPU hours. To expedite the training process, we leverage state-of-the-art techniques (e.g., [40–42]) that utilize a miniaturized NuScenes dataset to reduce the computational demand for communication rescores. As an example, Blanch et al. [42] demonstrates the use of a mini-dataset for hyperparameter optimization. Similarly, each model generated by neural architecture search (NAS) is trained on a standard mini-subset of the NuScenes dataset [5]. This technique reduces the evaluation time for each model to nearly 1 h, making the process approximately 12 times faster.

Figure 2 elaborates the TrajectoryNAS state diagram. The TrajectoryNAS workflow consists of three phases: Phase 1, exploration, where the metaheuristic algorithm suggests new architectures, and each architecture is trained using a mini dataset to compare with other suggested architectures. In Phase 2, the architecture with the highest accuracy on the mini dataset is retrained using the full-size dataset. Finally, in Phase 3, the fully trained model is deployed on hardware and tested with the test dataset to report the final results.

TrajectoryNAS is a one-stage trajectory prediction [21,25]. The model takes a sequence of Lidar data captured from the scene, which integrates a robust 3D backbone with cutting-edge neural architecture search (NAS) to refine map-view feature extraction from LiDAR point clouds. This innovative architecture further evolves by automating the design of multi-2D CNN detection heads, specifically tailored for future object detection and trajectory prediction. By detecting objects across multiple future timesteps and accurately projecting their movements back to the current moment, TrajectoryNAS stands out for its precision in trajectory prediction. This system not only anticipates the dynamic positioning of objects but also adjusts its computational strategies in real-time, ensuring a high degree of accuracy and efficiency in processing. The inclusion of NAS allows for continuous improvement of the detection and prediction heads, making TrajectoryNAS a highly adaptive and forward-thinking solution in the realm of autonomous navigation and surveillance technologies. TrajectoryNAS employs a hybrid optimization strategy to minimize optimization costs. The process is divided into two phases. The first phase, called exploration, involves the algorithm exploring various neural architecture designs to identify the optimal design. This phase is time-consuming as the algorithm must evaluate a wide range of parameters within the search space elaborated in Section 3.1. During the exploration phase, it is crucial to establish a comparative accuracy metric that can evaluate different architectures and determine the relative optimal design. To reduce processing time, the exploration phase utilizes a subset of the Nuscenes dataset [5], which contains significantly less data but maintains a distribution similar to the full dataset. To ensure that the selected model performs efficiently on the complete dataset, we use the full dataset in the second phase, known as exploitation, where we report the final accuracy of the designed model.

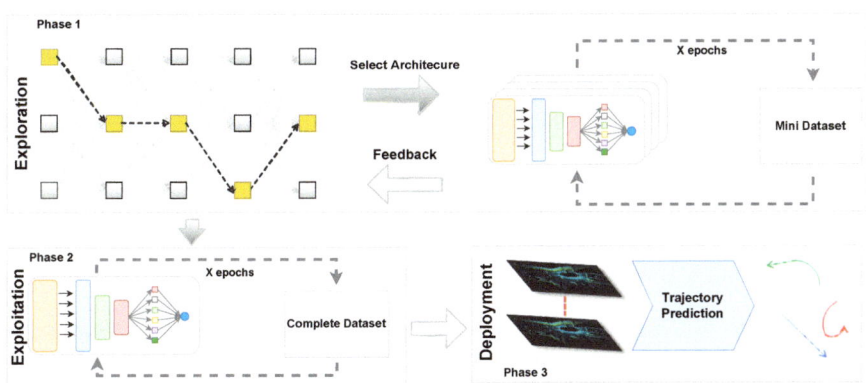

Figure 2. TrajectoryNAS state diagram. A model generated from the search space. The generated model trains using the mini dataset. The results are sent back to search space to generate a new model. The best final model is fully trained using the original dataset.

3.1. Search Space

TrajectoryNAS search space is demonstrated in Figure 3.

The TrajctryNAS architecture stands out as a solution for object detection and trajectory prediction, particularly in scenarios like autonomous driving, where understanding dynamic environments is paramount. It skillfully merges spatial and temporal object analyses, predicting not only the present state but also future trajectories.

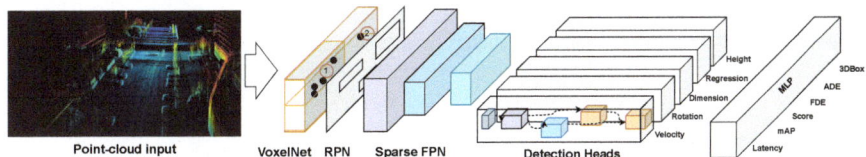

Figure 3. The overview of TrajcetoryNAS process.

3.1.1. 3D Object Detection with VoxelNet

Modern 3D object detection methods [17,43,44] utilize a 3D encoder that converts the point cloud into regular bins. A point-based network [45] then extracts features from all the points within each bin. The 3D encoder subsequently pools these features to form its primary feature representation. Most of the computational workload is handled by the backbone network, which operates exclusively on these quantized and pooled feature representations. The output of the backbone network is a map-view feature map

$M \in \mathbb{R}^{W \times L \times F}$ with width W, length L, and F channels in a map-view reference frame. The width and height are directly related to the resolution of the individual voxel bins and the stride of the backbone network. Common backbone architectures include VoxelNet [46,47] and PointPillars [43]. This work employs VoxelNet as the backbone network.

VoxelNet is a novel approach for 3D object detection from LiDAR data and comprises three functional blocks:

Feature Learning Network: This network processes raw LiDAR data by dividing the point cloud into 3D voxels. A crucial component is the voxel feature encoding (VFE) layer, which transforms each group of points within a voxel into a unified feature representation. By stacking multiple VFE layers, the network learns complex features that capture local 3D shape information within the point cloud.

Convolutional Middle Layers: After the feature learning network generates a volumetric representation with encoded features, these features are further processed by 3D convolutional layers. These layers aggregate local voxel features, transforming the point-cloud data into a richer and more informative high-dimensional representation.

Region Proposal Network (RPN): The final stage utilizes an RPN [48] to generate 3D object detections. The input to the RPN is the feature map provided by the convolutional middle layers. The network consists of three blocks of fully convolutional layers, with batch normalization (BN) and ReLU operations applied after each layer. The output of each block is up-sampled to a fixed size and concatenated to construct a high-resolution feature map.

3.1.2. Trajectory Prediction

TrajectoryNAS detects objects in both the current and future frames, projecting future detections back to the reference frame. We hypothesize that detecting objects in future frames requires the network to learn forecasted feature representations, as suggested by Peri et al. [25]. The network uses features extracted from the feature extraction module (VoxelNet) to predict features for the next timestep $(t + 1)$. After each prediction, a detection module refines the results. Initially, the extracted features are used for object detection in the current frame. Simultaneously, a copy of these features is passed to the prediction network to forecast features for the next timestep. This process is repeated iteratively, with predicted features being used for subsequent detection modules, until both the features and object detections are obtained for the final timestep.

Each detection module contains five parallel prediction heads, each responsible for a specific aspect of the object's state: velocity, rotation, dimension, regression (bounding box refinement), and height. These heads work in concert to provide a comprehensive description of an object's current position and orientation at time t.

To link objects across different frames, our network detects objects in both the current and future frames and predicts offsets to associate them back in time, assuming constant velocity between frames. Trajectory construction involves aligning all trajectories with the objects detected in the current LiDAR scan. Each detected object in the future frame (i) is matched to the previous frame (j) using the constant velocity equation. The distance between the detected object at time j and all other detected objects is calculated, and the closest object is then selected.

Such an approach allows TrajectryNAS to not only navigate but also anticipate complex dynamic behaviors, making it an invaluable asset in fields where predicting future states is crucial for proactive decision-making. This architecture's ability to foresee the direction and movement of objects enriches scene understanding and enhances planning for autonomous systems, offering a comprehensive and forward-looking perspective on environmental dynamics.

The TrajectoryNAS system automatically designs the region proposal network (RPN) and the prediction heads using the aforementioned layers. It explores an expansive space of 2^{300} potential architectures to identify an optimal balance between speed and accuracy.

This approach enables the selection of a highly efficient and accurate architecture tailored for specific applications.

3.2. Search Algorithm

To improve the accuracy of trajectory prediction while reducing network inference time, we employ the multi-objective simulated annealing (MOSA) algorithm, as described in [49]. The search algorithm optimizes the trajectory prediction in the design time and before training the model. The reason for using MOSA is its simplicity and its superior ability to explore a wide range of candidates compared to gradient-based algorithms. MOSA is also capable of finding global optima due to its effective exploration-exploitation balance. These attributes make MOSA a robust choice for optimizing complex, multi-objective problems such as trajectory prediction. MOSA selects candidates based on the probability of $min(1, exp(-\Delta/T))$, where Δ is the energy difference between present and newly generated candidates, and T is the regulating parameter for annealing temperature. Initially, T starts from a large value (T_{Max}) and gradually decreases to a small value (T_{Min}). Setting T_{Max} to a large value allows for exploration of non-optimal choices, while T_{Min} being small gives the maximum selection chance to optimal candidates (exploitation).

To achieve this optimization, we use a multi-objective energy function (Equation (1)).

The energy function (E) is the product of the network *latency* (t) and the weighted mean average precision of the predicted future place of the object and its actual place (*mAP*), weighted average displacement (*ADE*) error, and weighted final displacement error (*FDE*).

$$E = Latency \times mAP^\alpha \times ADE^\beta \times FDE^\gamma \tag{1}$$

where α, β, and γ are weights of *mAP*, *ADE*, and *FDE*, respectively. We do not use any proxy, such as Floating-Point-Operations-per-Second (FLOPs), for inference time estimation. Instead, we run the network directly on the target hardware (NVIDIA® RTX A4000 were sourced from NVIDIA Corporation, which is headquartered in Santa Clara, CA, USA) to measure the exact inference time. Algorithm 1 is a complete description of the TrajectoryNAS flow.

Algorithm 1 TrajecoryNAS

1: **procedure** EXPLORATION
2:    $M \leftarrow$ Mini-Dataset
3:    $A_{init} \leftarrow$ InitialArchitecture
4:    $T_{Max} \leftarrow$ MaximumTemperature
5:    $T_{Min} \leftarrow$ MinimumTemperature
6:    $T_{Factor} \leftarrow -Log(T_{Max}, T_{Min})$
7:    $A_{best} \leftarrow A_{Init}$
8:    $A_{current} \leftarrow A_{Init}$
9:    train($A_{current}, M$)
10:   **for** each iteration i from 1 to MaxIterations **do**
11:      $A_{new} \leftarrow$ GenerateNeighbor($A_{current}$)
12:      train(A_{new}, M)
13:      $\Delta E \leftarrow E(A_{new}) - E(A_{current})$
14:      **if** $\Delta E < 0$ **then**
15:         $A_{current} \leftarrow A_{new}$
16:      **else**
17:         $r \leftarrow$ Random number in $[0, 1]$
18:         **if** $r < min(1, exp(-\Delta/T))$ **then**
19:            $A_{current} \leftarrow A_{new}$
20:         **end if**
21:      **end if**

Algorithm 1 *Cont.*

```
22:        if E(A_current) < E(A_best) then
23:            A_best ← A_current
24:        end if
25:        T ← T_Max × Exp(T_Factor × (i/MaxIterations))
26:    end for
27:    return A_best
28: end procedure
29: procedure EXPLOITATION(A_best)
30:    C ← Complete-Dataset
31:    train(A_best, C)
32:    accuracy ← vaidate(A_best, C)
33:    return accuracy
34: end procedure
```

4. Experimental Setup

We demonstrate the effectiveness of our approach on a large-scale real-world driving dataset. We focus on modular metrics for detection and prediction, as well as system metrics for end-to-end perception and prediction.

4.1. Dataset

Our experimental analysis was performed on the nuScenes [5] dataset, which contains 1000 log snippets, each lasting 20 s. We utilized two officially released divisions of the dataset: the Mini and Trainval splits. The Mini split, which consists of 10 scenes, is a subset of the Trainval split. The Trainval split contains 700 scenes for training purposes and 150 scenes for validation. Additionally, the test split, containing 150 scenes, is designated for challenges and lacks object annotations.

4.2. Evaluation Metrics

We follow the detection and prediction metrics defined by [25] to have a fair comparison with other state-of-the-art. Specifically, we use average precision (AP_{det}) for detection and future average precision (AP_f) for trajectory prediction.

Detection Average Precision (AP_{det}): AP_{det} is defined as the area under the precision-recall curve [50], commonly averaged over multiple spatial overlap thresholds [51]. To compute AP, we first determine the set of true positives (TP) and false positives (FP) to evaluate precision and recall.

Future Average Precision (AP_f): future Average Precision (AP_f) is a metric used to evaluate the accuracy of future trajectory predictions anchored to detected objects in the current frame (t_{obs}). It penalizes incorrect future predictions (false predictions) and missed detection (missed predictions). A true positive (TP) requires a positive match both at the current timestamp (t_{obs}) and the final timestep (t_{obs} + T), Otherwise, a prediction is considered to be a false positive (FP). A successful match in the current frame is determined based on distance thresholds of 0.5, 1, 2, 4 m for the current frame and 1, 2, 4, 8 m for the final timestep [25]. AP_f considers all detections and penalizes missed predictions, typically measured by the miss rate.

We have defined three subclasses: static cars, linearly moving cars, and non-linearly moving cars [25], and we report AP_f and AP_{det} for these three classes. Subsequently, we evaluate the mean average precision for the future (mAP_f) as follows: $mAP_f = 1/3 \times (AP_f^{lin.} + AP_f^{non-lin.} + AP_f^{stat.})$. Similarly, mAP_{det} is evaluated as the average AP_{det} over the three subclasses. Subclass labels are determined based on the trajectory (whether ground truth or predicted). First, we calculate the intersection over union (IoU) between the bounding boxes at the first and last timestep. If the IoU is greater than 0, the trajectory is labeled as static. Next, we use the velocity from the first timestep to project a target box.

If the IoU between this target box and the last timestep box is greater than 0, the trajectory is labeled as linear. Trajectories that do not fit either category are labeled as non-linear.

mAP_f and mAP_{det} provide a more realistic evaluation by jointly assessing detection and prediction accuracy. They penalize both missed predictions and false predictions, ensuring that only predictions correctly matched to detected objects are considered true positives [25]. This joint evaluation embraces the inherent multi-future nature of prediction and is robust against imbalanced data scenarios, such as the high proportion of stationary cars in the nuScenes dataset.

4.3. Configuration Setup

For this study, Table 1 provides a brief overview of the configuration setup.

Table 1. Summarizing hardware specification, train, and search parameters.

Train/Test Hardware Device	Specification
GPU	NVIDIA® RTX A4000
GPU Compiler	CUDA v11.7 & cuDNN v8.2.0
DL Framework	PyTorch v1.9.1
Training and Search Parameters	**Value**
Full-Training Epochs	20
Batch Size	1
Learning Rate	5×10^{-4}
Optimizer	Adam
T_{Max}/T_{Min}	2500/2.5

5. Results

5.1. Trajectory Prediction Performance

As presented in Tables 2 and 3, the comparison of car and pedestrian trajectory prediction results demonstrates that TrajectoryNAS outperforms other state-of-the-art trajectory prediction methods in numerous parameters for car trajectory prediction and the majority of parameters for pedestrian trajectory prediction. Notably, the latency of TrajectoryNAS is comparable to that of Fast and Furious [21] and better than FutureDet [25], while TrajectoryNAS provides superior future average precision (AP_f) across all conditions for both linear and non-linear trajectories of cars and pedestrians. Future average precision (AP_f) is a novel trajectory prediction performance metric proposed by FutureDet [25] and proved to be more precise in demonstrating trajectory performance in comparison to previous metrics.

For cars, while Fast and Furious and FutureDet offer a marginal improvement in specific aspects when compared with TrajectoryNAS, TrajectoryNAS significantly surpasses the state-of-the-art in most parameters. This is evidenced by its top performance in average precision for static, linear, and non-linear trajectories, as well as its mean average precision (mAP), both for single ($K = 1$) and multiple ($K = 5$) predictions. Specifically, TrajectoryNAS achieves the highest detection accuracy and future average precision in almost all scenarios, highlighting its robustness and efficacy in car trajectory prediction.

Similarly, for pedestrian trajectory prediction, TrajectoryNAS demonstrates outstanding performance, particularly in accurately predicting linear and non-linear movements. It not only achieves the highest average precision scores across various scenarios but also maintains competitive latency, underscoring its effectiveness in real-time applications.

In conclusion, TrajectoryNAS advances the field of trajectory prediction by offering a highly accurate and efficient model. Its ability to provide better future average precision under different conditions for both cars and pedestrians, coupled with its comparable latency to leading models, positions TrajectoryNAS as a superior choice for trajectory prediction in dynamic environments.

Table 2. Comparison TrajcetoryNAS and state-of-the-art trajectory prediction model on cars according to accuracy and latency metrics.

Method	Time (ms)	K = 1								K = 5							
		$AP^{stat.}$		$AP^{lin.}$		$AP^{non-lin.}$		mAP		$AP^{stat.}$		$AP^{lin.}$		$AP^{non-lin.}$		mAP	
		$AP_{det.}$	AP_f	$AP_{det.}$	AP_f	$AP_{det.}$	AP_f	$AP_{det.}$	AP_f	$AP_{det.}$	AP_f	$AP_{det.}$	AP_f	$AP_{det.}$	AP_f	$AP_{det.}$	AP_f
Detection + Constant Velocity	21	70.3	66.0	65.8	21.2	90.0	6.5	75.4	31.12	70.3	66.0	65.8	21.2	90.0	6.5	75.4	31.2
Detection + Forecast [21]	20	69.1	64.7	66.1	22.2	86.3	7.5	73.8	31.5	69.1	64.7	66.1	22.2	86.3	7.5	73.8	31.5
FutureDet [25]	24	70.0	65.5	62.9	24.9	91.8	10.1	74.9	33.5	70.1	67.3	62.9	27.7	91.7	11.7	74.9	35.6
TrajectoryNAS (ours)	22	71.0	65.6	63.8	26	91.2	10.3	75	34	71	67.4	63.8	29.2	91.1	12.1	75.3	36.2

Table 3. Comparison TrajcetoryNAS and state-of-the-art trajectory prediction model on pedestrian according to accuracy and latency metrics.

Method	Time (ms)	K = 1								K = 5							
		$AP^{stat.}$		$AP^{lin.}$		$AP^{non-lin.}$		mAP		$AP^{stat.}$		$AP^{lin.}$		$AP^{non-lin.}$		mAP	
		$AP_{det.}$	AP_f	$AP_{det.}$	AP_f	$AP_{det.}$	AP_f	$AP_{det.}$	AP_f	$AP_{det.}$	AP_f	$AP_{det.}$	AP_f	$AP_{det.}$	AP_f	$AP_{det.}$	AP_f
Detection + Constant Velocity	21	55.1	33.3	73.5	27.8	96.9	12.4	75.2	25.5	55.1	33.3	73.5	27.8	96.9	12.4	75.2	24.5
Detection + Forecast [21]	20	53.7	35.0	73.9	30.8	97.2	13.3	74.9	26.4	53.7	35.0	73.9	30.8	97.2	13.3	74.9	26.4
FutureDet [25]	24	53.1	33.3	72.4	32.6	95.2	14.7	73.6	26.9	53.1	35.1	72.4	34.0	95.2	15.0	73.6	28.0
TrajectoryNAS (ours)	22	55.8	37.1	77.9	39.9	95.2	17.7	76.3	31.3	55.8	38.6	77.9	40.9	95.2	17.9	76.3	32.5

5.2. Analysing Search Methods

Figure 4 presents a detailed comparison of the energy function reduction (as defined in Equation (1)) during the search process employed by the TrajectoryNAS algorithm against those of random search and local search methods. This comparative analysis clearly demonstrates the limitations of both local search and random search techniques in effectively identifying the most optimal solution. Specifically, the best outcome identified through Random Search, characterized by an energy value of e = 0.19 as per Equation (1), was achieved in iteration 52. Similarly, Local Search's most effective solution registered an energy value of e = 0.186, and this result was obtained in iteration 50.

Despite these efforts, both methods fall significantly short when compared to the capabilities of the TrajectoryNAS algorithm. TrajectoryNAS not only surpasses these traditional search methodologies in efficiently navigating towards more optimal solutions but also showcases its superiority by discovering an exceptionally lower energy value of 0.113. This landmark achievement was realized in iteration 108, underlining the algorithm's advanced optimization prowess. Notably, the energy value associated with the best solution found by TrajectoryNAS is nearly half that of the best solutions unearthed by both random search and local search. This stark contrast underscores the advanced and sophisticated nature of TrajectoryNAS in exploring and exploiting the search space to find significantly more efficient solutions, thereby establishing a new benchmark in the quest for optimization within this context.

TrajectoryNAS overcomes Latent Acceptance Hill-Climbing (LAHC), a high-performance meta-heuristic algorithm as described by [52]. Figure 4 illustrates that LAHC becomes trapped in a local minimum and fails to escape to locate the global minimum. Consequently, the performance of LAHC is inferior to both local search and random search algorithms.

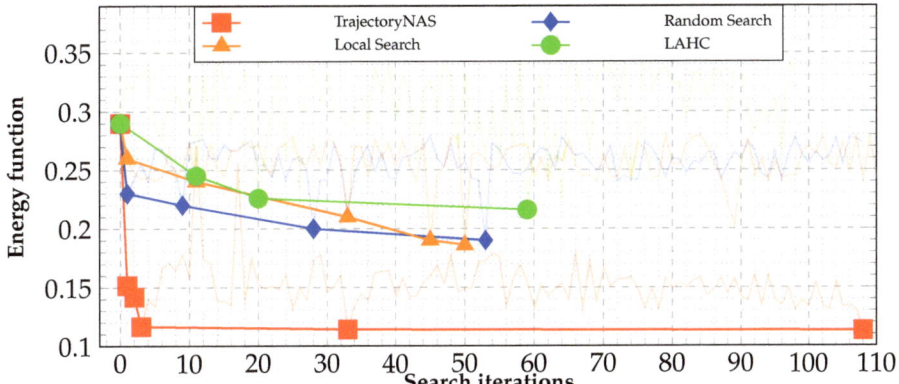

Figure 4. TrajectoryNAS optimization curve.

5.3. Visual Demonstration

The visual results of TrajectoryNAS are shown in Figure 5. As is evident, the results for both linear and non-linear activities for both cars and pedestrians closely match what occurs in the future. TrajectoryNAS is highly accurate in determining static and dynamic objects, and it rarely draws dynamic lines for static objects.

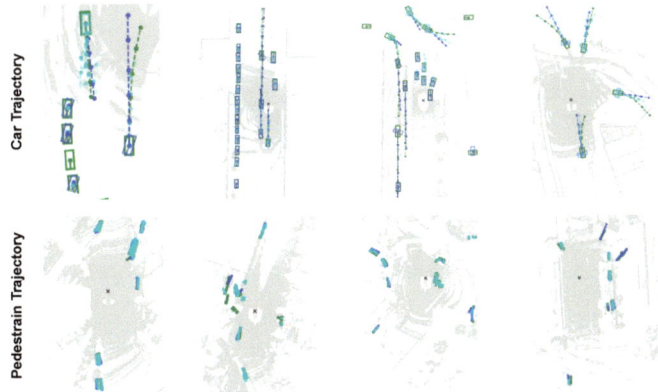

Figure 5. The visual demonstration of TrajectoryNAS; the first row is the trajectory prediction for cars, and the second row is the trajectory prediction for the pedestrian. Green lines are ground-truth. Blue lines are trajectory prediction with highest probability. Cyan lines are trajectory predictions with the highest probability.

6. Conclusions

Trajectory prediction is one of the most important components of autonomous driving systems. A well-designed trajectory prediction model can accurately predict the trajectories of surrounding objects near the ego vehicle within an acceptable inference time, helping to prevent collisions by ensuring the ego vehicle avoids crossing their paths. State-of-the-art trajectory prediction models suffer from their handcrafted design, which leads to suboptimal accuracy and latency.

To resolve this problem, we propose TrajectoryNAS, a neural architecture search approach tailored for trajectory prediction applications, which designs accurate and low-latency trajectory prediction models using metaheuristic algorithms. Our empirical studies demonstrate that TrajectoryNAS achieves a minimum of 4.8% higher accuracy in predicting the trajectories of objects with non-linear paths. This highlights its effectiveness in predicting the trajectories of objects with more freedom of movement than vehicles, such as pedestrians. Our future work involves enhancing TrajectoryNAS to support novel deep learning approaches, such as vision transformers (ViTs), which have been well-demonstrated in meeting autonomous driving requirements, such as long-range perception.

Author Contributions: Conceptualization, A.A.S.; Methodology, A.Z.; Validation, A.Z.; Data curation, A.A.S.; Writing—original draft, A.A.S. and A.Z.; Writing—review & editing, M.D.; Supervision, M.D. All authors have read and agreed to the published version of the manuscript.

Funding: This research was funded by the European Union and Estonian Research Council via project TEM-TA138, and the Swedish Innovation Agency VINNOVA project AutoDeep. The computations were enabled by resources provided by the National Academic Infrastructure for Supercomputing in Sweden (NAISS), partially funded by the Swedish Research Council through grant agreement no. 2022-06725.

Data Availability Statement: Data available on request due to restrictions (e.g., privacy, legal or ethical reasons)

Conflicts of Interest: The authors declare no conflicts of interest.

References

1. Liang, M.; Yang, B.; Zeng, W.; Chen, Y.; Hu, R.; Casas, S.; Urtasun, R. Pnpnet: End-to-end perception and prediction with tracking in the loop. In Proceedings of the IEEE/CVF Conference on Computer Vision and Pattern Recognition, Seattle, WA, USA, 13–19 June 2020; pp. 11553–11562.
2. Li, L.L.; Yang, B.; Liang, M.; Zeng, W.; Ren, M.; Segal, S.; Urtasun, R. End-to-end contextual perception and prediction with interaction transformer. In Proceedings of the 2020 IEEE/RSJ International Conference on Intelligent Robots and Systems (IROS), Las Vegas, NV, USA, 25–29 October 2020; pp. 5784–5791.
3. Marchetti, F.; Becattini, F.; Seidenari, L.; Del Bimbo, A. Multiple trajectory prediction of moving agents with memory augmented networks. *IEEE Trans. Pattern Anal. Mach. Intell.* **2020**, *45*, 6688–6702. [CrossRef]
4. Charroud, A.; El Moutaouakil, K.; Palade, V.; Yahyaouy, A.; Onyekpe, U.; Eyo, E.U. Localization and Mapping for Self-Driving Vehicles: A Survey. *Machines* **2024**, *12*, 118. [CrossRef]
5. Caesar, H.; Bankiti, V.; Lang, A.H.; Vora, S.; Liong, V.E.; Xu, Q.; Krishnan, A.; Pan, Y.; Baldan, G.; Beijbom, O. nuScenes: A multimodal dataset for autonomous driving. *arXiv* **2019**, arXiv:1903.11027.
6. Chang, M.F.; Lambert, J.W.; Sangkloy, P.; Singh, J.; Bak, S.; Hartnett, A.; Wang, D.; Carr, P.; Lucey, S.; Ramanan, D.; et al. Argoverse: 3D Tracking and Forecasting with Rich Maps. In Proceedings of the Conference on Computer Vision and Pattern Recognition (CVPR), Long Beach, CA, USA, 15–20 June 2019.
7. Leon, F.; Gavrilescu, M. A review of tracking and trajectory prediction methods for autonomous driving. *Mathematics* **2021**, *9*, 660. [CrossRef]
8. Phan-Minh, T.; Grigore, E.C.; Boulton, F.A.; Beijbom, O.; Wolff, E.M. Covernet: Multimodal behavior prediction using trajectory sets. In Proceedings of the IEEE/CVF Conference on Computer Vision and Pattern Recognition, Seattle, WA, USA, 13–19 June 2020; pp. 14074–14083.
9. Gao, J.; Sun, C.; Zhao, H.; Shen, Y.; Anguelov, D.; Li, C.; Schmid, C. Vectornet: Encoding hd maps and agent dynamics from vectorized representation. In Proceedings of the IEEE/CVF Conference on Computer Vision and Pattern Recognition, Seattle, WA, USA, 13–19 June 2020; pp. 11525–11533.
10. Liang, M.; Yang, B.; Hu, R.; Chen, Y.; Liao, R.; Feng, S.; Urtasun, R. Learning lane graph representations for motion forecasting. In *Proceedings of the Computer Vision–ECCV 2020: 16th European Conference, Glasgow, UK, 23–28 August 2020*; Proceedings, Part II 16; Springer: Berlin, Germany, 2020; pp. 541–556.
11. Ye, M.; Cao, T.; Chen, Q. Tpcn: Temporal point cloud networks for motion forecasting. In Proceedings of the IEEE/CVF Conference on Computer Vision and Pattern Recognition, Nashville, TN, USA, 20–25 June 2021; pp. 11318–11327.
12. Han, K.; Wang, Y.; Chen, H.; Chen, X.; Guo, J.; Liu, Z.; Tang, Y.; Xiao, A.; Xu, C.; Xu, Y.; et al. A survey on vision transformer. *IEEE Trans. Pattern Anal. Mach. Intell.* **2022**, *45*, 87–110. [CrossRef] [PubMed]
13. Yuan, Y.; Weng, X.; Ou, Y.; Kitani, K.M. Agentformer: Agent-aware transformers for socio-temporal multi-agent forecasting. In Proceedings of the IEEE/CVF International Conference on Computer Vision, Montreal, BC, Canada, 11–17 October 2021; pp. 9813–9823.
14. Khandelwal, S.; Qi, W.; Singh, J.; Hartnett, A.; Ramanan, D. What-if motion prediction for autonomous driving. *arXiv* **2020**, arXiv:2008.10587.
15. Weng, X.; Ivanovic, B.; Kitani, K.; Pavone, M. Whose track is it anyway? Improving robustness to tracking errors with affinity-based trajectory prediction. In Proceedings of the IEEE/CVF Conference on Computer Vision and Pattern Recognition, New Orleans, LA, USA, 18–24 June 2022; pp. 6573–6582.
16. Wang, S.; Sun, Y.; Liu, C.; Liu, M. Pointtracknet: An end-to-end network for 3-d object detection and tracking from point clouds. *IEEE Robot. Autom. Lett.* **2020**, *5*, 3206–3212. [CrossRef]

17. Yin, T.; Zhou, X.; Krahenbuhl, P. Center-based 3d object detection and tracking. In Proceedings of the IEEE/CVF Conference on Computer Vision and Pattern Recognition, Nashville, TN, USA, 20–25 June 2021; pp. 11784–11793.
18. Li, X.; Guivant, J.E. Efficient and Accurate Object Detection With Simultaneous Classification and Tracking Under Limited Computing Power. *IEEE Trans. Intell. Transp. Syst.* **2023**, *24*, 5740–5751. [CrossRef]
19. Simon, M.; Amende, K.; Kraus, A.; Honer, J.; Samann, T.; Kaulbersch, H.; Milz, S.; Michael Gross, H. Complexer-yolo: Real-time 3d object detection and tracking on semantic point clouds. In Proceedings of the IEEE/CVF Conference on Computer Vision and Pattern Recognition Workshops, Long Beach, CA, USA, 16–17 June 2019.
20. Weng, X.; Yuan, Y.; Kitani, K. PTP: Parallelized tracking and prediction with graph neural networks and diversity sampling. *IEEE Robot. Autom. Lett.* **2021**, *6*, 4640–4647. [CrossRef]
21. Luo, W.; Yang, B.; Urtasun, R. Fast and furious: Real time end-to-end 3d detection, tracking and motion forecasting with a single convolutional net. In Proceedings of the IEEE Conference on Computer Vision and Pattern Recognition, Salt Lake City, UT, USA, 18–23 June 2018; pp. 3569–3577.
22. Casas, S.; Luo, W.; Urtasun, R. Intentnet: Learning to predict intention from raw sensor data. In Proceedings of the Conference on Robot Learning PMLR, Zurich, Switzerland, 29–31 October 2018; pp. 947–956.
23. Zeng, W.; Luo, W.; Suo, S.; Sadat, A.; Yang, B.; Casas, S.; Urtasun, R. End-to-end interpretable neural motion planner. In Proceedings of the IEEE/CVF Conference on Computer Vision and Pattern Recognition, Long Beach, CA, USA, 15–20 June 2019; pp. 8660–8669.
24. Weng, X.; Wang, J.; Levine, S.; Kitani, K.; Rhinehart, N. Inverting the pose forecasting pipeline with SPF2: Sequential pointcloud forecasting for sequential pose forecasting. In Proceedings of the Conference on Robot Learning PMLR, Atlanta, GA, USA, 6–9 November 2021; pp. 11–20.
25. Peri, N.; Luiten, J.; Li, M.; Ošep, A.; Leal-Taixé, L.; Ramanan, D. Forecasting from lidar via future object detection. In Proceedings of the IEEE/CVF Conference on Computer Vision and Pattern Recognition, New Orleans, LA, USA, 18–24 June 2022; pp. 17202–17211.
26. He, X.; Zhao, K.; Chu, X. AutoML: A Survey of the State-of-the-Art. *Knowl.-Based Syst.* **2021**, *212*, 106622. [CrossRef]
27. Elsken, T.; Metzen, J.H.; Hutter, F. Neural architecture search: A survey. *J. Mach. Learn. Res.* **2019**, *20*, 1997–2017.
28. Zoph, B.; Le, Q.V. Neural architecture search with reinforcement learning. *arXiv* **2016**, arXiv:1611.01578.
29. Hsu, C.H.; Chang, S.H.; Liang, J.H.; Chou, H.P.; Liu, C.H.; Chang, S.C.; Pan, J.Y.; Chen, Y.T.; Wei, W.; Juan, D.C. Monas: Multi-objective neural architecture search using reinforcement learning. *arXiv* **2018**, arXiv:1806.10332.
30. Loni, M.; Sinaei, S.; Zoljodi, A.; Daneshtalab, M.; Sjödin, M. DeepMaker: A multi-objective optimization framework for deep neural networks in embedded systems. *Microprocess. Microsyst.* **2020**, *73*, 102989. [CrossRef]
31. Loni, M.; Zoljodi, A.; Sinaei, S.; Daneshtalab, M.; Sjödin, M. Neuropower: Designing energy efficient convolutional neural network architecture for embedded systems. In Proceedings of the International Conference on Artificial Neural Networks, Munich, Germany, 17–19 September 2019; pp. 208–222.
32. Liu, C.; Chen, L.C.; Schroff, F.; Adam, H.; Hua, W.; Yuille, A.L.; Fei-Fei, L. Auto-deeplab: Hierarchical neural architecture search for semantic image segmentation. In Proceedings of the IEEE/CVF Conference on Computer Vision and Pattern Recognition, Long Beach, CA, USA, 15–20 June 2019; pp. 82–92.
33. Liu, H.; Simonyan, K.; Yang, Y. Darts: Differentiable architecture search. *arXiv* **2018**, arXiv:1806.09055.
34. Loni, M.; Mousavi, H.; Riazati, M.; Daneshtalab, M.; Sjödin, M. TAS:Ternarized Neural Architecture Search for Resource-Constrained Edge Devices. In Proceedings of the Design, Automation & Test in Europe Conference & Exhibition DATE'22, Antwerp, Belgium, 14 March 2022.
35. Cai, H.; Gan, C.; Wang, T.; Zhang, Z.; Han, S. Once-for-all: Train one network and specialize it for efficient deployment. *arXiv* **2019**, arXiv:1908.09791.
36. Dong, X.; Liu, L.; Musial, K.; Gabrys, B. NATS-Bench: Benchmarking NAS Algorithms for Architecture Topology and Size. *IEEE Trans. Pattern Anal. Mach. Intell.* **2021**, *44*, 3634–3646. [CrossRef]
37. Loni, M.; Zoljodi, A.; Maier, D.; Majd, A.; Daneshtalab, M.; Sjödin, M.; Juurlink, B.; Akbari, R. DenseDisp: Resource-Aware Disparity Map Estimation by Compressing Siamese Neural Architecture. In Proceedings of the 2020 IEEE Congress on Evolutionary Computation (CEC), Glasgow, UK, 19–24 July 2020; pp. 1–8. [CrossRef]
38. Xu, H.; Wang, S.; Cai, X.; Zhang, W.; Liang, X.; Li, Z. Curvelane-nas: Unifying lane-sensitive architecture search and adaptive point blending. In *Proceedings of the Computer Vision—ECCV 2020: 16th European Conference, Glasgow, UK, 23–28 August 2020; Proceedings, Part XV 16*; Springer: Berlin, Germany, 2020; pp. 689–704.
39. Loni, M.; Zoljodi, A.; Majd, A.; Ahn, B.H.; Daneshtalab, M.; Sjödin, M.; Esmaeilzadeh, H. FastStereoNet: A Fast Neural Architecture Search for Improving the Inference of Disparity Estimation on Resource-Limited Platforms. *IEEE Trans. Syst. Man Cybern. Syst.* **2021**, *52*, 5222–5234. [CrossRef]
40. Xie, S.; Li, Z.; Wang, Z.; Xie, C. On the adversarial robustness of camera-based 3d object detection. *arXiv* **2023**, arXiv:2301.10766.
41. Kälble, J.; Wirges, S.; Tatarchenko, M.; Ilg, E. Accurate Training Data for Occupancy Map Prediction in Automated Driving Using Evidence Theory. In Proceedings of the IEEE/CVF Conference on Computer Vision and Pattern Recognition, Seattle, WA, USA, 17–21 June 2024; pp. 5281–5290.
42. Blanch, M.R.; Li, Z.; Escalera, S.; Nasrollahi, K. LiDAR-Assisted 3D Human Detection for Video Surveillance. In Proceedings of the IEEE/CVF Winter Conference on Applications of Computer Vision, Waikoloa, HI, USA, 1–6 January 2024; pp. 123–131.

43. Lang, A.H.; Vora, S.; Caesar, H.; Zhou, L.; Yang, J.; Beijbom, O. Pointpillars: Fast encoders for object detection from point clouds. In Proceedings of the IEEE/CVF Conference on Computer Vision and Pattern Recognition, Long Beach, CA, USA, 15–20 June 2019; pp. 12697–12705.
44. He, C.; Zeng, H.; Huang, J.; Hua, X.S.; Zhang, L. Structure aware single-stage 3d object detection from point cloud. In Proceedings of the IEEE/CVF Conference on Computer Vision and Pattern Recognition, Seattle, WA, USA, 13–19 June 2020; pp. 11873–11882.
45. Qi, C.R.; Su, H.; Mo, K.; Guibas, L.J. Pointnet: Deep learning on point sets for 3d classification and segmentation. In Proceedings of the IEEE Conference on Computer Vision and Pattern Recognition, Honolulu, HI, USA, 21–26 July 2017; pp. 652–660.
46. Yan, Y.; Mao, Y.; Li, B. Second: Sparsely embedded convolutional detection. *Sensors* **2018**, *18*, 3337. [CrossRef]
47. Zhou, Y.; Tuzel, O. Voxelnet: End-to-end learning for point cloud based 3d object detection. In Proceedings of the IEEE Conference on Computer Vision and Pattern Recognition, Salt Lake City, UT, USA, 18–23 June 2018; pp. 4490–4499.
48. Ren, S.; He, K.; Girshick, R.; Sun, J. Faster R-CNN: Towards real-time object detection with region proposal networks. *IEEE Trans. Pattern Anal. Mach. Intell.* **2016**, *39*, 1137–1149. [CrossRef] [PubMed]
49. Amine, K. Multiobjective simulated annealing: Principles and algorithm variants. *Adv. Oper. Res.* **2019**, *2019*, 8134674. [CrossRef]
50. Everingham, M.; Van Gool, L.; Williams, C.K.; Winn, J.; Zisserman, A. The pascal visual object classes (voc) challenge. *Int. J. Comput. Vis.* **2010**, *88*, 303–338. [CrossRef]
51. Lin, T.Y.; Maire, M.; Belongie, S.; Hays, J.; Perona, P.; Ramanan, D.; Dollár, P.; Zitnick, C.L. Microsoft coco: Common objects in context. In *Proceedings of the Computer Vision–ECCV 2014: 13th European Conference, Zurich, Switzerland, 6–12 September 2014*; Proceedings, Part V 13; Springer: Berlin, Germany, 2014; pp. 740–755.
52. Burke, E.K.; Bykov, Y. The late acceptance Hill-Climbing heuristic. *Eur. J. Oper. Res.* **2017**, *258*, 70–78. [CrossRef]

Disclaimer/Publisher's Note: The statements, opinions and data contained in all publications are solely those of the individual author(s) and contributor(s) and not of MDPI and/or the editor(s). MDPI and/or the editor(s) disclaim responsibility for any injury to people or property resulting from any ideas, methods, instructions or products referred to in the content.

Communication

Determining the Level of Threat in Maritime Navigation Based on the Detection of Small Floating Objects with Deep Neural Networks

Mirosław Łącki

Faculty of Navigation, Gdynia Maritime University, 81-225 Gdynia, Poland; m.lacki@wn.umg.edu.pl

Abstract: The article describes the use of deep neural networks to detect small floating objects located in a vessel's path. The research aimed to evaluate the performance of deep neural networks by classifying sea surface images and assigning the level of threat resulting from the detection of objects floating on the water, such as fishing nets, plastic debris, or buoys. Such a solution could function as a decision support system capable of detecting and informing the watch officer or helmsman about possible threats and reducing the risk of overlooking them at a critical moment. Several neural network structures were compared to find the most efficient solution, taking into account the speed and efficiency of network training and its performance during testing. Additional time measurements have been made to test the real-time capabilities of the system. The research results confirm that it is possible to create a practical lightweight detection system with convolutional neural networks that calculates safety level in real time.

Keywords: deep neural networks; detection and classification; safety of marine navigation; image processing; object detection techniques

Citation: Łącki, M. Determining the Level of Threat in Maritime Navigation Based on the Detection of Small Floating Objects with Deep Neural Networks. *Sensors* **2024**, *24*, 7505. https://doi.org/10.3390/s24237505

Academic Editors: Man Qi and Matteo Dunnhofer

Received: 29 September 2024
Revised: 17 November 2024
Accepted: 22 November 2024
Published: 25 November 2024

Copyright: © 2024 by the author. Licensee MDPI, Basel, Switzerland. This article is an open access article distributed under the terms and conditions of the Creative Commons Attribution (CC BY) license (https://creativecommons.org/licenses/by/4.0/).

1. Introduction

The safety of marine navigation directly impacts the protection of human lives, the environment, and valuable cargo. Key factors in maintaining navigational safety include accurate positioning, effective communication, understanding of environmental conditions, and navigational situation.

Threats to marine navigation are numerous and diverse. They range from natural hazards like severe weather, rough seas, and poor visibility to man-made dangers such as collisions with other vessels, grounding on reefs or sandbars, and accidents in busy ports or narrow channels. In addition to these traditional threats, the increasing presence of floating objects poses a significant risk. These hazards, often referred to as marine debris, include a wide range of items such as derelict fishing gear, rubber, textiles, metal, and various types of plastic waste. The accumulation of such debris in the oceans can lead to dangerous situations for vessels, especially in busy shipping lanes, coastal areas, and regions prone to natural disasters.

Plastic debris and other floating objects can cause severe damage to a vessel's hull, propellers, and rudders if collided with, potentially leading to accidents, grounding, or even sinking. These hazards can also obstruct critical navigation channels, disrupt traffic, and create challenges for search and rescue operations. Furthermore, smaller vessels and recreational boats are particularly vulnerable to these risks, as they may lack the robust detection and avoidance systems found on larger ships, that can integrate satellite imagery, oceanographic models, and reports from other vessels and from citizen science programs [1]. Thus, the safety of smaller vessels depends mostly on continuous visual observation.

Constantly watching for floating obstacles in the course of a vessel is a challenging and complex task. Several factors make this task difficult:

1. Visibility limitations: Floating objects, especially smaller ones like plastic debris or partially submerged containers, can be difficult to spot, particularly in poor-visibility conditions such as fog, rain, or darkness. Even in daylight, the vastness of the open ocean can make it nearly impossible to detect all potential hazards in a vessel's path.
2. Human limitations: Maintaining continuous vigilance for floating obstacles requires intense concentration and can be mentally and physically exhausting for the crew. Human error, fatigue, and distraction are significant factors that can reduce the effectiveness of manual monitoring.
3. Unpredictability of debris: The location and movement of floating debris are often unpredictable. Currents, tides, and winds can disperse debris across large areas, making it difficult to anticipate where these objects might be encountered. Additionally, debris can move rapidly, especially in rough seas, increasing the challenge of detecting and avoiding it.
4. Technology gaps: While radar and sonar systems can detect larger objects, they are less effective at identifying smaller or partially submerged debris. Some modern vessels are equipped with advanced optical sensors or infrared cameras, but these technologies also have limitations, particularly in adverse weather or lighting conditions, and require constant attention if not integrated with an automated detection system.

Given these challenges, relying solely on human watchkeepers or traditional navigation tools is not sufficient to ensure safety. This is where a system based on deep neural network architecture (DNN) may be useful. The proposed solution can provide automated alerts, as shown on Figure 1, and enhance a crew's ability to detect and avoid floating obstacles.

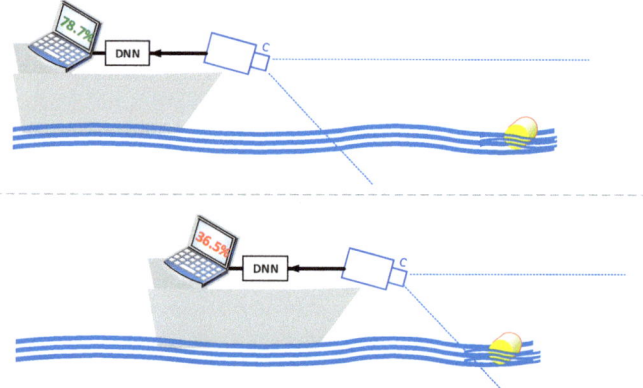

Figure 1. General proposal of a system calculating the level of safety regarding the distance from detected floating objects with usage of deep neural network.

There are two main DNN structures considered in this study: Fully Connected (Dense) Neural Network (NN) and Convolutional Neural Network (CNN). In Dense NN each neuron in one layer is connected to every neuron in the subsequent layer. Dense net-works typically require a large amount of data and computational resources to train effectively and are sensitive to overfitting. Standalone large Dense NN is not as efficient as smaller Dense NN combined into convolutional architecture [2].

Convolutional Neural Networks (CNNs) are a type of deep learning algorithm that have been widely used in image object recognition tasks [3–5]. CNNs are specifically designed to deal with the variability of two-dimensional shapes and have shown superior performance compared to other techniques. CNNs are a powerful tool for image object recognition and classification. They are capable of extracting meaningful features from images and have been successfully applied in various domains, including medical imaging, activity recognition, real-time remote sensing monitoring, and big data analysis.

According to reviews of deep learning (DL) methods, there are several hundred important articles describing concepts and applications [6]. An example of the successful application of CNNs in image object recognition is the Inception architecture proposed by Szegedy et al. [7]. This architecture achieved state-of-the-art performance in the ImageNet Large-Scale Visual Recognition Challenge 2014 (ILSVRC14), which is a benchmark dataset for image classification and detection tasks. Inception architecture utilizes multiple parallel convolutional layers with different filter sizes to capture features at different scales, allowing for more accurate recognition of objects in images. Another example is the use of CNNs in medical image recognition. Here CNNs were applied to train medical images, such as magnetic resonance imaging [8], invasive ductal carcinoma [9], and computed tomography images [10], and achieved higher recognition rates compared to traditional methods. Furthermore, CNNs have been applied in various fields beyond image recognition. For example, CNNs have been used in recognizing human activity using wearable sensors [11] or the seismic facies classification [12].

Another interesting approach to image classification solution is deep transfer learning (DTL). This is a machine learning approach that enables models trained on one task to be adapted and applied to a related but different task. It leverages the pre-trained knowledge from a source domain (such as images, text, or other data) to improve performance on a new target domain, especially when labeled data for the target domain are limited or costly to obtain. This approach overcomes the drawbacks of traditional machine learning, in which the training datasets are separated and used individually for each task. A complete survey with detailed classification of DTL models has been presented in [13].

The novelty of this study is in its ability to calculate threat level for a vessel regarding the influence of detected floating objects on the safety of navigation. The practical application may be created in the future as a navigational support system for watch officers or helmsmen, generating warnings easy to read as threat levels calculated by trained neural networks. Other studies in the field of small floating object detection focus mainly on proper and fast identification and classification [14], also in low-light conditions [15]; thus the goals of the other studies are slightly different. The contribution of this study is in the design and comparison of a few different network topologies and selection of the best ones that fit the 'lightweight' category.

The structure of the paper is as follows: After this introduction, there is Section 2 de-scribing materials and methods used in this study. The next chapter provides simulation results for 12 neural networks of different topologies. The results are discussed in Section 4, and Section 5 consists of a brief summary and conclusions.

2. Materials and Methods

The architecture of a CNN consists of multiple layers, including convolutional layers, pooling layers, and fully connected dense layers (Figure 2). Input image is divided into three separate color channel input layers.

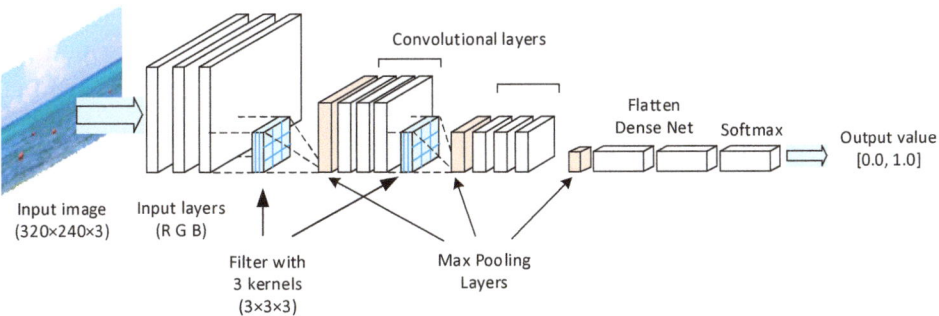

Figure 2. General architecture of CNN used in this study.

Next, the filters are applied to the input layers to extract local features. These filters, each 3 × 3 in size, are learned through the training process and can capture different patterns and textures in the image [7]. The pooling layers, introduced in 1990 by Yamaguchi et al. [16], downsample the feature maps to reduce the spatial dimensions and extract the most important features. The size of the pooling layers in this study is 2 × 2. The result of each convolutional layer is calculated by ReLu (rectified linear unit) activation function, which replaces each negative value by zero according to Equation (1):

$$f(x) = \max(0, x) \quad (1)$$

where x is the value from neurons of convolutional layer. The fully connected dense layers combine the extracted features and make safety calculations based on them [17]. The last dense network output is calculated with softmax function

$$s(x_i) = \frac{e^{x_i}}{\sum_{j=1}^{n} e^{x_j}} \quad (2)$$

which transforms a vector of real numbers into a vector of probabilities.

The images, required to train neural networks and validate them, were acquired from the following free image and video depositories: depositphotos.com, pexels.com, istockphoto.com, and stockvault.net. The examples of the images are shown on Figure 3.

net

net

sea

Figure 3. Examples of input images.

The training set contains 200 images divided into two categories:
1. Clear water surface (tag: sea)—this collection contains photos of a calm and a stormy sea, with different daylight conditions, sometimes with flying or floating birds;
2. Polluted water surface (tag: net)—this collection contains photos of different plastic debris, fishing gear, buoys, and similar small floating objects.

The validation set was divided into the same categories as the training set. It consists of 280 images, completely different from the training set to observe possible overfitting. Overfitting occurs when a model learns to perform very well on the training data but fails to generalize to new, unseen data. Overfitting often occurs when the model is too complex relative to the amount of training data. In CNNs this could mean having too many layers, filters, or parameters relative to the size and diversity of the training dataset. An overfitted model fails to make correct predictions on data that it hasn't encountered before. To prevent overfitting, the dropout layer has been added after the dense layer. Dropout randomly turns off neurons during training with probability 0.5 for each neuron.

Additional training parameters and limitations:
- 30 epochs of training;
- Each training batch consists of 30 input images;
- Each image is resized to 320 × 240 px
- The result is divided into two crisp categories, depending on network output value: not safe [0–0.5] and safe [0.5–1];

- Average accepted performance for each NN not less than 80% during training and validation;
- Size of the network not more than 1 GB, for lightweight property requirements.

3. Results

This section provides a description of the experimental results of 12 different DNN architectures and their interpretation.

Table 1 shows the basic parameters of each network. The numbers for the CNN layers are the numbers of filters, and the numbers for the Dense layers are the number of neurons.

Table 1. Comparison of the size and average accuracy of the tested neural networks.

Network No.	CNN1	CNN2	CNN3	Dense	Size [MB]	Avg Test Accuracy	Avg Validation Accuracy
1	-	-	-	8	21.1	0.504	0.719
2	-	-	-	256	664.5	0.549	0.546
3	-	-	-	512	1310	0.598	0.598
4	2	-	-	8	3.4	0.664	0.483
5	2	4	-	16	3.34	0.648	0.727
6	2	4	8	32	3.16	0.63	0.734
7	4	8	16	64	12.5	0.816	0.778
8	8	16	32	128	49.9	0.86	0.851
9	16	32	64	256	199.8	0.881	0.818
10	16	32	64	512	399.3	0.861	0.849
11	32	64	128	512	799.1	0.859	0.759
12	64	128	256	512	1560	0.858	0.716

The basic criteria for efficient network are an accuracy of at least 0.8 in both training and validation. The first four networks are unable to achieve good validation results and perform very poorly during training. Dense architecture requires many more neurons to operate effectively, which causes its size to significantly exceed the allowable limit of 1 GB.

Although network 9 achieved better average results during training, network 10 has better results during validation and a smaller difference in values between the average result of training and validation (Figure 4). The size of network no. 10 is also less than 400 MB; therefore in this study, network no. 10 was selected as the best candidate for further tests and timing measurements.

Networks 11 and 12 show overfitting symptoms, and therefore are probably too big for this task. Additionally, network 12 is more than 1 GB in size, what makes it not as lightweight as expected.

Another important parameter that can help evaluate neural network performance is loss value. Loss is a measure of how well the network's predictions match the actual target values. It quantifies the error between the predicted output and the true output during training. The goal of training a CNN is to minimize this loss value so that the network can make accurate predictions on new, unseen data. The loss function serves as a guide to adjust the weights and biases of the network. By calculating the loss after each forward pass, the network determines how far off its predictions are from the true values. Using this information, the network updates its parameters to reduce the error in future predictions.

In this research Binary Cross-Entropy (BCE), also known as Log Loss, has been used, with a formula as follows:

$$BCE = -(y \times \log(y_p) + (1-y) \times \log(1-y_p)) \tag{3}$$

where y is the true label, and y_p is the predicted label. This formula is applied for each training and validation step, and the total loss is usually averaged across all the examples in a batch. BCE is a commonly used loss function for binary classification tasks, where there are only two possible outcomes, as in this task the system classifies images as safe empty sea surface (labeled as "sea") or detected floating objects (labelled as "net"). BCE measures the difference between the true label and the predicted probability from the network. The examples of loss values are presented on Figure 5.

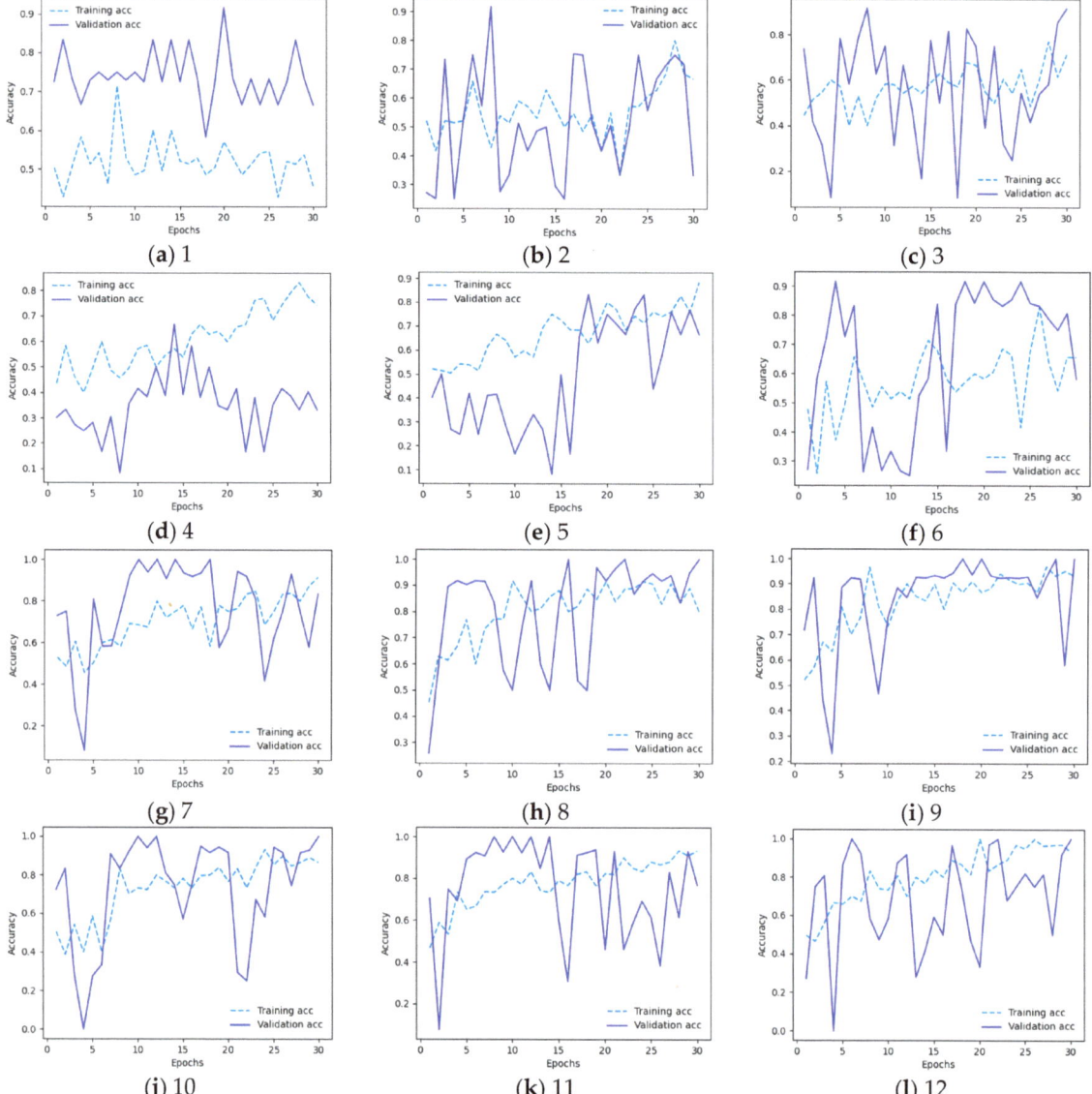

Figure 4. Training (dashed) and validation (solid) results for each network: (**a**) Network no 1 shows huge underfitting gap and is unable to learn effectively; (**b**–**f**) Networks 2–6 have big differences between training and validation values (**g**–**j**) Networks 7–10 perform well in comparison to other networks. (**k**,**l**) Networks 11 and 12 are also quite good, but due to overfitting occurrence and bigger size, they are not as good as smaller ones.

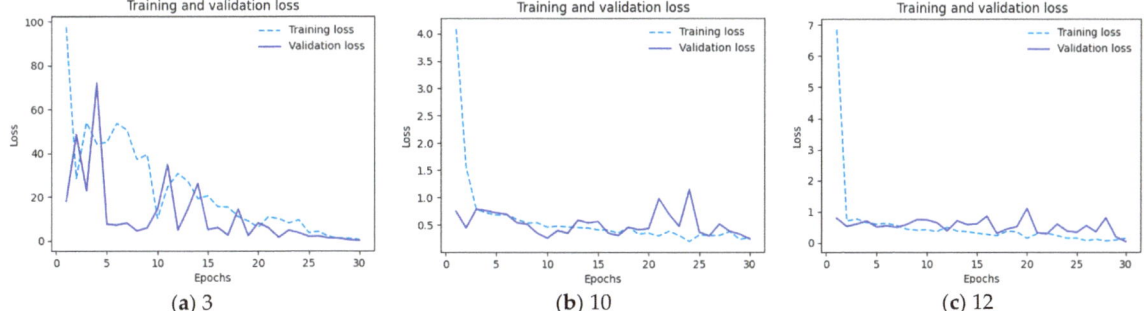

Figure 5. Examples of loss value for chosen networks: (**a**) Network no. 3 generates large loss values; (**b**) Network 10 has loss value about 0.5; (**c**) Network 12 has also values below 1, but there are some visible differences between training and validation values.

During training, the loss value is often monitored over each epoch or batch. A decreasing loss value typically indicates that the model is learning and improving its performance. Overfitting can be detected if the training loss continues to decrease while the validation loss starts to increase. A low loss does not always correspond to high accuracy, and vice versa, as they provide different insights into the model's performance.

Prediction examples shows that network no. 10 properly intentified clean sea surface and images with floating objects, as presented on Figure 6.

Figure 6. Examples of prediction results of network no. 10. Values in brackets closer to 1 indicate sea surface without floating objects.

Additional time measurements have shown that the system is capable of processing about 20 images per second, which is sufficient and even excessive, because the expected target refresh of the threat level value in the final application will be no more than 2 times per second.

In this system, neural network calculations are performed entirely on the CPU. In the future, it could be considered to compare the performance of the hardware with hardware solutions using GPU and FPGA. At the current stage, however, the speed of the system using only the CPU is sufficient.

Downloading and saving images from the camera strongly depends on the hardware and software used. Cameras connected directly to a computer by cable can generate even several dozen images per second. The bottleneck of the transfer here may save individual images to the disk. In the tested system, the wireless camera connected via Wi-Fi allowed for recording 5 images per second at a resolution of 640 × 486 pixels (96 dpi). This was also a sufficient result for the smooth operation of the system.

4. Discussion

The obtained results of the system calculations are satisfactory, although in the case of estimating the level of safety or threat during navigation, using one strict threshold (e.g., value 0.5) is impractical. Classifying a given navigation situation only in two stages, as safe (predicted tag: sea) or dangerous (predicted tag: net), as shown on Figure 7, used for training neural networks is too restrictive. Therefore, in further studies, estimation using fuzzy values defining the ranges of the safety level will be considered.

True: net, Pred: sea
[0.5925232]
(a)

True: net, Pred: net
[0.48333532]
(b)

Figure 7. Examples of different prediction results of similar images: (**a**) A result above 0.5 indicates relatively safe situation; (**b**) A result below 0.5 is treated as not safe.

Dividing the 0–1 range into several smaller intervals allows for a more accurate estimation of the threat level at a given moment, which will mainly depend on the distance from the detected object on the sea surface. In the current system, this can be realized at the system output in the end user interface, and from this point of view the system works properly.

As can be seen in Figure 8, the longer the distance to the object, the higher the level of safety, and in Figure 8a the situation can be described as relatively safe, because the value calculated by the system is 0.71. The second situation in Figure 8b would require the watchman to take some action, because the detected objects are much closer and the system calculated the level of safety at about 31%.

Three subsets of safety levels may be distinguished:

- 0.0 to 0.4—not safe, taking action required;
- 0.4 to 0.6—relatively safe, but require intensive monitoring;
- 0.6 to 1.0—safe, some small object may be floating in long distance.

The safety level (Figure 9) and the number of subsets strictly depend on the distance to the detected floating objects, and to the type of vessels and their maneuvering characteristics. Additionally, the environmental conditions should be taken into account, i.e., visibility, wave height, wind, and other weather parameters.

True: net, Pred: sea [0.710709] (a)

True: net, Pred: net [0.3137841] (b)

Figure 8. Comparison of network results depending on distance to floating objects. Range of values is a real value from 0 (not safe) to 1 (safe). Safety level in square brackets. Longer distance (**a**) qualifies situation as relatively safe (0.71). When vessel approaches a little closer (**b**) then safety level drops significantly to 0.31.

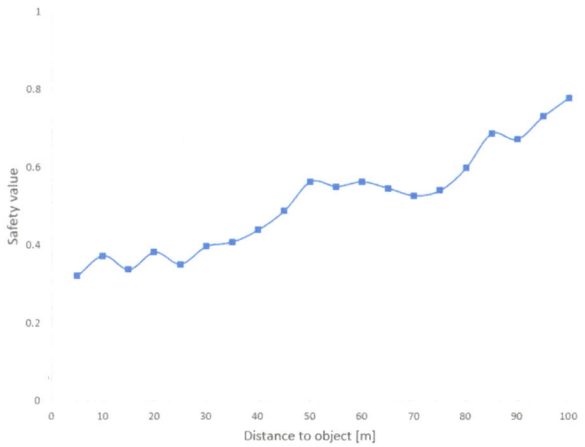

Figure 9. Safety level in relation to distance to floating object.

Additionally, the system can be expanded to more precisely identify and mark objects on the image. One of the good solutions enabling real-time identification and classification of objects is the YOLO algorithm presented in 2015 [18]. The example of practical application of this algorithm for small-object detection in an underwater environment is described in [19]. The algorithm is still being developed and improved with new solutions that allow users to cope with some of the shortcomings of previous versions [20]. It is also worth noticing that each addition of new piece of visual or numerical information on the system screen may reduce readability or even overwhelm the user with information. Therefore, such auxiliary systems should be legible, and any additional elements should be optional and can be temporarily turned off.

5. Conclusions

The safety of marine navigation relies on a combination of skilled seamanship and technological advancements, including the effective use of decision support systems. The proposed solution is a good candidate that meets the conditions to become an easy-to-use lightweight application of the decision support system. Such a system during navigation through sea areas may help to assess safety level regarding the presence of small floating objects on the surface, difficult to spot in time without additional sensors other than the watchful eyes of the observer on board. Further practical research and measurements are needed to more thoroughly verify and test different navigation scenarios, especially in rough weather, low-light conditions, and high traffic.

Funding: This study was funded from the statutory activities of Gdynia Maritime University, grant number WN/2024/PI/2.

Institutional Review Board Statement: Not applicable.

Informed Consent Statement: Not applicable.

Data Availability Statement: Data available at http://kpisk.umg.edu.pl/lacki/research/2024/ accessed on 21 November 2024

Conflicts of Interest: The author declares no conflicts of interest.

References

1. Fraisl, D.; See, L.; Bowers, R.; Seidu, O.; Fredua, K.B.; Bowser, A.; Meloche, M.; Weller, S.; Amaglo-Kobla, T.; Ghafari, D.; et al. The Contributions of Citizen Science to SDG Monitoring and Reporting on Marine Plastics. *Sustain. Sci.* **2023**, *18*, 2629–2647. [CrossRef]
2. Huang, G.; Liu, Z.; Pleiss, G.; Van Der Maaten, L.; Weinberger, K.Q. Convolutional Networks with Dense Connectivity. *IEEE Trans. Pattern Anal. Mach. Intell.* **2022**, *44*, 8704–8716. [CrossRef] [PubMed]
3. Moon, J.; Lim, S.; Lee, H.; Yu, S.; Lee, K.-B. Smart Count System Based on Object Detection Using Deep Learning. *Remote Sens.* **2022**, *14*, 3761. [CrossRef]
4. Bashir, S.M.A.; Wang, Y. Small Object Detection in Remote Sensing Images with Residual Feature Aggregation-Based Super-Resolution and Object Detector Network. *Remote Sens.* **2021**, *13*, 1854. [CrossRef]
5. Coleman, S.; Kerr, D.; Zhang, Y. Image Sensing and Processing with Convolutional Neural Networks. *Sensors* **2022**, *22*, 3612. [CrossRef] [PubMed]
6. Alzubaidi, L.; Zhang, J.; Humaidi, A.J.; Al-Dujaili, A.; Duan, Y.; Al-Shamma, O.; Santamaría, J.; Fadhel, M.A.; Al-Amidie, M.; Farhan, L. Review of Deep Learning: Concepts, CNN Architectures, Challenges, Applications, Future Directions. *J. Big Data* **2021**, *8*, 53. [CrossRef] [PubMed]
7. Szegedy, C.; Liu, W.; Jia, Y.; Sermanet, P.; Reed, S.; Anguelov, D.; Erhan, D.; Vanhoucke, V.; Rabinovich, A. Going Deeper with Convolutions. In Proceedings of the 2015 IEEE Conference on Computer Vision and Pattern Recognition (CVPR), Boston, MA, USA, 7–12 June 2015; pp. 1–9.
8. Zhang, H.; Zhang, W.; Shen, W.; Li, N.; Chen, Y.; Li, S.; Chen, B.; Guo, S.; Wang, Y. Automatic Segmentation of the Cardiac MR Images Based on Nested Fully Convolutional Dense Network with Dilated Convolution. *Biomed. Signal Process. Control* **2021**, *68*, 102684. [CrossRef]
9. Kumaraswamy, E.; Kumar, S.; Sharma, M. An Invasive Ductal Carcinomas Breast Cancer Grade Classification Using an Ensemble of Convolutional Neural Networks. *Diagnostics* **2023**, *13*, 1977. [CrossRef] [PubMed]
10. Manabe, K.; Asami, Y.; Yamada, T.; Sugimori, H. Improvement in the Convolutional Neural Network for Computed Tomography Images. *Appl. Sci.* **2021**, *11*, 1505. [CrossRef]
11. Moya Rueda, F.; Grzeszick, R.; Fink, G.A.; Feldhorst, S.; Ten Hompel, M. Convolutional Neural Networks for Human Activity Recognition Using Body-Worn Sensors. *Informatics* **2018**, *5*, 26. [CrossRef]
12. Abid, B.; Khan, B.M.; Memon, R.A. Seismic Facies Segmentation Using Ensemble of Convolutional Neural Networks. *Wirel. Commun. Mob. Comput.* **2022**, *2022*, 7762543. [CrossRef]
13. Yu, F.; Xiu, X.; Li, Y. A Survey on Deep Transfer Learning and Beyond. *Mathematics* **2022**, *10*, 3619. [CrossRef]
14. Kim, J.-H.; Kim, N.; Park, Y.W.; Won, C.S. Object Detection and Classification Based on YOLO-V5 with Improved Maritime Dataset. *J. Mar. Sci. Eng.* **2022**, *10*, 377. [CrossRef]
15. Emanuelsson, E.; Wang, L. Real-Time Characteristics of Marine Object Detection under Low Light Conditions: Marine Object Detection Using YOLO with near Infrared Camera. Master's Thesis, Chalmers University of Technology, Gothenburg, Sweden, 2020.
16. Yamaguchi, K.; Sakamoto, K.; Akabane, T.; Fujimoto, Y. A Neural Network for Speaker-Independent Isolated Word Recognition. In Proceedings of the First International Conference on Spoken Language Processing, Kobe, Japan, 18–22 November 1990; pp. 1077–1080.
17. Simonyan, K.; Zisserman, A. Very Deep Convolutional Networks for Large-Scale Image Recognition. *arXiv* **2015**, arXiv:1409.1556.
18. Redmon, J.; Divvala, S.; Girshick, R.; Farhadi, A. You Only Look Once: Unified, Real-Time Object Detection. In Proceedings of the IEEE Computer Society Conference on Computer Vision and Pattern Recognition, Las Vegas, NV, USA, 27–30 June 2016.

19. Zhang, M.; Xu, S.; Song, W.; He, Q.; Wei, Q. Lightweight Underwater Object Detection Based on YOLO v4 and Multi-Scale Attentional Feature Fusion. *Remote Sens.* **2021**, *13*, 4706. [CrossRef]
20. Terven, J.; Córdova-Esparza, D.-M.; Romero-González, J.-A. A Comprehensive Review of YOLO Architectures in Computer Vision: From YOLOv1 to YOLOv8 and YOLO-NAS. *Mach. Learn. Knowl. Extr.* **2023**, *5*, 1680–1716. [CrossRef]

Disclaimer/Publisher's Note: The statements, opinions and data contained in all publications are solely those of the individual author(s) and contributor(s) and not of MDPI and/or the editor(s). MDPI and/or the editor(s) disclaim responsibility for any injury to people or property resulting from any ideas, methods, instructions or products referred to in the content.

MDPI AG
Grosspeteranlage 5
4052 Basel
Switzerland
Tel.: +41 61 683 77 34

Sensors Editorial Office
E-mail: sensors@mdpi.com
www.mdpi.com/journal/sensors

Disclaimer/Publisher's Note: The title and front matter of this reprint are at the discretion of the Guest Editors. The publisher is not responsible for their content or any associated concerns. The statements, opinions and data contained in all individual articles are solely those of the individual Editors and contributors and not of MDPI. MDPI disclaims responsibility for any injury to people or property resulting from any ideas, methods, instructions or products referred to in the content.